H. Planck   M. Dauner   M. Renardy (Eds.)

Medical Textiles for Implantation

With 198 Figures and 41 Tables

Springer-Verlag
Berlin Heidelberg New York
London Paris Tokyo
Hong Kong Barcelona

Proceedings of the 3rd International ITV Conference on Biomaterials, Stuttgart,
June 14-16, 1989

Dr. Ing. Heinrich Planck
Dipl. Ing. Martin Dauner
Dipl. Biol. Monika Renardy

Institut für Textil- und Verfahrenstechnik
Körschtalstraße 26
W-7306 Denkendorf
FRG

ISBN-13:978-3-642-75804-1   e-ISBN-13:978-3-642-75802-7
DOI: 10.1007/978-3-642-75802-7

2125/3145-543210 - Printed on acid-free paper

PREFACE

Triennial the Division of Biomedical Engineering of the Institute of Textile Technology and Chemical Engineering, Denkendorf, is organizing conferences on specific topics in the field of polymeric materials for use in the biomedical areas. The aim is to bring together scientists from all over the world working on this specific topic, to present the newest state of the art and to discuss their problems in a more concentrated atmosphere and at last to create and intensivate their cooperation.

Following two conferences on "Polyurethanes in Biomedical Engineering" (1983 and 1986), the Institute of Textile Technology and Chemical Engineering set a theme, which is very closely related to its own task: "Medical Textiles for Implantation".

As technical materials, textiles can be classified in two fields of application:
- first, textiles used for highly flexible, strong, but only tension load bearing systems, e.g. tows;
- second, textiles manufactured to flat shaped devices to separate two regions more or less semipermeable, e.g. clothing;
- a combination of both are reinforced systems like tubular fabrics e.g.; here pressure load will be transformed to tensile load, the separation may be performed by a coating.

In the biological systems the classification can be used in the same manner:
- Tension load bearing structures are ligaments and tendons, semipermeable separation is realized by cell membranes as well as by cell layers, for example the skin.
- The combination of both of the principles can be found for example in arteries and the trachea.
  While collagen builds up the load bearing fibers, bidirectional oriented smooth muscle cells transform pressure in the blood vessels to tensile load.

Starting with the more basic contributions to the use of textiles and polymers in biomedical applications, these proceedings are continued with load bearing suture materials and ligament prostheses. Vascular and tracheal prostheses combine both functions; the coating may have synthetic or autologous origin. Finally separation by a technical structure is required in bioartificial organs as well as in the artificial skin.

The papers are completed by selected remarks taken from the discussions. The round table discussion is referred, too, to give a more general scope.

The editors wish to thank:
- The speakers for the excellent papers and the participants for the active discussion.
- The following companies for their financial support of the conference:

        Karl Otto Braun KG, Wolfstein

        Daimler Benz AG, Stuttgart

        Dornier GmbH, Friedrichshafen

        Enka AG, Wuppertal

        Carl Freudenberg, Weinheim

        Hoechst AG, Frankfurt

        Karl Mayer GmbH, Obertshausen
- Martina Maier for her cooperation in organizing the conference and in typing and preparing the manuscripts in their final form.

Institute of Textile Technology	M. Dauner
and Chemical Engineering,	M. Renardy
Denkendorf, Germany	H. Planck

CONTENTS

I. GENERAL ASPECTS

1. General Aspects in the Use of Medical Textiles for
    Implantation   1
    Planck, H.

2. In Vitro Evaluation of the Effects of Macrophages and
    Synthetic Materials on the Fibroblast Proliferation
    and Collagen Synthesis   17
    Lim, J.O., Chu, C.C., Appel, M.

3. Contributions to Biocompatibility during Hemodialysis   35
    Seyfert, U.T., Helmling, E., Grothe-Pfautsch, U.,
    Albert, F.W., Pindur, G., Wenzel, E.

4. Macroporous Textile and Microporous Nonwoven Vascular
    Prostheses: Histological Aspects of Cellular Ingrowth
    into the Structure   61
    Jerusalem, C.R., Hess, F.

5. Cytotoxicity Tests in the Biological Evaluation of
    Medical Devices   77
    Müller-Lierheim, W.

6. Fact Database Mediplast. Polymers in Medicine   85
    Planck, H., Umutyan, G., Konle, S., Kunberger, A.

7. The Central Register for Side Effects of Biomaterials   93
    Seyfert, U.T., Bohnert, G., Pindur, G., Bambauer, R.,
    Jutzler, G., Kiesewetter, H., Wenzel, E

II. LIGAMENTS AND TENDONS

8. Influence of Processing Parameters for Artificial
    Ligaments   99
    Dauner, M., Planck, H., Syré, I., Dittel, K.-K.

9. Treatment of Ligamentous Instabilities of the Knee-Joint
    under the Special Aspect of Prosthetic Ligament Recon-
    struction   111
    Dittel, K.-K., Dauner, M., Planck, H., Syré, I.

10. Histology of Aramide Cords (Kevlar {R}) Used as a
    Cruciate Knee Ligament Substitute in the Sheep   123
    Jerusalem, C.R., Dauner, M., Planck, H., Dittel, K.-K.

11. Mechanical Properties of Various Ligament Prostheses   129
    Dürselen, L., Claes, L.

III. VASCULAR PROSTHESES

12. Dilatability and Stretching Characteristics of Poly-
    ester Arterial Prostheses. Evaluation of the Elastic
    Behaviour   137
    Debille, E., Guidoin, R., Charara, J., Torché, D.,
    Bernier-Cardou, M., Marceau, D., Boyer, D., Chaput, C.,
    Dadgar, L., Cardou, A.

13. Polyurethanes and Their Cytocompatibility for Cell
    Seeding   187
    Gerlach, J., Schauwecker, H.H., Planck, H.

14. Cellular and Cytoskeletal Response of Vascular Cells
    to Mechanical Stimulation   193
    Dartsch, P.C., Betz, E.

15. Perigraft Reaction of Vascular Prosthetic Grafts and
    Therapeutic Management   219
    Pallua, C., Meinecke, H.J., Hepp, W.

IV. SUTURES

16. The Historical Development of Sutures Comparing the
    Manufacturing Process, Handling Characteristics and
    Biocompatibility   223
    Kniepkamp, H.

17. Test Methods for Surgical Sutures   231
    Planck, H., Weber, O., Elser, C., Renardy, M., Mayr, K.,
    Milwich, M.

18. The Effect of Bacteria on the Degradation of Absorbable
    Sutures   243
    Elser, C., Renardy, M., Planck, H.

V. BIOARTIFICIAL ORGANS

19. Bioartificial Endocrine Pancreas. Prerequisites and
    Current Status   247
    Federlin, K., Zekorn, T.

20. Improvement of Insulin Diffusion by Protein Coating
    of Membranes for Use in a Bioartificial Pancreas   253
    Zekorn, T., Siebers, U., Bretzel, R.G., Renardy, M.,
    Planck, H., Federlin, K.

21. Polymeric Membranes for Use in a Bioartificial Diffusion
    Pancreas: Insulin Diffusion in Vitro   259
    Renardy, M., Zschocke, P., Planck, H., Trauter, J.,
    Zekorn, T., Siebers, U., Federlin, K.

22. Morphological and Functional Studies on Implanted
    Diffusion Membranes   265
    Siebers, U., Zekorn, T., Bretzel, R.G., Sturm, R.,
    Planck, H., Renardy, M., Trauter, J., Zschocke, P.,
    Federlin, K.

23. First Experiences on Epithelialisation of an Expanded
    PTFE Tracheal Prosthesis   273
    Gerhardt, H.J., Kaschke, O., Wenzel, M., Böhm, F.,
    Haake, K.

24. Proliferation and Differentiation of Human Ciliated
    Epithelia on PTFE-Prostheses   285
    Böhm, F., Wenzel, M., Gerhardt, H.J.

VI. ARTIFICIAL SKIN AND PATCHES

25. Controlled Microporosity: A Key Design Principle for
    Artificial Skin   293
    Hinrichs, W.L.J., Kuit, J., Feijen, J., Lommen,
    E.J.C.M.P., Wildevuur, Ch.R.H.

26. A New Kind of Collagen Membrane to be Used as Long
    Term Skin Substitute   305
    Hafemann, B., Sauren, B., Hettich, R.

27. The Gore-Tex Surgical Membrane for Temporary Skin
    Closure after Complicated Cardiac Operations   319
    Mestres, C.A., Pomar, J.L., Barriuso, C., Ninot, S.,
    Mulet, J.

28. Pathological Analysis of the Explanted Gore-Tex Surgi-
    cal Membrane after Surgical Implantation   323
    Mestres, C.A., Pomar, J.L.

29. Bioabsorbable Non-Woven Fabric for Surgery   329
    Nakamura, T., Shimizu, Y., Watanabe, S., Shiraki, K.,
    Hyon, S.-H., Suzuki, M., Shimamoto, T., Ikada, Y.

30. Water Vapor and Oxygen Permeabilities of Polyether-
    urethanes   333
    Karakelle, M., Taller, R.A.

ROUNDTABLE DISCUSSION AT THE END OF THE CONFERENCE   343

KEYWORDS   351

AUTHOR INDEX   353

GENERAL ASPECTS IN THE USE OF MEDICAL TEXTILES FOR IMPLANTATION

Planck, H.

Institut für Textil- und Verfahrenstechnik, D-7306 Denkendorf

1. Introduction

Textile structures play a major role in biomedical engineering in general, and especially at devices to be implanted. In the early fifties, the use of textiles for the replacement of blood vessels opened a new area for medical textiles. Voorhes found 1952 /1,2/ in animal tests, that porous woven fabrics were covered by an endothelium-like cell-layer. He concluded, that the ingrowth of fibroblasts coming from the surrounding tissue, would be possible, forming a base for the growth of endothelium at the inner surface of vascular grafts.

Today, textiles have a large scale of applications. The most important applications are listed in table 1.

Table 1: Application of textile structures as implants

Application	Device
- abdominal wall	- patch, zipper
- artery	- patch, prosthesis
- biohybrid organ	- membrane
- bone	- osteosynthese-plate, pin
- dura	- patch
- heart valve	- sewing-ring
- heart wall	- patch
- hipjoint	- acetabulum
- ligament	- augmentation, prosthesis
- tendon	- prosthesis
- trachea	- prosthesis
- vein	- prosthesis
- wound	- suture

Today the main applications are patches for the heart/vessel-system, prostheses for blood vessels as well as ligaments. Very important also is the use of resorbable or nonresorbable braided sutures. All these devices are made of or are containing textile structures.

H. Planck M. Dauner M. Renardy (Eds.)
Medical Textiles for Implantation
© Springer-Verlag Berlin Heidelberg 1990

2. Functions of implanted textiles

Textiles can fulfil different functions as implanted materials /15/:

One function is the replacement of tissue. This function is necessary, when no natural tissue is available, when the new tissue has not the required properties like stability or extensibility, or when the penetration of tissue would not allow the device to function over a long period of implantation. This criterion for instance is valid for vascular grafts with a diamater less than 6 mm, where penetrating tissue reduces the free lumen of the device. According to the law of Hagen-Poiseuille the blood flow is reduced with the 4th power of the reduction of the diameter. Here the used textile structures are woven or knitted fabrics.

In bioartificial organs (blood vessels or skin), polymeric structures were seeded with homologeous resident cells. The immune reaction to the polymeric material can be reduced by the biological surface.

For the replacement of ligaments and tendons, textile structures have to transfer load from bone to bone or from muscle to bone. High strength materials with required elastic extensibility allow a minimal cross-section of the prostheses. For ligament-prostheses braids, woven fabrics or warp knits are used. For tendon replacement, only braids or twisted yarn bundles are applicated.

The other function can be the tissue support, where the textile structure reinforces the new tissue, which penetrates the pores of the textile. This function is required at large calibre vascular grafts, arterial patches, reinforcement of the abdominal wall, as well as for augmentation devices for ligaments. The used structures are mainly warp knitted or woven. Ligament-augmentions are woven fabric or braided structures.

The use of textiles for implantation started with the use of sutures in form of twisted yarns. The function of those auxilary devices is the temporary or longterm fixation of tissue or bone. Today, sutures are braided to get a smooth surface and higher flexibility. Zippers are used, to have a reopenal passage through the abdominal wall for desinfection of the peritoneal cavity after a strong bacterial infection. In such a device, woven fabrics or warp knits are used. Skin-passages for implantable catheters or leads are made from non-wovens or warp knits.

Textile structures are used also for reinforcing polymeric materials, to get higher strength, higher bending stiffness or improved damping properties. For example, we are working on the fiber reinforcement of UHMWPE to get an improved creep resistance /27/. Osteosynthesis-plates are reinforced to get higher strength and higher stiffness. Polymeric plates without fibers have not enough strength. As unidirectional reinforcement fiber-bundles are used or woven fabrics. Resorbable pins for bone fixation consist of braids embedded in a polymer matrix.
Membranes were taken as a base for those biohybrid organs, which consist of polymeric membranes combined with natural cells. The membrane is the basic material, on which cell layers can be developed. Other membranes are used as immunoisolation of foreign cells (hollow fiber membranes or membrane pockets).

In most cases, membranes are not made of textiles directly. They are produced by phase-separation technics /3/, where polymers are solidified in contact with nonsolvents. Flat membranes must be supported by textiles like woven fabrics or non-wovens, to get enough strength.

3. Textile definitions

Before the main differences in the textile structures will be discussed, some textile definitions should be explained.

3.1. Yarn-properties

The thickness of a yarn is defined as the yarn titer. The titer is calculated as weight per length. Dtex means the calculate weight of a yarn of 10.000 m length. For denier, which is still common in the United States, the standard length is 9.000 m.

The classification of a yarn, for example 108 dtex f 54, means, that a 10.000 m yarn weigh 108 grams and that the yarn consists out of 54 filaments. Then the titer t of the individual filament can be calculated as

$$t = \frac{T}{f}$$

where
T = total titer of the yarn
f = number of filaments.

The single titer t in the yarn example above is 2 dtex, a typical value for a fine polyester yarn. The diameter d of one single filament can be calculated from the following equation:

$$d = 2 \cdot 10^{-2} \sqrt{\frac{t}{\pi \cdot \rho}} \quad [mm]$$

where
t = single titer [dtex]
ρ = density of the material [g/cm^3]

The strength of a yarn is tested with tensile tester according to DIN 853.834. To compare the strength of several different materials, a titer-specific strength value cN/tex is calculated by dividing the absolut strength value through the titer of the yarn.

3.2. Yarnproduction

Standard yarns are mostly produced by dry or melt spinning technics /30,31/. High modulus yarns are made by wet or gel-spinning technics.

The strength will be achieved by drawing the yarn under heat. The stress-strain- as well as the shrinkage- behaviour of the yarn is influenced by the total draw-ratio and the heat setting of the yarn.
Drawing processes influence the crystallinity of the polymer in the yarn. The ratio of crystallinity on the other hand is responsible also for the degradation rate of resorbable materials. With a same molecular weight high crystalline yarns show a slower degradation rate in vivo than medium or low crystalline yarns.

The most important properties of the yarn used for implants are listed in table 2.

Table 2: Mechanical Properties of Fibers used in Medical Textiles
/ITV Fasertafel (1986) TPI (in parts)/

	Density [g/cm³]	Linear Tensile Strength (1) [cN/tex]	Elongation to Break [%]	Loop Strength % of (1) [%]	Knot Strength % of (1) [%]	Glass Transition Temperature [°C]
PETP	1.38	70	15	70	50	85
PA 6.6.	1.14	80	25	90	50	90
PP	0.90	48	25	90	-	-10
PVAL	1.26	66	10	70	80	130
Para-Aramid	1.45	170	4	80	25	300
HMPE	0.97	190	3	-	15	-
C-Fiber	1.87	156	1	0	0	-

3.3. Porosity and poresize of textiles

For medical textiles, the porosity and the poresize are very important, because they have a great influence on the in vivo behaviour of a textile implant. The porosity can be defined as penetration rate of a medium, while poresize is a dimension.

Buxton /4, 5/ has standardized the porosity by measuring the permeability of water under constant pressure, through a defined square unit per time. But there are problems related to this method. Surface interactions between solvent and material will influence the results. Measuring a hydrophobic material, water will only pass through large pores. Also swelling of the material will alter the results. This method is therefore not reproducable over longer periods of time. Other methods using air or solvents, are alternatives, but no real solutions to determine this very important property. Also the number of pores per square unit, counted under a microscope, or the relation of measured density to the theoretic density, has been proposed as a definition of the porosity /17/.

The same difficulties exist in defining the pore size. The measurement of the pore size of textile structures shows great problems, because the pores have no ideal round cross section. Even though a round cross section is mostly the base for the calculation of the pore size. The pores of textile structures are influenced by the fiber layers. There are no tubelike pores; it is a complicated capillary system. A direct measurement of the pore dimensions is not possible. One way to determine the pore size is to measure the light or dark areas of a surface under a microscope /17/. In velours surfaces, the velours can alter such a measurement by covering the pores in the third dimension. But at the moment, there is no better simple method known. For small pores, like in a membrane, the poresize can be measured by the bubble point method, which shows a very good reproducability /32/.

4. Textile structures

Textiles are made of fibers. Fibers are in general comparable to the fibrilles in collagenous materials. Some textile stuctures are directly comparable to collagen with regard to fiber orientation as well as to mechanical properties.
All kinds of textile structures are used for medical textiles. The technologies to produce them will be explained and discussed, as well as the main differences between the structures, their properties and their main applications.

4.1. Woven Fabrics

Besides twisted yarn for sutures, first woven fabrics were used for medical applications in the early fifties for arterial prostheses.

A woven fabric has yarns in two perpendicular directions: the warp yarn in the longitudinal and the weft yarn in the transversal direction (Fig. 1). The stress-strain-behaviour of a fabric in warp and weft direction is shown in figure 2. The yarns are trapped in the fabric structure, so that there is only a little increase in fabric elongation, compared with the yarn elongation itself. The yarn stress-strain-curves are also shown (interrupted lines).

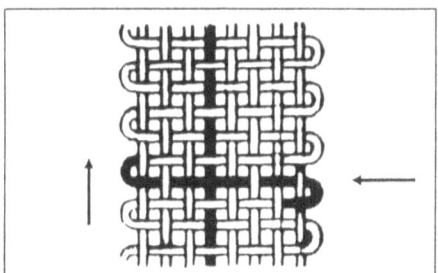

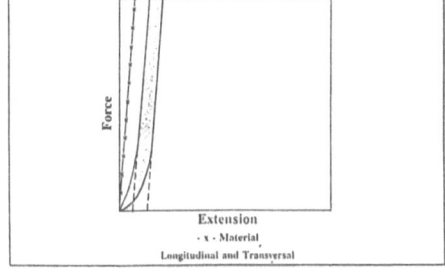

Fig. 1: Structure of Woven Fabric   Fig. 2: Stress-Strain-Behaviour of Woven Fabric

The technic, to produce standard woven fabrics, is simple and well known.

More complicated is the technic to produce woven velours, which allows a better adherence of tissue to the prosthetic wall. The loop yarn, for instance a texturized yarn, is less frequently interlaced to the fabric base than the base yarn. The textured yarn forms the velours loops during shrinkage under heat or solvent if a high shrinkage yarn is used for the base /6,33/.

A woven fabric has a tight structure. Unporous to microporous structures can be produced through pattern, shrinkage of yarn or the weft and warp density. The surface can be smooth as well as rough in form of velours. Bifurcation for vascular grafts can be made by using a jaquard-loom /7 - 9/.

A tight fabric can also be of disadvantage. Woven fabrics are normally stiff and inflexible. They have no extensibility, if no elastic material, like texturized yarn or elastic polymers like polyurethane, are used. The selvedges tend to unravel, when stressed.

Woven fabrics were used for large diameter arterial prostheses near the heart, for ligament-replacement or augmentation, as well as for the support of membranes and other reinforced materials.

4.2. Braids

Braided structures are similar to woven fabrics (Fig. 3).
Here are also two yarn systems in two directions. But in the braiding technology, there is a larger variability in the range of crossing angle between the yarn systems, from about 20 to 170 degrees. The extensibility of a braid is influenced by the pattern of the braid, as well as by the braid angle. The stress-strain-behaviour is shown in figure 4.

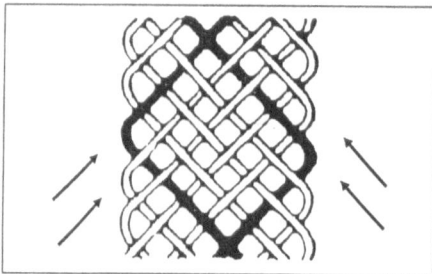
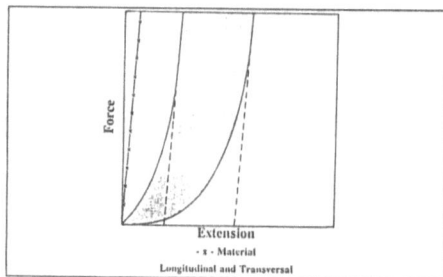

Fig. 3: Structure of Braid   Fig. 4: Stress-Strain-Behaviour of Braid

The technic, to produce braids, is very simple and well known. There are two carrier-systems, one is moved clockwise, the other counter-clockwise. The carriers transport the yarn bobbins in a fixed circular way, by crossing each the other (see Cpt. 8).

The possibility of high load transfer with defined reversible extensibility is of advantage for load bearing devices. The stress-strain-behaviour and the bending stiffness can be varied in a wide range. Thus, braids with for example 9 % elongation at break can be made of high modulus yarns, which have only 3 % elongation at break by themselves /24/.
The only disadvantage of braided structures is the tendency to unravel. Only straight tubes and tapes are produced.

Braids are used for sutures, ligament prostheses and augmentation devices.

4.3. Knitted structures

In textile technology, there are two different ways, to make knitted-structures: Weft knitting and warp knitting. A mesh is formed by the loop head and the loop shanks.

There is one general problem related to all knitted structures: they elongate under stress. The large hysteresis of knitted structures is not only due to the hysteresis of the yarn itself, but mainly to the straightening of the loops and the possiblity of the movement of the yarn relative to the individual loops. The mobility of the yarn increases with increasing porosity, that means with increasing loop length /6/.

4.3.1 Weft knit

In a weft knit, each row is produced by one yarn. Therefore the fiber direction is perpendicular to the fabric direction (Fig. 5).
This has a direct influence on the stress-strain-behaviour, which is demonstrated in figure 6. First, under low stress the fabric can be largely elongated. The deformation of the meshes is nearly reversible. Then the stress will be applied to the yarn itself. Penetrating tissue will reduce this extensibility, will make the knitted structure stiff. This means, that the extensibility in longitudinal direction is also reduced.

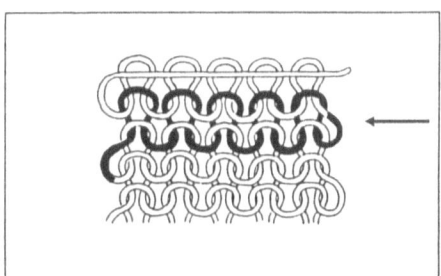

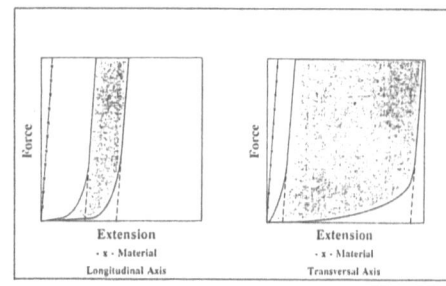

Fig. 5: Structure of Weft Knit   Fig. 6: Stress-Strain-Behaviour of Weft Knit

Each loop is formed by one latch needle from one yarn, one loop after the other. Figure 7 shows a SEM of a typical weft knitted structure in a single jersey pattern, with the loop head and the two shanks.

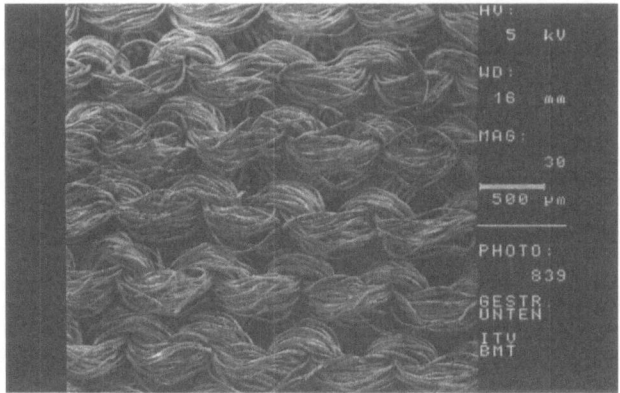

Fig. 7: Structure of Weft Knit, SEM

The advantage of such a knitted structure is the high extensibility and flexibility, if there is no ingrowth of tissue. The porosity and the poresize can be influenced in a wide range. Velours structures can also be achieved /34/.

Bifurcated tubes can be produced by transfer techniques, shifting loops from one needle to another. Such techniques are used to make finger-gloves for instance.

The disadvantage of a knitted structure is the low ladder proof and the deformation of the loops under dynamic load, which leads to a set extension of the structure. Therefore, weft knit structures are not often used for implants.

4.3.2. Warp knits

Warp knitted structures are also made from loops, which are made out of the warp yarn (Fig. 8).
The fibers are in fabric direction, which leads to a different stress-strain-behaviour (Fig. 9).

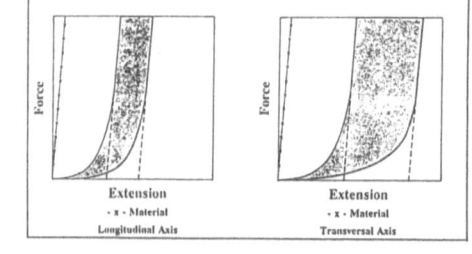

Fig. 8: Structure of Warp Knit   Fig. 9: Stress-Strain-Behaviour of Warp Knit

The properties of a warp knit fabric can be adjusted to the requirements in a wide range. Contrary to the weft knitting process, the latch needles are mounted on bars, and they are moved all at the same time.

The warp yarns are positioned by guide bars to the needles of the raschel machine, and loops are formed interlacing the warp yarns next to each other. Figure 10 shows the SEM of a warp knit fabric.

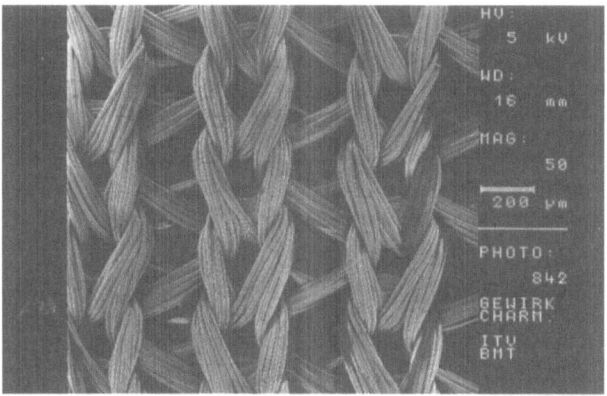

Fig. 10: Warp Knitted Vascular Prosthesis, SEM

Bifurcations are produced by using a raschel-machine with additional guide bars, which forms one large or two small tubes. The advantage of the warp knitting technology is, that one can adjust the extensibility of the fabric, thus getting fabrics with high or medium porosity, with varying poresize. It is also possible to make smooth surfaces or velours structures.

The only disadvantage is the high effort needed, and the sensitivity of this technology. Nevertheless, warp knits were used in a large scale for arterial prostheses, patches, ligaments and skin-passages.

4.4. Non-wovens

Nonwoven structures are very similar to collagen based structures (Fig. 11). They can be isotropic or non-isotropic, depending on the technique used. The possibility to influence the stress-strain-behaviour is low (Fig. 12).

Only non-wovens made from elastomeric materials like polyurethanes have a high extensibility.

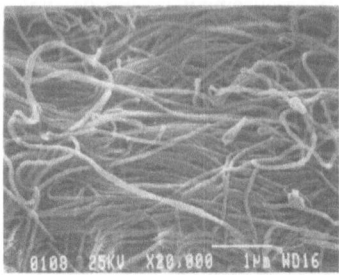
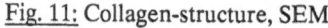

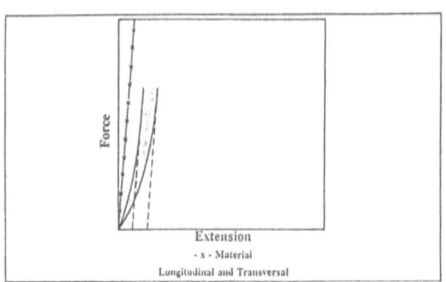

Fig. 11: Collagen-structure, SEM   Fig. 12: Stress-Strain-Behaviour of Non-woven
Structure

We distinguish between dry laid and wet laid non wovens. The common technique to produce
PETP or PTFE non-wovens is the dry laid technique, where the card is used. This non-
wovens are made from staple fibers. If the fibers forming the felt are interlaced by needling, a
non-isotropic structure will be achieved. The fibers will be punched through the felt by the
needles.

For spunlaced non-wovens, air- or water-jets are used to form the structure. The stability is
given by fiber entanglement.

On the other hand, if felts are bonded together by an adhesive, non-wovens with a high fiber
orientation can be produced. For spunbonded non-wovens, multifilament yarns are formed
directly after the extrusion process to a non-woven structure. The still sticky fibers are
bonded at the fiber surfaces by cohesion.

Bifurcations can be made from non-wovens, sewing three tubes together.

In medical textiles, there are two other technics to produce non-woven structures with high
extensibility, made from elastomeric materials like polyurethanes.

In an electrostatic process, a polymer solution is injected to a electrostatic field /28/, and
fibers are formed along the electrostatic fieldlines and transported to a rotating mandrel,
where the fibers are collected. The fiber orientation is very high.

In our labs we have developed a spraying method to produce elastic non-wovens /25-27/. A
polymer solution is sprayed through a nozzle (Fig. 13).

Fibers are formed, which are transported by air to a rotating mandrel, where they are bonded
cohesively together by the remaining solvent. The main fiber orientation can be preset by the
angle of the airstream relative to the axis of the mandrel. Fiber layers with different fiber
orientation can also be achieved. The fiber orientation has an influence on the stress-strain-
behaviour of the non-woven, from which implants can be made.

A typical structure is shown in figure 14.

The advantage of non-wovens is that their structure is similar to the structure of collagen. The
porosity of those structures are depending on the technique used and can be varied in a wide
range, from medium to microporosity. The bending stiffness is depending on the material
used, as well as on the package density.

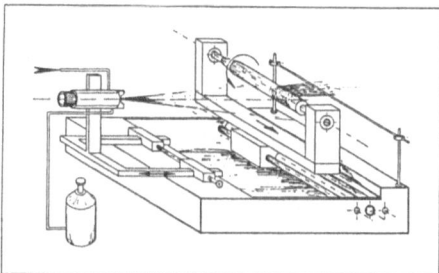
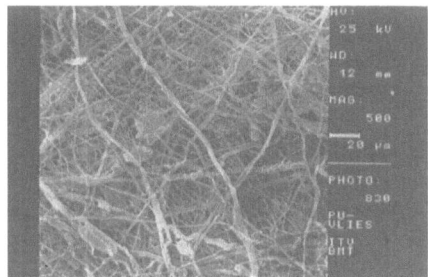

Fig. 13: Spraying Technic for Non-wovens   Fig. 14: Sprayed Non-woven Vascular
                                                      Prosthesis, SEM

Needled non-wovens have a low extensibility, a low flexibility, and, if they are thick, a high bending stiffness. The two other non-wovens don't have these disadvantages, but they are more expensive.

Nonwoven structures are used for small calibre vascular grafts, skin passages and patches as well as for reinforcement of plastic materials and membranes.

Microporous PTFE-membranes have also a fibrillated structure and should therefore be considered as a non-woven structure. Goretex or other microporous PTFE-structures are produced by pie-extrusion-technics /10,12/. Hereby, a dispersion of PTFE-particles and oil is extruded to a tube. After sintering, the tube will be extended. The bonded contact points of the particles form by this stretching a knot-fibril-structure. The length of the fibrils and the poresize depend on the draw ratio /11,12/.

Bifurcations are very difficult to make by this technique.

5. Possible variations

In textile technology, technical parameters can be varied in a wide range, to get required properties like surface structure, stress-strain behaviour, poresize or porosity.

In general, the material used, the fiber properties and the structure of the textile fabric will have a major influence on these parameters. The cross-section of the fiber, for instance round or trilobal, the fabric thickness and the shrinkage are other factors, also the bending stiffness.

To reduce the porosity of textiles, it is common to shrink the material under heat or in contact with solvents. If low-shrink and high-shrink yarns are brought together in one fabric, surfaces like velours can be formed. Very often texturized yarns are used /33/.

In the weaving, the pattern as well as the warp and weft density can be varied to change the porosity. Higher warp and weft density of course will lead to a lower porosity. The warp or weft density can also be varied by changing the titer of the yarn.

The possible modifications in braiding will be discussed later (see Cpt 8).

In the weft and warp knitting, the porosity of the fabrics can be influenced by chosing the fineness of the machine (number of needles per inch). But also the pattern, especially in the warp knitting technology, has an important influence on the required properties. Higher yarn tensions or lower take-off speeds, influence the fabric porosity.

With regard to non-wovens, the fiber orientation has an impact on the stress-strain-behaviour. The porosity is mainly influenced by the fiber thickness and the non-woven density. Normally a higher density is achieved by a higher compacting. With the electrostatic, as well as with the fiber spraying technique, the package density will be influenced by the polymers and solvents used, by the spraying parameters, and by the rotating speed of the mandrel.

6. In vivo behaviour of textile structures

For medical application of textile structures, the in vivo behaviour is of main interest.

Under nonpathological situations the in vivo behaviour depends on the interaction of

>   product   - material-properties,
>   - textile structure,
>   - manufacturing processes,

and

>   body   - environment around the textile structure,
>   - immune reaction of the body.

The surface chemistry and its homogeneity are important with regard to the reaction of the body /13,14,21,22/. Protein adsorption starts immediately after contact of the blood to the polymeric surface. Electrical charges and the hydrophily or hydrophoby of a surface are important for the protein adsorption. Hydrophobic surfaces react more thrombogenic than hydrophilic surfaces, because they adsorb fibrogen and gamma-globulin, while hydrophilic surfaces adsorb anti-thromobogenic albumin /23/. The quantity of adsorbed proteins is also depending on the structure. Smooth surfaces adsorb less proteins than rough surfaces /29,30/. This can be explained by the free surface area of the structure. Therefore, velours surfaces adsorb more proteins than plain fabric surfaces, because of their high specific surface.

The manufacturing process can change the surface properties. Contact to solvents during cleaning can crack the surface or strain the polarity or micro-roughness. Sterilisation procedures with heat or radiation can cause oxidation of the surface /13,14/.

The textile structure /15,18,19/, but also the fiber titer and form as well as the titer of the yarn can have an influence on the adherence or penetration of connective tissue /20/. Tissue will penetrate into open structures. Very fine or heavily twisted yarns give a tight structure, into which no tissue can penetrate. For a good penetration, the poresize should be around 30 μm to 50 μm.

The surface of the textile has an influence on the formation of tissue or cell-layers. A velours causes a stronger tissue formation than a smooth surface /15/.

Material and structure can also lead to an encapsulation of the implant by the connective tissue or its penetration into pores of the textile structure, which is then a base for a good fixation of the implant, as well as for cell-layer formation.
On the other hand, if tissue can grow into pores or penetrate through an implanted textile wall, the stress-strain-behaviour will be changed. Basically, the penetrating connective tissue will reduce the extensibility /29/. The increase in the Youngs modulus and the reduction of the extensibility is depending on the textile structure, the material used and its porosity, but also on the morphology and quantity of penetrating tissue. The same parameters influence the elastic behaviour of the implant.

Penetrating tissue reduces also the extensibility of knitted stuctures by filling the loops and fix them. In braids however, when structure stays under high load, there is no penetration since there is no space between the yarns. On the other hand, if the load is reduced, tissue can penetrate in open spaces under expanding the cross section of the implant, related with a reduction of the implant length. This reduces the extensibility of the braid. How big the change is, depends on the construction. For instance, elongation of braids made of monofilaments with high bending stiffness is in general reversible. If the braids are made from flexible yarns and/or in tight construction (high braid-angle), the penetrating tissue reduces the extensibility, but the elastic elongation and the elastic recovery will be increased.

The site of the implanted textile has also an important influence of the in vivo behaviour. The reaction can be quite different. In a muscle or in the peritoneum, there is a much greater contact to blood or body fluids respectively than in the connective tissue. The body fluids can also have different ions, or enzyme concentrations, which can cause different degradation rates on the polymer. And last but not least, the local stress-situation around an implant, for instance load induced micromotions or flow conditions, can influence the in vivo behaviour of that implant.

The reaction of the organism is generally understood as a foreign body reaction in many different ways. First the protein adsorption, which defines the body reaction as well as the diffusion of proteins: When a textile structure comes in contact with blood, the adsorbed proteins will determine more or less the thrombus formation if in a second stage, fibrin deposition occurs. For the development of a new endothelium, the lysis of the first fibrin layer at the inside of a vascular graft for example is the first step, then a monolayer of endothelial cells has to be built up inside, which is achieved by the ingrowth of fibroblasts through the fabric /1,20/. In the contact with connective tissue, these protein-layers play a major role with regard to adhesion or penetration of tissue, cell spreading or cell adhesion.

Developing textile implants, one has always to consider the fact, that following the operation and because of immune reaction, a more or less strong scar tissue will be formed around the implant. This leads to a more or less high reduction of elasticity of the implant due to the poor elasticity of the scar tissue. The reduction depends on the thickness of this scar tissue, which is influenced mainly by the material (surface chemistry, interface-physics), but also by the surface properties of the textile (roughness, poresize, porosity), and by the stress situation around the implant. High bending stiffness, motion or friction between implant and surrounding tissue lead to thicker scar tissue formation or to bone lysis /15,24/.

Degradation phenomena can also be caused by immune reaction, as body fluids change their composition. Enzymes might react with the synthetic material, or foreign giant cells might cause a superoxidation, which then leads to a degradation. Vascular prostheses have shown postoperative aneurysma, as well as dilatation /6/. In several applications, degradation and calicification were observed.

Some influences, that the properties of the textile structure have on the in vivo behaviour of implants, can be summarized as

 - an influence of the poresize on the patency rate of small-calibre vascular grafts (Fig. 15).
   We have learned from PTFE-vascular grafts, as well as from grafts, we have developed, that an optimum patency rate will be found for structure with poresizes from 30 to 40 μm /16,20/.

 - higher porosity of the textile structure eases the penetration of tissue /20/.
   The more structurized the surface is, it means the more rough, the better is the adhesion of tissue (Fig. 16).

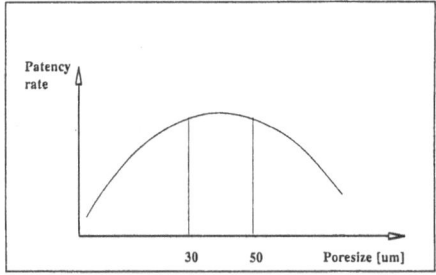

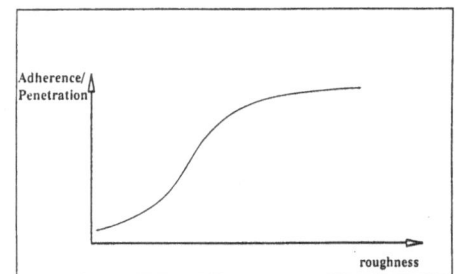

Fig. 15: Poresize, Influence to Patency Rates

Fig. 16: Surface Roughness or Influence to Adherence/Penetration of Tissue

 - smooth surfaces of catheters show no tissue adhesion at all. Velours surfaces improve tissue adhesion. From our own experiments with vascular grafts and also from our cell seeding trials we know, that cells will only grow on structurized surfaces. On totally smooth surfaces or around large pores no cell-formation can be observed.

- the formation of scar tissue depends on the bending stiffness of the implant (Fig. 17). But there is only a tissue increase up to a certain limit. Above that, no significant changes will be found.

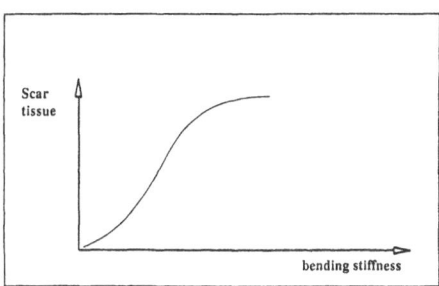

Fig. 17: Influence of Bending Stiffness to Scar Tissue

7. Summary

If one wants to develop a textile implant, it is very important, not only to define the requirements of the product, but also to check in advance on possible reactions of the implant with the human body. Textile structures have to be characterized by structure, material, fiber surface and degradation behaviour. Because of interaction with the body, in vivo the mechanical and/or biological behaviour of the implant might change. This has to be considered constructing the implant. Since there is a large scale of different textile technologies with a lot of modifications available, the required properties of a device can be achieved in most cases. But this needs interdisciplinary team work, work that has to be done jointly by engineers, chemists, textile technologists and physicians, each using his specific know how.

REFERENCES:
/1/ Sauvage, L.: Dacron Arterial Grafts: Comparative Structures and Basis for Successful Use of Current Prostheses
in: Kambic et al: Vascular Graft Update
ASTM STP 898 (1986), 16 - 24
/2/ Voorhees, A. et al:
Annals of Surgery, 135 (1952) 3, 332 ff
/3/ Strathmann, H.: Trennung von molekularen Mischungen mit Hilfe synthetischer Membranen
Steinkopf-Verlag, Darmstadt (1979)
/4/ Buxton, B. et al.: Practical Considerations in Fabric Vascular Grafts
Am. J. Surg. 125 (1973), 288 ff
/5/ Guidoin, R. et al: Textile Arterial Prostheses: Is Water Permeability Equivalent to Porosity?
J. Biomed. Mat. Res. 21 (1987), 65 - 87

/6/ Harrision, P.W.: Vascular Grafts: Textile Structures and Their Performance
The Textile Institute: Textile Progress Vol. 15 (1986) 3, 1-33

/7/ USP 2.845.959 (1958)

/8/ USP 2.978.787 (1961)

/9/ USP 3.096.560 (1963)

/10/ Lenz, J. et al: Neuartige Methoden zur Herstellung von Polytetrafluorethylen-Fasern und -Filamenten, deren Eigenschaften und Anwendung
Chemiefaser / Textilindustrie (1978), 50 ff

/11/ USP 3 953 566

/12/ DOS 27 02 513 (1977)

/13/ Planck, H. et al.: Polyurethanes in Biomedical Engineering I
Elsevier Publisher (1984)

/14/ Planck, H. et al.: Polyurethanes in Biomedical Engineering II
Elsevier Publisher (1987)

/15/ Planck, H.: Kunststoffe und Elastomere in der Medizin
Kohlhammer-Verlag Stuttgart (in Druck)

/16/ Body, K.L. et al: The effects of pore size and endothelial cell seeding upon the performance of small diameter e-PTFE- vascular grafts under controlled flow conditions
J. Biomed. Mat. Res. 22 (1988), 163 - 177

/17/ Pourdeyhimi, B.: Porosity of surgical mesh fabrics: New technology
J. Biomed. Mat. Res. 23 A1 (1989), 145 - 152

/18/ Wolf, P. et al: In Vitro Thrombogenicity Test of Materials Used in Arterial Prostheses
Haemostasis 13 (1983), 113 - 118

/19/ Didisheim, P. et al: Relative Role of Surface Chemistry and Surface Texture in Blood-Material Interactions
Trans. Am. Soc. Artif. Intern. Organs (1983), 169 - 176

/20/ Hoffman, A.S.: Medical Application of Polymeric Fibers
J. Appl. Polym. Science.: Appl. Polym. Symp. 31 (1977),
313 - 334

/21/ Hastings, G.W. et al.: Macromolecular Biomaterials
CRC-Press, Boca Raton (1983)

/22/ Sevastianov, V.I. et al.:
Trans. 3rd World Biomat Congr. Kyoto (1988), 91

/23/ Lelah, M.D. et al.: Polyurethane in Medicine
CRC Press (1986)

/24/ "Entwicklung von Bandprothesen als Kreuzbandersatz"
Institut für Textil- und Verfahrenstechnik Denkendorf
Denkendorfer Forschungsberichte (1989)

/25/ Planck, H.: Entwicklung einer textilen Arterienprothese mit faserförmiger Struktur
Dissertation Universität Stuttgart (1980)

/26/ DE 28 06 030

/27/ DE 28 57 925

/28/ DOS 25 34 935

/29/ Dauner, M.; Planck, H.: Publication in preparation

/30/ Mayer, M.: Faserextrusion
in: Handbuch der Kunststoffextrusionstechnik II
Hanser-Verlag München (1987)

/31/ Lückert, H.: Fibers: 3. General Production Technology
in: Ullmann's Encyclopedia of Industrial Chemistry
VCH Verlagsgesellschaft (1987)

/32/ Coulter Electronics GmbH: Nachschlagehandbuch für das Coulter Porometer
Coulter Electronics GmbH, Krefeld (1983)

/33/ Hoffman, H.L.: Fabrication and Testing of Polyester Artificial Grafts
in: Kambic et al: Vascular Graft Update
ASTM STP 898 (1986), 71 - 81

/34/ DOS 24 61 370 (1974)

IN VITRO EVALUATION OF THE EFFECTS
OF MACROPHAGES AND SYNTHETIC MATERIALS
ON THE FIBROBLAST PROLIFERATION AND
COLLAGEN SYNTHESIS

Jeong OK Lim , C. C. Chu, Max Appel*

Department of Textiles & Apparel, Martha Van Rensselaer Hall,
Cornell University, Ithaca, NY 14853-4401

I. INTRODUCTION

The essential role of macrophages in the biological defense system has been
well illustrated in many studies using foreign materials of either biological or
synthetic nature.[1-5] The presence of synthetic foreign materials whether because
of necessity like modern surgical devices for the reconstruction of injured or
diseased tissues/organs (e.g., vascular grafts, surgical mesh fabrics, artificial
knee joint, etc.), or because of incidences like silica asbestos particles,
eventually lead to a chronic inflammatory response followed by fibrous capsule
formation around the synthetic foreign materials. A recent study of the ultrastructure of large accumulated cells immediately adjacent to synthetic foreign
materials (i.e., a biodegradable wound closure material) indicated that
macrophages played a major role in the phagocytosis of wound debris, and
subsequent degradation, invasion, and phagocytosis of the synthetic foreign
material.[6]

There is a close relationship between macrophage functions and the
biological elimination and defense mechanisms against synthetic foreign
materials. A series of in vitro and in vivo studies of the interface between
macrophages and foreign materials indicated that various types of foreign
materials ranging from synthetic polymers like nylon, polyethylene to glass and
inorganic and organic powders resulted in different extents of macrophage
adherence; upon adherence, macrophages would be activated to participate in many
functions ranging from fibrinolytic activity to the release of growth factors.[7-14]

Upon stimulation by the presence of foreign materials, macrophages produce
many factors which regulate cell-cell interaction and cell-mediated interactions
for modulating inflammatory and immunological response of the tissue. The
replication of many types of cells is mediated by macrophage's release of growth
factors. Examples are fibroblasts, vascular smooth muscle cells, lymphocytes
participating in an immune reaction, and endothelial cells.[15,16] Among the many
mediators released by activated macrophages, fibroblast growth regulatory
molecules are considered to be the most important in the tissue fibrosis

H. Planck M. Dauner M. Renardy (Eds.)
Medical Textiles for Implantation
© Springer-Verlag Berlin Heidelberg 1990

formation around the synthetic foreign materials.[1,4,15] Interleukin 1 (IL-1), released by macrophages, has been identified as an important mediator to regulate fibroblast growth and proliferation,[15,17] and the activation of T lymphocytes.[18] Interleukin 1 can not only induce the collagen synthesis of fibroblasts by stimulating fibroblast activity but also reduce collagen content by stimulating fibroblast collagenase production.[19] These findings and the reported feedback inhibition of immunological activity by prostaglandins and glucocorticosteroids following IL-1 release[20-22] illustrate the complexity of the role of the macrophage in biological defense and elimination mechanism.

Most of the studies of macrophage-related tissue fibrosis have been focused on pulmonary fibrosis, an interstitial lung disorder induced by silica or asbestos foreign particles and characterized by an irreversible progressive fibrotic thickening of alveolar walls.[23-25] However, the use of synthetic biomaterials to examine macrophage's fibrogenic activity has received relatively little attention,[26-27] even though synthetic biomaterials have been used for the repair of injured or diseased tissues for the last few decades. Indirect evidences observed in vivo have shown that different types of synthetic biomaterials elicit different degrees of fibrous tissue formation. Whether the differing extent of fibrous tissue formation is due to differential stimulation of the macrophage fibrogenic activity by the various synthetic biomaterials is not fully known.

The objective of this study was to evaluate the effect of 13 synthetic materials differing in chemical structure and geometrical configuration on the macrophage fibrogenic activity.

II. MATERIALS AND METHODS

A. Synthetic Substrates

Thirteen polymers and fabrics were chosen for this study, and they differed in chemical constituents and geometrical configuration. As shown in Table 1, the chemical composition of the testing substrates included Polyethylene terephthalate, Polyolefins, Polysulfone, Polystyrene (tissue culture plate), Silicone, and Silver; while the geometrical configuration of the substrates ranged from films, rubber, to woven, knitted, non-woven, and felt fabrics. These synthetic materials were cut into circular discs of 16mm diameter. The film and some non-woven samples were treated with CCl_4 extraction to remove any residual chemicals which may be harmful to cells. There was no need to clean the fabric samples because they were originally used as surgical fabrics.

All synthetic substrates were characterized in terms of weight per cm^2, thickness, density, and construction pattern in the case of fabrics. Their surface morphology was characterized by Jeol 35CF scanning electron microscope

and surface area by Quantsorb surface area analyzer model QS-9.[28] The total
surface area of a sample, S_t, was determined by Eq. 1.

$$S_t = X_m N A/M \qquad (1)$$

where X_m = Weight of monolayer adsorbate (N_2 gas)

N = Avogadro number (6.023×10^{23})

A = Cross sectional area of adsorbate molecule
($16.2 \times 10^{-20} m^2$ for N_2 gas)

M = adsorbate molecular weight (28 for N_2 gas)

The specific surface area was determined by Eq. 2.

$$S = S_t/\text{Weight of sample} \qquad (2)$$

The synthetic substrates were sterilized by ethylene oxide before use.

B. Biological Materials

1. <u>Macrophage Suspensions</u>: Alveolar macrophages were obtained from the
lung tissues of pethoges-free Desgle dogs of 1-2 months old. The dogs' lungs
were minced and stirred in McCoy-L15 cell medium (Gibco Labs, Grand Island, NY)
at room temperature for one hour to mechanically separate macrophages from lung
tissues. The solution was filtered through a cheesecloth, centrifugated at 1200
rpm for ten minutes, and resuspended in McCoy L-15 medium. The centrifugation
and resuspension were repeated twice. The final pellets were resuspended in
McCoy L-15 supplemented with 10% Fetal Bovine Serum (FBS) to make a final
macrophage suspension of 10^6/ml.

2. <u>Fibroblast Suspension</u>: Fresh fibroblasts were obtained from the
remaining dogs' lung pieces after the removal of macrophages. They were
trypsinized and filtered through a cheesecloth. The filtered solution was
centrifugated and resuspended in McCoy L-15 cell medium three times. At the end
of the third centrifugation, the pellets were suspended in McCoy L-15
supplemented with 10% FBS. The resulting fibroblast suspension had $2-2.5 \times 10^5$
cells/ml.

3. <u>Artificial Stimulation of Macrophages</u>: The macrophage suspensions
obtained in II.B.1. were used to culture onto different synthetic substrates for
the purpose of activating macrophages to release soluble fibroblast growth
factor.

The sterilized synthetic substrates were placed on the bottom of each well
(Corning plastic 24-well tissue culture plates) filled with 0.5 ml of McCoy-L15
cell medium supplemented with 10% FBS. One ml of the macrophage suspension
prepared in II.B.1. was overlaid onto the synthetic substrate. Three types of
control groups were used and their compositions are given in Table 2. A
comparison between the sample and control One groups would reveal the ability of
synthetic substrates in general to activate macrophage's fibrogenic activity. A

comparison between the sample and control Two groups would indicate the extent that these synthetic substrates affected macrophage's activation relative to tissue culture plates. A comparison between control Two and Three groups would show the ability of a tissue culture plate to activate macrophage fibrogenic activity.

All sample and control tissue culture plates were incubated in 5% CO_2 at 36°C for two hours. This duration of incubation was chosen to achieve the optimal release of fibroblast growth factor (FGF) by macrophages. Lemaire et al. reported that the maximum release of FGF by alveolar macrophages was observed within one hour of contact with substrates,[25] while Bitterman et al. reported that a maximum release of FGF by macrophages occurred between three and four hours of incubation with substrates.[1]

At the end of two hours incubation, the supernatants were collected, centrifuged (to remove debris), and were ready for transferring onto fibroblast monolayer cell culture as described below.

4. Transferring of Fibroblast Growth Factor onto Fibroblast Monolayer: Fibroblast suspensions prepared in II.B.2. were cultured on tissue culture multiwell plates without synthetic substrates for four days at 36°C in 5% CO_2. At the end of the fourth day, the supernatant medium of the fibroblast monolayer was removed and 0.1ml of the centrifuged supernatant from II.B.3. along with 1ml of either fresh McCoy L-15 cell medium supplemented with 1% FBS or [3H]-thymidine cell medium, depending on what was to be measured, were introduced and incubated for two days in 5% CO_2 at 36°C. The macrophage's fibrogenic activity on fibroblasts was evaluated by the measurement of fibroblast proliferation and their collagen synthesis as described below.

C. Determination of Fibroblast Proliferation

The extent of fibroblasts DNA synthesis was used to indicate the magnitude of their proliferation. [3H] - thymidine was obtained from duPont with 6.7 Ci/mmole specific activity, and 0.15 ml of this isotope mixed with 50ml of McCoy L-15 cell medium supplemented with 1% FBS. The final solution had 0.5 μCi/ml. One ml of this isotope solution was overlaid onto fibroblast monolayer before the introduction of 0.1 ml of the activated macrophage supernatant. The final volume of each well was 1.1 ml. The culture plates were incubated for two days, at 30°C, 5% CO_2. At the end of two days, the supernatants were removed, and the plates were rinsed with McCoy L-15 medium twice followed by 2% SDS to detach adhered fibroblasts. 0.1 ml of the detached fibroblast suspension was collected in a glass vial, mixed with 10 ml of ready-to-use liquid scintillation counting coctail (Kodak), and counted by Beckman/LS 9800 Series Liquid Scintillation System.

D. Determination of Hydroxyproline (HOP) Content

Regular 1 ml McCoy L-15 cell medium supplemented with 1% FBS was overlaid
onto fibroblast monolayer followed by 0.1 ml of the activated macrophage
supernatant to make a final volume of 1.1 ml. The fibroblast culture plates were
then incubated for two days at 5% CO_2, 36°C. At the end of the second day,
supernatants of the culture plates were collected and centrifuged to remove solid
debris. The resulting supernatants were used to determine the amount of collagen
synthesized by fibroblasts. A standard calibration curve of absorbance (at 560
mn) vs HOP concentration was constructed by using HOP standard stock solutions.
The HOP contents of the testing substrates were calculated from the standard
calibration curve. The detailed procedures for HOP determination have been given
elsewhere.[29,30]

E. Gross Morphological Observation

The synthetic substrates cultured with macrophages for two hours were
removed from the wells, rinsed twice with phosphate buffer, fixed with methanol,
and stained with Giemsa Trypan Blue. Photographs of the stained substrates were
taken.

F. Statistical Analysis

The general dependence of fibroblast DNA synthesis and HOP contents (as
dependent variables) on macrophage and material factors (as independent
variables) was examined by two-way analysis of variance. Student t-test was used
to determine whether the differences in DNA synthesis or HOP contents between
macrophage treated and untreated synthetic substrates were statistically
significant. The significance level chosen was 0.05 and the significance
probability, p-value, was calculated to determine the plausibility of the
hypothesis. The difference in dependent variables between the sample and control
groups was considered significant when $p < 0.05$ and very significant when $p <
0.01$.

III. RESULTS AND DISCUSSION

A. Characteristics of Synthetic Substrates

As shown in Table 3 and Figure 1, a wide range of material characteristics
of the synthetic substrates was observed. Among these characteristics, surface
area and morphology were suspected to have a greater influence on macrophages'
response. The 13 synthetic substrates could be classified into three groups, as
shown in Table 4, according to their surface area. Films and silicone rubber had
the smallest surface area (less than 100 cm^2), followed by woven and knitted
fabrics, silver-coated nylon non-woven, spun-bonded polyester; while the two felt
fabrics, Tyvek and Sauvage Filamentous Dacron fabrics, had the highest surface
area. Among the fabric substrates, silver-coated nylon fabric had the lowest

surface area. These differences in surface area of the substrates were also evident in their SEM micrographs as shown in Figure 1. Those synthetic substrates with smooth and nonporous morphology were also those with low surface area.

B. Fibroblast Proliferation

The results of fibroblast proliferation in terms of [^{3}H]-thymidine labeled DNA synthesis are given in Table 5. A two-way analysis of variance (ANOVA) was performed for examining whether there was any statistically significant difference in fibroblast proliferation between the presence and absence of macrophages or between different substrates, and the data were given in Table 6. The Table demonstrates that there was a very strong, significant difference ($p <$ 0.00001) in the degree of fibroblast proliferation due to either the presence of macrophages or different synthetic substrates.

The magnitudes of macrophage induced fibroblast proliferation were found to closely depend on the chemical structure and geometrical configuration of the synthetic substrates. As shown in Figure 2, synthetic substrates of fabric configuration, in general, activated macrophage fibrogenic activity more than the substrates of film configuration. Among the fabric substrates, silver-coated non-woven nylon fabric showed the highest macrophage induced fibroblast proliferation followed by woven Dacron, Tyvek, spunbonded polyester, knitted Dacron, Teflon felt, Sauvage filamentous Dacron, Dacron felt, and polypropylene mesh.

Among the four film substrates having very similar surface area (about 47-50 cm^2) and morphology (Polyester, Polysulfone, Polyethylene, and tissue culture plates), PET appeared to be the most active toward macrophage fibrogenic activity followed by Polysulfone, Polyethylene, and Polystyrene (tissue culture plates). There was virtually no difference in fibroblast DNA synthesis between macrophage treated and untreated tissue culture plates which suggests that, for reasons not clear at the present time, Polystyrene does not activate macrophage fibrogenic activity under the testing condition.

A statistical analysis of the difference in fibroblast proliferation between each macrophage treated and untreated substrate revealed that 6 out of 13 tested substrates (silver-coated non-woven nylon, woven Dacron, knitted Dacron, Teflon felt, Tyvek, and spunbonded polyester), showed statistically significant increases in fibroblast proliferation at $p < 0.05$.

Silicone rubber was the only substrate among the 14 tested materials (including tissue culture plates) to exhibit macrophage inhibitory effect on fibroblast proliferation. The reason is not clear at the present time. Similar inhibitory effect was reported by Schmidt et al. _in vitro_ culture of human peripheral blood derived monocytes with diamond dust of 1-2 μm particle size at the concentrations greater that 50 μg/ml.[17] Diamond dust of less than 0.5 μm, however, showed a 1-3 fold increase in fibroblast proliferation activity.

In the absence of macrophages (controls 1 and 3), the effect of synthetic substrates on fibroblast proliferation activity was much less pronounced as evident in Figure 3, except a few materials (i.e., silicone rubber, PET film, Sauvage filamentous Dacron, and Tyvek), most of them exhibited similar fibroblast proliferation activities as tissue culture plates. Teflon felt and Polypropylene mesh were the only two substrates that showed lower fibroblast proliferation activities than the tissue culture plate.

C. Hydroxyproline Contents

A two-way analysis of variance (Table 7) indicated that HOP contents were very significantly different between the presence and absence of macrophages at p < 0.00001. Similarly, HOP contents were significantly different among the synthetic substrates at p < 0.0001. However, there was no significant interaction between macrophage presence and materials. This suggests that the difference in HOP contents due to macrophage factor bore no relationship to the difference in the HOP contents due to the nature of materials.

Figure 4 illustrates the differences in HOP contents of each synthetic substrate due to macrophage factor. Except a few exceptions, fabric substrates showed higher macrophage induced HOP contents than film substrates. Among the 14 substrates tested (including tissue culture plates), silver-coated nonwoven nylon fabric exhibited the highest macrophage induced HOP contents followed by woven Dacron, knitted Dacron, Tyvek, Teflon felt, and film substrates. Polypropylene mesh, Dacron felt, spunbonded polyester and silicone rubber, however, had the lowest macrophage induced HOP contents among the 13 synthetic substrates, but their HOP contents were still higher than the tissue culture plates.

A statistical analysis of the macrophage induced HOP contents of each synthetic substrate at p < 0.05 indicated that silver-coated nonwoven nylon and Tyvek showed the most significant increases in HOP contents when treated with macrophages than the corresponding untreated ones. PET and Polysulfone films, woven Dacron, Sauvage filamentous Dacron, Teflon and Dacron felts, and spunbonded polyester also showed statistically significant increases in HOP content when treated with macrophages. Only polyethylene film, silicone rubber, knitted Dacron, and Polypropylene mesh did not have statistically significant macrophage induced HOP contents at p < 0.05.

D. Correlation of Fibroblast Proliferation and Collagen Synthesis

It is a common belief that an increase in fibroblast proliferation would result in a corresponding increase in HOP content. This relationship was examined in this study.

As shown in Figure 5, a correlation coefficient of 0.83 was found between the macrophage induced fibroblast DNA synthesis and HOP contents of the 13 synthetic substrates. This suggests that if a synthetic substrate is able to activate macrophages to promote subsequent fibroblast proliferation, there is a very good probability (83%) that this same substrate can also activate macrophages to promote fibroblast collagen synthesis. This correlation between fibroblast DNA synthesis and HOP contents was further strengthened when only four synthetic substrates which exhibited statistically significant increase in both DNA synthesis and HOP contents (silver-coated nonwoven nylon, woven Dacron, Tyvek, and spunbonded polyester) were used. The correlation coefficient reached to 0.93 as shown in Figure 6. This supports the findings that if there is more fibroblast proliferation due to macrophages, there is correspondingly more collagen synthesis.

IV. REFERENCES

1 Bitterman PB, Rennard SK, Hunninghake GW, Crystal RG (1982) Human alveolar macrophage growth factor for fibroblasts Regulation and partial characterization, J Clin Invest, 70(4):806-22

2 Wahl SM, Wahl LM, McCarthy JB, Chedid L (1979) Macrophage activation by mycobacterial water soluble compounds and synthetic muramyl dipeptide, J Immunol, 122(6):2226-31

3 Heppleston AG, Styles JA (1967) Activity of a macrophage factor in collagen formation by silica, Nature, 214:521-522

4 Anderson JM, Miller KM (1984) Biomaterial biocompatibility and the macrophage, Biomaterials, 5(1):5-10

5 Van Furth R (1975) Mononuclear Phagocytes in Immunity, Infection, and Pathology, Blackwell Sci Publ, Oxford

6 Matlaya BJ, Salthouse TN (1983) Ultrastructural observations of cells at the interface of a biodegradable polymer: Polyglactin 910, J Biomed Mater Res, 17(1):185-97

7 Miller K, Handfield RI, Kagan E (1978) The effect of different mineral dusts on the mechanism of phagocytosis: A scanning electron microscope study, Environ Res, 15, 139-154

8 Leake ES, Wright MJ (1979) Variations on the form of attachment of rabbit alveolar macrophages to various substrata as observed by scanning electron microscopy, J Reticuloendothel Soc, 25, 417-441

9 Rasp FL, Clawson CC, Hoidal JR, and Repine JE (1979) Quantitation and scanning electron microscopic comparison of human alveolar macrophage and polymorphonuclear leukocyte adherence to nylon fibers in vitro, J Reticuloendothel Soc, 25, 101-109

10 Berhard B, Manolescu N, Simonescu N, Ciocnitu V (1983) Changes concerning the external structure of peritoneal macrophages, due to the DQ-12 standard dust," Environ Res, 31, 256-265

11 Leake ES, Wright MJ, Myrvik QN (1975) Differences in surface morphology of alveolar macrophages attached to glass and to millipore filters: A scanning electron microscope study, J Reticuloendothel Soc, 17, 370-379

12 Spilizewski KL, Marchant RE, Anderson JM, Hiltner A (1987) In vivo leukocyte interactions with the NHLBI-DTB primary reference materials: polyethylene and silica-free polydimethylsiloxane, Biomaterials, 8, 12-17

13 Sipe JD (1985) Interleukin 1 as a key factor in the acute-phase response, in The Acute-Phase Response to Injury and Infection, Chapter 2, Gordon and Koj (Eds), Elsevier Science Publishers, The Netherlands, 23-35

14 Dinarello CA (1984) Interleukin 1, Rev Infect Dis, 6, 51-95

15 Leibovich SJ, Ross RA (1976) A macrophage dependent factor that stimulates the proliferation of fibroblasts in vitro, Am J Pathol, 84(3):501-14

16 Martin BM, Gimbrone MA, Unanue ER, Cotran RS (1981) Stimulation of nonlymphoid mesenchymal cell proliferation by a macrophage-derived growth factor, J Immunol, 126(4):1510-5

17 Schmidt JA, Mizel SB, Cohen D, Green I (1982) Interleukin 1, a potential regulator of fibroblast proliferation, J Immunol, 128(5):2177-82

18 Klaus GGB (1987) <u>Lymphocytes - A Practical Approach</u>, IRL Press, Washington, DC, pp 211-212

19 Postlethwaite AE, Lachman LB, Mainardi CL, Kang AH (1983) Interleukin 1 stimulation of collagenase production by cultured fibroblasts, J Exp Med, 157(2):801-6

20 Besedovsky HO, del Ray A, Sorkin E, Dinarello CA (1986) Immunoregulatory Feedback Between Interleukin-1 and Glucocorticoid Hormones, Science <u>233</u>:652-654

21 Gery I (1982) Production and Assay of Interleukin-1 (IL-1), In: Isolation, Characterization, and Utilization of T Lymphocyte Clones, Academic Press, p 41

22 Kenney JS, Masada MP, Eugui EM, Delustro BM, Mulkins MA, Allison AC (1987) Monoclonal Antibodies to Human Recombinant Interleukin 1 (IL 1): Quantitation of IL 1 and Inhibition of Biological Activity, J Immunol, <u>138</u>:4236

23 Spencer H (1977) Pathology of the Lung, W.B. Sanders, Philadelphia

24 Bitterman PB, Adelberg S, Crystal RG (1983) Mechanisms of pulmonary fibrosis spontaneous release of the alveolar macrophage derived growth factor in the interstitial lung disorders, J Clin Invest, 72(5):1801-13

25 Lemaire I, Beaudoin H, Masse S, Grondin C (1986) Alveolar macrophage stimulation of lung fibroblast growth in asbestos - included pulmonary fibrosis, Am J Pathol, 122(2):205-11

26 Miller KM, Anderson JM (1988) Human monocyte/macrophage activation and Interleukin 1 generation by biomedical polymers, J Biomed Mater Res, 22:713

27 Stea S, Ciapetti G, Pizzoferrato A (1984) The release of macrophage-dependent fibroblast stimulating activity (MFS) from macrophages treated with powdered biomaterials, Biointeractions - Materials/Interactions, City University, London, January 4-6, p 11

28 Gregg S, Sing K (1967) <u>Adsorption, Surface Area and Porosity</u>, Academic Press, New York

29 Bergman I, Loxley R (1963) Two improved and simplified methods for the spectrophotometric determination of hydroxyproline, Analytic Chem, 35(12):1961-65

30 Switzer BR, Summer GK (1971) Improved method for hydroxyproline analysis in tissue hydrolyzates, Analytical Biochem, 39:487

Table 1

Classification of Testing Materials according to Chemical Structure

Chemical Structure	Materials
Polyester (PET) (Polyethylene terephthalate)	-Mylar^R 400D polyester film -Reemay^R spunbonded polyester -USCI, DeBakey^R Dacron woven fabric -Meadox Cooley Graft^R Dacron knitted fabric -USCI Sauvage^{TM} filamentous Dacron fabric -Meadox Dacron felt
Polysulfone	-UDEL^R polysulfone film
Olefin	-Marlex^R polyethylene film -Tyvek^R spunbonded olefin -Marlex^R polypropylene mesh -Meadox Teflon^R (polytetrafluoroethylene) felt
Silicone	-Silastic^R elastomer
Polyamide	-Silver-coated nonwoven nylon

Table 2

Components of Sample and Control Groups

Classification	Components
Sample	Substrate (#1 to #13) + Macrophage suspension (1ml) + Cell Medium (0.5 ml)
Control 1 Group	Substrate (#1 to #13) + Cell medium (1.5 ml)
Control 2	Macrophage suspension (1 ml) + Cell medium (0.5 ml) on Tissue Culture Plate
Control 3	Cell medium (1.5 ml) on Tissue Culture Plate

Table 3

Characterization of Testing Materials

# MATERIALS	WEIGHT g/cm²	THICKNESS cm	DENSITY g/cm³	WARP #/cm	WEFT #/cm	SURFACE cm²/disc	AREA cm²/T.S.S*
1. polyester film	0.012	0.009	1.33			79.1	39
2. polysulfone film	0.031	0.03	1.03			200	96
3. polyethylene film	0.009	0.0093	0.964			81	40
4. Silicone rubber	0.018	0.03	0.6			86.6	41.7
5. Tyvek	0.014	0.016	0.875			407	200
6. Spunbonded Polyester	0.008	0.023	0.347			169	100.2
7. Silver-Coated Nylon	0.015	0.043	0.35			218.5	99.1
8. Woven Dacron fabric	0.018	0.025	0.72	26	37	351	170
9. Knitted Dacron fabric	0.030	0.043	0.69	20	29	542	258.1
10. Filamentous Dacron fabric	0.022	0.07	0.31			296.3	136.1
11. polypropylene mesh	0.011	0.05	0.22				
12. Dacron felt	0.036	0.13	0.27			1463	629
13. Teflon felt	0.15	0.16	0.94			1992	830

*T.S.S (Top surface of substrate)
The top surface area of substrate was calculated subtracting the lower surface of disc and circumference area (which are considered not to react with cells especially macrophages during experiment) from the total surface area of substrate disc.

Table 4

Classification of Materials According to Surface Area

Group 1 (< 100cm²)	Group 2 (250 to 350 cm²)	Group 3 (400 to 860 cm²)
Polysulfone film	Woven Dacron fabric	Teflon felt
Silicone rubber	Knitted Dacron fabric	Dacron felt
Polyethylene film	Spunbonded Polyester	Tyvek
Polyester film	Silver Coated Nylon	Filamentous Dacron

Table 5

Fibroblast DNA Synthesis

	Sample Group[a] (CPM/Well)	Control 1 Group[b] (CPM/Well)
1. Polyester film	16860 ± 930	13810 ± 1460
2. Polysulfone film	14625 ± 1325	12335 ± 1030
3. Polyethylene film	12085 ± 445	11350 ± 950
4. Tyvek	17655 ± 365	13060 ± 300
5. Silicone rubber	13345 ± 1005	15580 ± 660
6. Woven Dacron fabric	17695 ± 365	10870 ± 685
7. Knitted Dacron fabric	16935 ± 675	12445 ± 1010
8. Filamentous Dacron fabric	16165 ± 1585	13275 ± 400
9. Silver-coated nylon	18660 ± 960	12045 ± 1270
10. Polypropylene mesh	10340 ± 895	8485 ± 265
11. Dacron felt	13980 ± 845	11490 ± 1420
12. Teflon felt	15785 ± 780	9630 ± 1880
13. Spunbonded polyester	16950 ± 1430	11300 ± 1300
Control 2		11025 ± 1000
Control 3		11065 ± 840

Final serum concentration in both sample and control groups: 1%

a. Macrophage suspension + cell medium on testing substrates

b. Control 1 - Cell medium on testing substrates
   Control 2 - Macrophage suspension + cell medium on tissue culture plates
   Control 3 - Cell medium on tissue culture plates

Table 6

Summary of Two-way ANOVA of Fibroblast DNA Synthesis

SOURCE	DEGREE OF FREEDOM	SUM OF SQUARES	MEAN SQUARES	F	P
MATERIAL	13	252320656	19409282	5.211	P<0.00001**
MACROPHAGE PRESENCE	1	222423920	222423920	59.722	P<0.00001**
INTERACTION	13	135578032	10429079	2.800	P<0.0227*

*Significant
**Strongly significant

Table 7

Summary of Two-way ANOVA of Fibroblast Hydroxyproline Content

SOURCE	DEGREE OF FREEDOM	SUM OF SQUARE	MEAN SQUARE	F	P
Material	13	74.53	5.73	4.34	P<0.0001**
Macrophage Presence	1	143.52	143.52	108.72	P<0.00001**
Interaction	13	14.69	1.13	0.85	P<0.6074

**Strongly significant

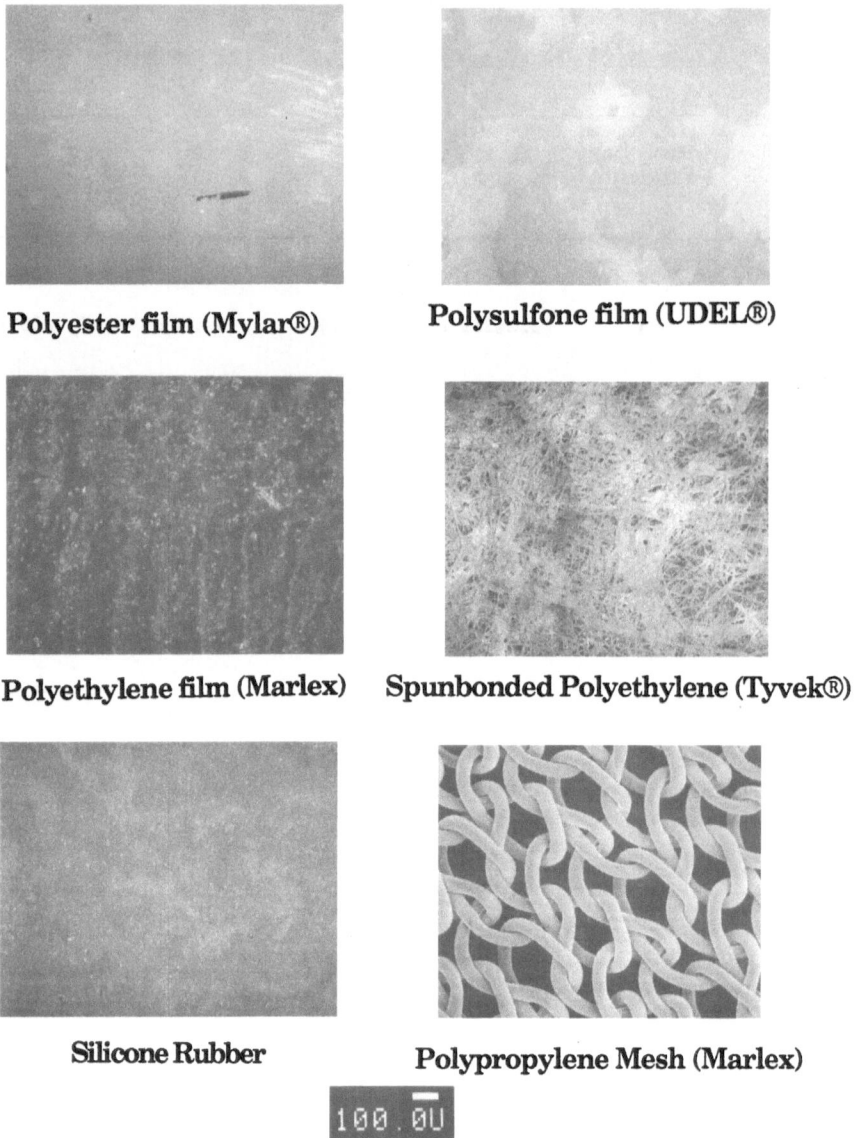

Figure 1. Scanning Electron Micrographs of 13 Testing Substrates.

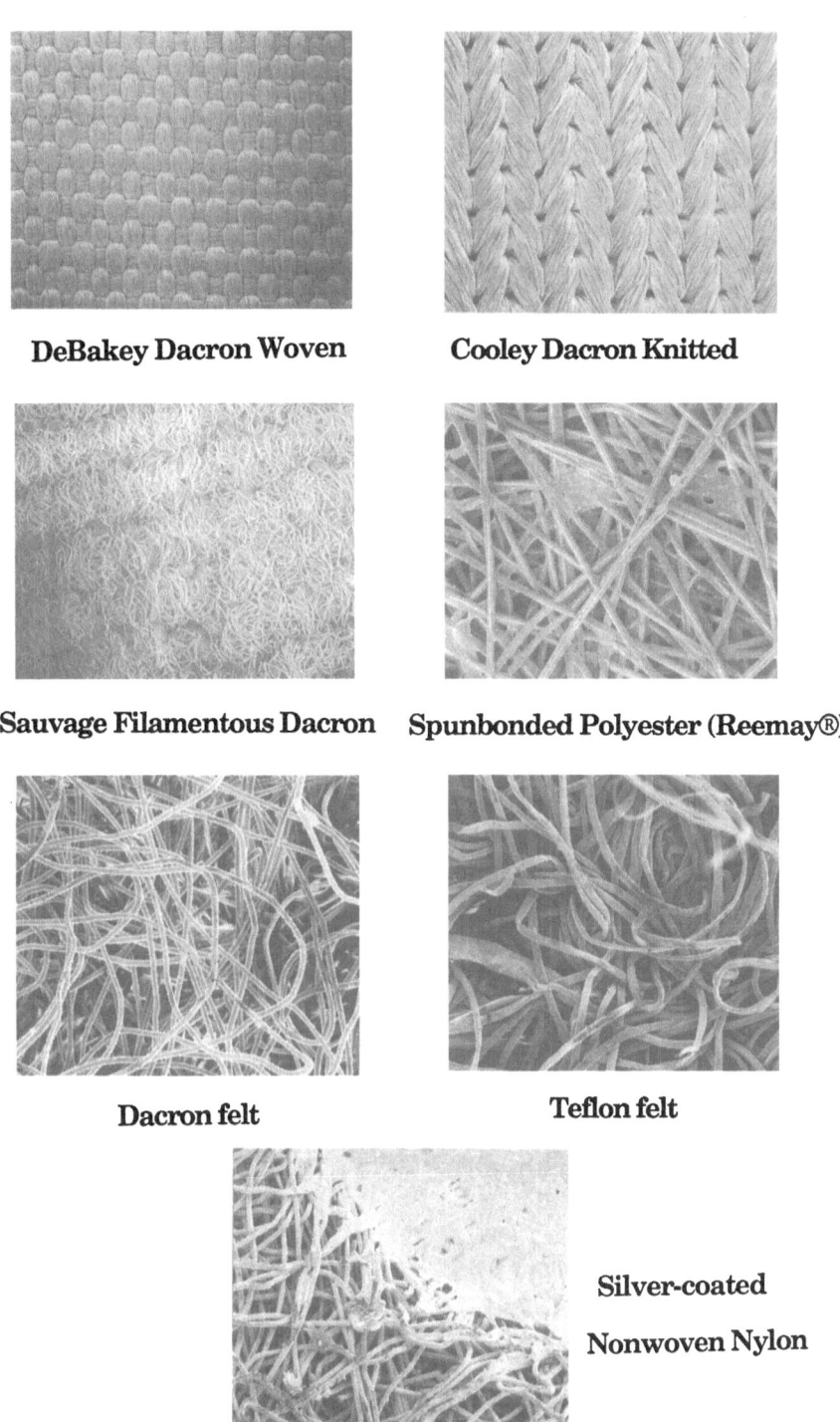

DeBakey Dacron Woven

Cooley Dacron Knitted

Sauvage Filamentous Dacron

Spunbonded Polyester (Reemay®)

Dacron felt

Teflon felt

Silver-coated
Nonwoven Nylon

Figure 1. Scanning Electron Micrographs of 13 Testing Substrates.

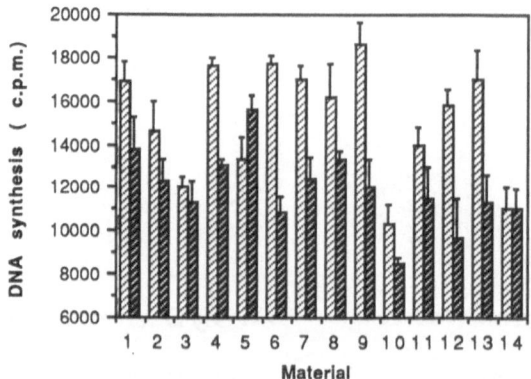

Figure 2. The Effect of Macrophages on Fibroblast DNA Synthesis on Testing Substrates. ▨ with Macrophages, ■ without Macrophages. 1-Polyester film, 2-Polysulfone film, 3-Polyethylene film,4-Spunbonded Polyethylene (Tyvek®), 5-Silicone rubber, 6-Woven Dacron fabric, 7-Knitted Dacron fabric, 8-Filamentous Dacron fabric, 9-Silver coated nonwoven Nylon, 10-Polypropylene mesh, 11-Dacron felt, 12-Teflon felt, 13- Spunbonded Polyester (Reemay®), 14-Controls 2 & 3 (Polystyrene).

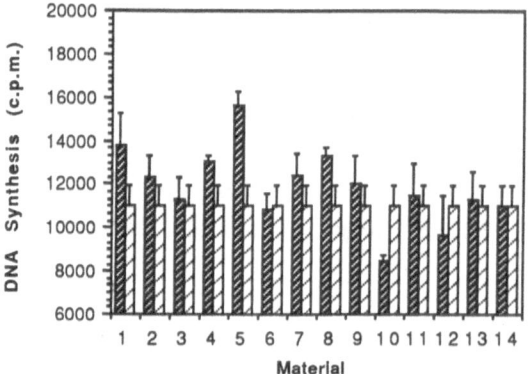

Figure 3. The Effect of Testing Substrates on Fibroblast DNA Synthesis (absence of macrophages). ▨ Control 3, ■ without Macrophages. 1-Polyester film, 2-Polysulfone film, 3-Polyethylene film,4-Spunbonded Polyethylene (Tyvek ®), 5-Silicone rubber, 6-Woven Dacron fabric, 7-Knitted Dacron fabric, 8-Filamentous Dacron fabric, 9-Silver coated nonwoven Nylon, 10-Polypropylene mesh, 11-Dacron felt, 12-Teflon felt, 13- Spunbonded Polyester (Reemay ®), 14-Controls 2 & 3 (Polystyrene).

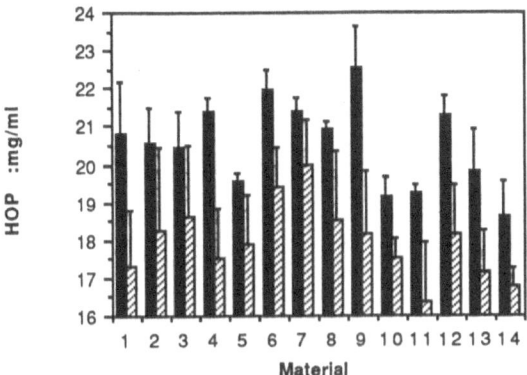

Figure 4. The Effect of Macrophages on Fibroblast Hydroxyline Contents on Testing Substrates. ■ with Macrophages, ▨ without Macrophages. 1-Polyester film, 2-Polysulfone film, 3-Polyethylene film, 4-Spunbonded Polyethylene (Tyvek®), 5-Silicone rubber, 6-Woven Dacron fabric, 7-Kiniited Dacron fabric, 8-Filamentous Dacron fabric, 9-Silver coated nonwoven Nylon, 10-Polypropylene mesh, 11-Dacron felt, 12-Teflon felt, 13-Spunbonded Polyester (Reemay®), 14-Controls 2 & 3 (Polystryrene).

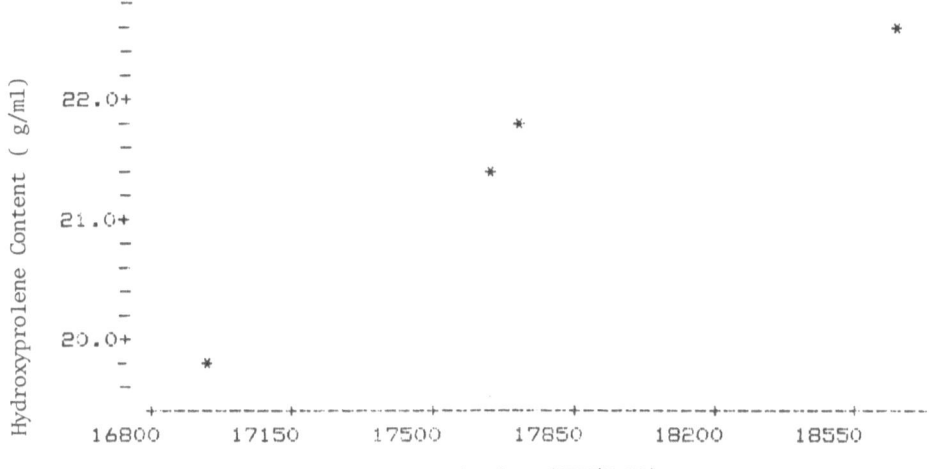

Figure 5. Correlation between Fibroblast DNA Synthesis and Hydroxyprolene Contents (in the Presence of Macrophages) of Four Synthetic Substrates Showing Strongly Significant Macrophage Fibrogenic Activity. The Four Substrates: Woven Dacron Fabric, Silver Coated Nonwoven Nylon, Spunbonded Polyethylene (Tyvek), and Spunbonded Polyester (Reemay). The Correlation Cofficient is 0.929.

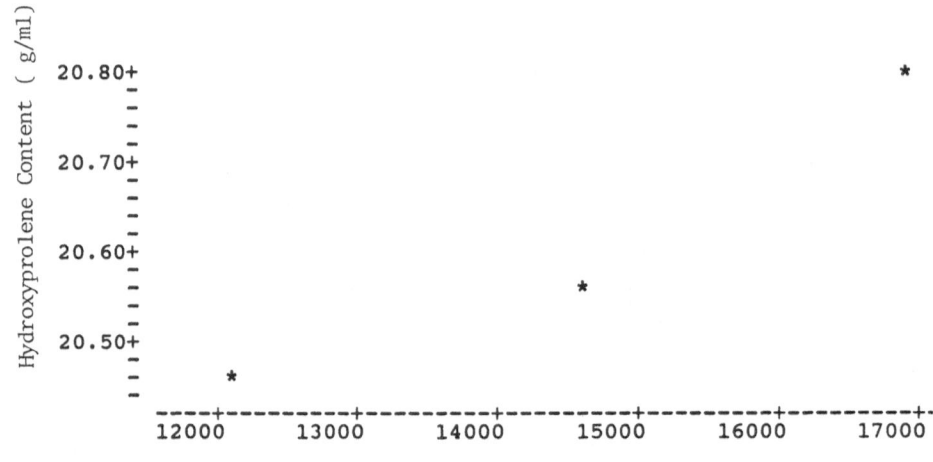

Figure 6. Correlation between Fibroblast DNA Synthesis and Hydroxyprolene Content (In the Presence of Macrophages) of Three Film Substrates: Polyester, Polysulfone, and Polyethylene. The Correlation Cofficient is 0.966.

SELECTED DISCUSSION REMARKS:

Gogolewski: Could you relate the effect of macrophages' activity to the chemistry of the polymer chain? Is it possible that the macrophages' activities had to do with the release of additives from the polymer?

Chu: We have extracted all the synthetic materials before their study in cell culture and we believe extraction has removed all the additives and the impurities. You can see at the three film materials (PES, PSU, PE) that they affect the macrophages' fibrogenic activity differently (fig. 2, columns 1 - 3). We don't know how it relates exactly to the chemical structure. May be it is due to the hydrophobicity or hydrophilicity.

Turnquist: I have a question on the silver coated nylon material. Past history indicates that nylons would be degraded in vivo.

Chu: Originally this commercial material was not intended for medical usage. We are interested in the conductive silver-metallic coating rather than in the nylon substrate. We intended to test how the conductive coating affects macrophages' fibrogenic activity and the secretion of fibroblast growth factor. And we use a similar type of material as antimicrobial suture and the in vitro studies are very promising.

CONTRIBUTIONS TO BIOCOMPATIBILITY DURING HEMODIALYSIS

U.T. Seyfert, E. Helmling, U. Grothe-Pfautsch, F.W. Albert,G. Pindur, E. Wenzel

Abt. für Klinische Hämostaseologie und Transfusionsmedizin, Unikliniken, D - 6650 Homburg/Saar, FRG

ABSTRACT

Thrombogenicity is the property of a foreign surface to induce clotting processes or formation of aggregates after contact with blood. Today different drugs are available which stop these reactions. The substances we use like conventional and low molecular weight heparin, prostazyclin and prostazyclinanaloga have different action profiles.
Effective inhibition of platelet activation, however, does not necessarily grant clotting-free haemodialysis.
Induction of clotting through contact of blood with foreign surfaces during Cuprophane haemodialysis has to be considered as multifactorial phenomenon. Leucocytes play an important role in the induction and advancement of these clotting processes.
Beside the sort of anticoagulation patient's prethrombotic state, rheological factors as well as physicochemical properties of foreign membranes decisively influence thrombogenicity.
A new synthetic membrane (AN 69) inactivates sufficiently the clotting in an extracorporal system and allow a reduction in dosages of heparin.
Studies of bioincompatibility reactions have shown increase in interleukin-2 concentrations.
Platelet activation, platelet turnover, disturbances of endothelium, fibrinolysis activation and granulocyte activation and complement drop are reproducible parameters of our described interaction model. They also permit to compare different haemodialysis membranes.
Keywords: biocompatibility - hemodialysis - interaction - coagulation system - immune defense system

H. Planck M. Dauner M. Renardy (Eds.)
Medical Textiles for Implantation
© Springer-Verlag Berlin Heidelberg 1990

INTRODUCTION

During the first dialysis, which Georg Haas in Gießen performed about 1915, clotting of blood in contact with foreign surfaces used in the course of dialysis has already been a central problem. When heparin was detected by Mc Lean in 1916, it seemed to be the "ideal" substance for avoiding clotting.

Optimal anticoagulaton which eliminates the bleeding risk during haemodialysis in patients with a bleeding risk and which simultaneously stops clotting in the extracorporal circulation has not yet been achieved.

Inspite of progresses as regards development of membranes for dialysis, interactions between foreign surfaces and the clotting system continue to be a problem.

It is known that haemodialysis using Cuprophane membranes causes activation of complements, a passing decrease in leucocyte count and activation of platelets. Leucocytes in contact with foreign surfaces are considered to play a decisive role in the induction of coagulation (2).

AIM OF THE STUDY

1. Studies of antithrombotic and anticoagulatory actions of different substances during haemodialysis

2. Studies of bioincompatibility of different dialysis membranes during therapy with conventional heparin.

PATIENTS, MATERIALS AND METHODS

1. PATIENTS

10 patients (6 female, 4 male) undergoing maintenance haemo-dialysis (haemodialysis duration between 0,5 and 10 years) were studied. 8 out of these 10 patients suffered from a pyelonephritis. In one case a renal artery stenosis and a poly-cystic renal disease led to endstage kidney failure.

Criteria for inclusion and exclusion are presented in tables 1 and 2.

Tab.1: Criteria for inclusion

- Age between 20-65 years
- Continuous dialysis because of chronic renal insuffi-
  ciency and stable haemodialysis conditions
- Hb   > 5,5 g/l
- Platelet count > 100 000/mm^3
- Quick   > 70%
- aPTT, Thrombin time, fibrinogen concentration within
  normal range

Tab 2.: Criteria for exclusion

- Severe cardiovascular and liver diseases
- arterial hypotension (systolic blood pressure < 90 mmHg)
- autonomous insufficiency
- Serum-Kalium > 7 nmol/l
- Hb   < 5,5 g/l
- Medicines which influence platelet aggregation.
- Erythropoetin therapy.

2. DRUGS

In the studies conventional heparin-UH (Organon Technika), low
molecular weight heparin-LMWH (Fragmin-Kabi, München),
Prostazyclin - natrium (Epoprostenol - Wellcome, London) and
Prostazyclinanalogon CG 4203 (Grünenthal, Aachen) were used.

3. TRIAL

The study includes patients with 5 consecutive haemodialysis
sessions on the whole, where dialysis time and 30 min follow-up
postdialysis phase serves as control period. This was a controlled, sequential, single-blind, cross-over study.

Tab. 3.

Dialysis number	bolus	dosages during haemodialysis
1	50 U/kg per body weight	Heparin 1000 U/h
2	(500 U i.v.)	CG 4203 (25 ng/kg/min) and Heparin 200 U/h
3	-	CG 4203 (35 ng/kg/min)
4	-	Prostazyclin-Natrium (5 ng/kg/min)
5	low molecular weight heparin LMWH 35 anti-Xa-units/kg body weight	15 anti-Xa-units/ kg body weight

4. MODALITIES OF HAEMODIALYSIS

Haemodialysis was performed with a dialysis apparatus of Drake Willock, Modell-# 4215/16 with a capillary dialysator of Fresenius MTS C 1,3 m^2, (Cuprophane membrane). Dialysis was performed by a Cimino-fistula or a femoral shunt against bicarbonate (SW 56 BC-Schiwa, Glahndorf).

During the different dialyses blood flow was kept at 150 - 200 ml/min, dialysis flow at 500 ml/min and the ultrafiltration rate was kept constant. Weight loss during haemodialysis was 1,5 +/- 0,5 kg.

In the second part of the study a polyacrylonitrile-membrane (AN 69-Hospal, Nürnberg) was used beside the Cuprophane membrane. The dialysis with a highflux-dialysator was performed by a Monitral S dialysis apparatus (Hospal).

Anticoagulation was performed with conventional heparin (bolus 50 U heparin/kg body weight, intravenously given. During haemodialysis heparin 1000 U/h was administered.)

5. MEASUREMENT PARAMETERS, VALUES AND EVALUATION TIME

Haematological parameters:

Before, 10, 30, 150, 300 min (corresponds to end of HD) after start of dialysis and 30 min after end of dialysis haematocrit, haemoglobin, number of erythrocytes and leucocytes were determined (Hemolog 8/90-Technikon).

Haemostaseological parameters:

At the above mentioned points in time Quick (Dade), activated partial thromboplastin time (APTT), thrombin time (TT), reptilase time, antithrombin III-activity (Boehringer Mannheim), thrombincoagulase time (Boehringer Mannheim), protein C-concentration, plasminogen activity, antiplasmin activity (all Boehringer Mannheim), FD dimer ELISA (Boehringer Mannheim), heparin concentrations (Heptest-Labor), anti-Xa-concentration (Kabi München) platelet factor 4 concentration (Boehringer Mannheim), factor VIII:C (Behring Marburg) were determined. Thrombocyte count was determined by a 147 C thrombocyte analyzer (Dilab. Karlsruhe).

Interleukin-II-concentration (Gen-Tech. Munich) was determined during haemodialysis with Cuprophane and polyacrylonitrile hemodialysis.

ADP-induced platelet aggregation was determined according to the method described by Born (5) and it was recorded with a fully automated four channel aggregometer, Model PAP 4 (Biodata, Colora Frankfurt). Final concentration of platelet aggregation inducing ADP was 4×10^{-6} mol/l. Finally a linear continuous dilution was performed till the threshold of complete desaggregation was reached. The mean value of the inhibition of platelet aggregation maximal amplitude compared to the value at entry for the different concentrations was calculated and indicated in percentage numbers.

Heparinized blood samples were taken to determine hemorheological parameters like viscosity (Rheomat, Fresenius) and erythrocyte aggregation index (Myrene, FRG).

6. SCREENING ASSESSMENT

Macroscopic assessment of capillaries was as follows:
They were described as "clear" when in the diameter < 50 red
fibres were found, as "stripped" when the number > 50, as
"clotted" in complete clotting in the capillary system.

7. SCANNING ELECTRON MICROSCOPIC EXAMINATION

After dialysis capillaries were cleaned in a recirculation first
with Sorenson phosphate buffer (1-15 molar, pH 7,0), 10 min.
second with fixation solution (4% glutaraldehyde in phosphate
buffer pH 7,0) and again with Sorenson-phosphate buffer for 10
minutes. Scanning electron microscopic images were made by the
Institut für Med. Physik, Universität Münster.

8. SIDE EFFECTS

Patients were asked about the following symptoms and these were
recorded. Flush, palmar erythema, headache, nausea, pruritus,
vomiting, abdominal pain, angina, erythema. Drop in blood pressure
could be kept under control by volume substitution with
physiologic saline (max 500 ml/dialysis).

9. EX VIVO - SYSTEMS

Human native blood (heparin concentrations 0.5, 1.0, 1.5 U/ml) is
recirculated in a tube system using a Cuprophane vs. AN 69-
membrane (3 trials) across two roller pumps at a constant speed
(200 ml/min) and PF-4-concentration were determined.

10. STATISTICAL EVALUATION

After calculation of the arithmetic means and standard deviations
in the different dialysis groups, the following statistical
methods were used for controlling the statements:
1. t-test for paired numbers
2. linear regression analysis. Significance level was fixed at a
   probability of 1%.
3. Graphic representation was done by Chartmaster C I 3-51904-610
A. Decision Resources, INC, Westport, USA.

RESULTS

All laboratory results were represented corrected according to haematocrit.

Mean values of the ADP induced platelet aggregation in dialysis patients did not differ from that in normal subjects (Fig. 1)

Thrombocyte number, plasma clotting factors, antithrombin III, protein C, plasminogen, antiplasmin were not significantly influenced by the different antithrombotic substances used during haemodialysis (data not shown).

In all cases of haemodialysis a significant drop ($p<0.001$) in leucocyte number was observed within the first 30 minutes after beginning of dialysis and independently of the antithrombotic substances used. (Fig. 2) There were also not significant differences between conventional heparin and low molecular weight heparin.

Independent of the antithrombotic substance used ADP-induced platelet aggregation was remarkably parallel in behaviour to that of the leucocyte number during haemodialysis. No differences between conventional and low molecular heparin were observed (Fig. 3-6).

Within the first 30 minutes of haemodialysis a significant drop in ADP-induced platelet aggregation ($p < 0.001$) was observed for the use of all antithrombotic substances. This reduction is followed by a continuous elevation of ADP-induced platelet aggregation till the end of dialysis. (Fig. 7). No differences between conventional heparin and low molecular heparin was found.

Prostazyclin in the dosage used has the strongest antiaggregatory potency (23 %). Conventional and low molecular weight heparin and CG 4203 do not differ significantly in their inhibiting of platelet aggregation. Antiaggregatory potency of heparin was 35%, that of CG 4203 was 34,5%. Even the combination of low doses of heparin with low doses of CG 4203 has no advantage over the separate use of both substances. Synergistic effect has not been observed (antiaggregatory potency was 33%).

Effective inhibition of thrombocyte aggregation in the dosages used does not grant a dialysis free of clotting processes.

Differences in antifactor Xa acitivity were found when using conventional or low molecular weight heparin (Fig. 8).

Scanning electron microscopic examinations show characteristic findings for heparin (UH, LMWH), prostazyclin and CG 4203.

The morphological images of filters in heparin dialysis examined by scanning electron microscopic means showed fibrin lined conglomerates consisting of erythrocytes, leucocytes and thrombocytes. The accumulation of activated thrombocytes (pseudopodia formation, emptied granula) was remarkable. No differences between conventional and LMW heparin were found (Fig. 9).

Dialysis with prostazyclin produces a different picture. Thrombocytes and fibrin deposits are missing in the sections; filter membrane is covered with erythrocytes. (Fig. 10).

In contrast membrane surface is covered with fibrin fibres where singular thrombocytes, erythrocytes and granulocytes are found when using the prostazyclin analogon CG 4203. (Fig. 11)

In the second part of this study blood compatibility and in particular the thrombogenicity of a Cuprophane membrane and a fully synthetic polyacrylonitrile membrane were investigated. Anticoagulation was always brought on with conventional heparin.

There was no statistically significant initial drop in leucocyte number during haemodialysis with a polyacrylonitrile membrane (AN 69) observed (Fig. 12).

There were also no differences between the two membranes concerning the behaviour of ADP induced platelet aggregation (data not shown).

Statistically significant changes concerning clotting parameters, in particular, inhibitors were not observed except an increase in factor VIII:C ($P<0.001$) during haemodialysis with Cuprophane membrane (Fig. 13). An elevation in factor VIII: C-activity has to be taken as an indication of endothelial disturbances.

F-D-dimers are cross-linked fibrin fibres resulting from plasmin-induced cleavage products of factor XIII. During AN 69 haemodialysis in comparison to Cuprophane-dialysis a significant increase in cleavage products ($p < 0,001$) was observed (Fig. 14).

As indicator for the turn-over of platelets the platelet factor 4 concentration was determined. During AN 69-haemodialysis elevated platelet factor-4-concentrations which have to be considered as induced by heparin as our studies of PTT and heparin concentration (0.67 during AN vs 0.5 U/ml during Cuprophanee) indicate (p < 0,05).

Our ex vivo model could also confirm a heparin induced platelet factor-4 exliberation depending on concentrations (heparin concentration 0.5, 1.0, 1.5 IU/ml, Fig. 15). The elevation was only mitigated during Cuprophane haemodialysis.

Scanning electron microscopic studies of membranes showed the typical morphological image of a Cuprophane filter during heparin anticoagulation. The polyacrylonitrile membrane had deposits of erythrocytes and of singular monocytes, yet there were no fibrin linked conglomerates (Fig. 16).

Except of an increased initial value in comparison to that in normal subjects (IL-2 concentration < 0.5 IU/ml) there were no significant changes in interleukin 2-concentration (tab. 4) during haemodialysis. We observed 3-6 fold increases in IL-2 concentrations during a bioincompatibility accident (data not shown).

Tab. 4: Course of interleukin 2 concentrations during hemodialysis. No significant differences between Cuprophanee vs. AN 69 hemodialysis.

Interleukin-II-concentration (IU/ml) Normal < 0.5 IU/ml	1.37 ± 1.27	1 ± 0.6	1.6 ± 0.9	2 ± 2.8 n.s.

Furtheron we examined the influence of chronic renal failure and different membranes during haemodialysis on haemorheological parameters. There was a significant increase in plasma viscosity and fibrinogen in patients (Fig. 17) with endstage kidney failure compared to normals (p < 0.001). At different times during haemodialysis there were concordant changes of hematocrit, viscosity and erythrocyte aggregation index (r= 0.8) due to the hemofiltration rate (weight loss 1.2 - 2.9 kg) in our patients

(Fig. 18). Therefore we clearly could demonstrate the close relationship between the hemofiltration rate and an increase of viscosity, but different membranes cannot be discriminated by these special haemorheological parameters.

DISCUSSION

The drugs used (conventional heparin, low molecular weight heparin, prostazyclin and a prostazyclin analogon) have different action profiles which have varied effects on the parameters studied. The wanted antithrombotic effect of heparin is based on the inhibition of plasmatic clotting by interaction with anti-thrombin III. Conventional heparin proved its value as an anti-thrombotic substance for routine use during haemodialysis. Beside its good antithrombotic efficiency it has the advantage to be easily monitored and that there is the possibility: of a rapid antagonisation by protamin.

Low molecular weight heparin (11) has a high antithrombotic (anti-Xa) activity and a comparably lower anticoagulatory effect. Laboratory monitoring and comparability of different low molecular weight heparins without having international standards is quite problematic. The frequency of thrombotic events in filters resulted in our patient group a definite preference for heparin dialysis. Serious metabolic and clinical side effects were also not observed.

The antiaggregatory potential during prostazyclin dialysis was convincing (17,18) since the incidence of adverse effects occuring during dialysis with the mentioned dosages was low. Reduction in blood pressure was easy to compensate by substituting a maximum of 500 ml physiological saline. Prostazyclin led to an increase of the cAMP concentration in the platelets, to a decrease of the intracellular calcium concentration and to a reduced action of platelet's fibrinogen receptors. Thus, prostazyclin inhibits structural changes and aggregation of platelets. Scanning electron microscopic pictures of a dialysis membrane free of platelets and fibrin is quite impressive.

The antiaggregatory action and the scanning electron microscopic picture during haemodialysis with CG 4203 are comparable to that with heparin and show fibrin linked conglomerates of leucocytes, thrombocytes and erythrocytes. Compared to heparin this substance has no advantages during a 5 hour dialysis. The synergistic effect of heparin and the prostazyclin analogon was not observed. The use of higher dosages seems not to be practicable due to considerable adverse effects.

Effective inhibition of platelet activation does not grant a haemodialysis free of clotting.

Activation of complements, drop in leucocyte number and thrombogenicity of the Cuprophane membrane could not be eliminated by using these substances.

Induction of clotting processes through contact of blood with foreign surfaces during Cuprophane membrane haemodialysis has to be considered as a multi-factorial phenomenon.

The first interaction between blood and foreign surfaces is the absorption of plasma proteins at the surface. This contact with surfaces changes the action of different blood components and may result in an activation of enzymatic activities and a reduction of cell components. When proteins are deposited at the surface, blood cells like platelets, leucocytes and erythrocytes are rapidly absorbed (Fig. 19).

This process influences physical properties of surfaces, blood flow, and, in particular, geometry of flow and the properties of blood cells and of clotting processes (Fig. 20).

In our opinion, in this special interaction model leucocytes e.g. monocytes play an important role as far as induction and perpetuation of these clotting processes are concerned. Thrombocyte activation works not primarily as clotting priming agent but is a reactive phenomenon at the Cuprophane membrane. Imagining that the foreign surface in contact with blood acquires a property typical of antigens and, thus, initiate a complement activation (14,15), a reaction chain of clotting processes taking place in the extracorporal dialysis circulation can be constructed.

Summarizing it can be said that beside the kind of anti-coagulation, the prethrombotic state in patients, rheological factors (geometry or flow behaviour) and physicochemical properties of the foreign surface are decisive parameters of thrombogenicity. Therefore, measures are needed to improve bioincompatibility of foreign surfaces with the human organism.

We undertook investigations with a new synthetic polyacrylonitrile membrane. The use of this membrane shows in comparison with the Cuprophane membrane no statistically significant initial drop in leucocyte number (13). A mitigated course of activation of complement and of granulocytes is observed.

Behaviour of platelet aggregation does not differ from that found during Cuprophane haemodialysis. It has to be thought of a catalytic activation, e.g. by PAF-release (platelet activating factor) and secretion of neutral proteases (3,6,7,8) when reduction in leucocyte number does not occur. Endothelial releasing factors may also cause such a cascade although endothial alterations judged by factor VIII release when using the new polyacrylonitrile membrane could not be objectivized. Scanning electron microscopic pictures show deposits of erythrocytes and monocytes at the polyacrylonitrile membrane. Activated thrombocytes could not be detected.

The new synthetic polyacrylonitrile membrane inhibits sufficiently the clotting process in an extracorporal system and, thus, allows a reduction of the heparin dosage (higher concentrations of heparin). In consequence, there is a reduced bleeding risk.

It was interesting to see that there is an increase in F-D-dimers during dialysis with polyacrylonitrile membrane which seems to be protective.

The behaviour of platelet 4 concentrations (10,16) cannot be considered as thrombophilic factor but as induced by heparin. In accordance with the literature platelet factor 4 can be removed from endothelium by using high concentrations of heparin.

Furthermore, we found indications for a reduced concentration of beta-thromboglobulin during AN 69-haemodialysis compared to that with Cuprophane membrane which implies that platelet activation does not occur.

Interleukin 2 (4,9) is a lymphokine which is formed after activation of T-helper lymphocytes and therefore it is important for development and control of immune reactions.

Except of elevated basal values in our patients compared to normal subjects there were no statistically significant changes in interleukin 2 concentrations during haemodialysis with Cuprophane membrane vs. AN 69 membrane. However, it was remarkable that interleukin II concentrations were increased, when bioincompatibility reactions like pruritus and high levels of ETO antibodies occur. When replacing Cuprophane membrane by a polyacrylonitrile membrane the increase of interleukin-II concentrations was stopped. In 2 patients who responded to polyacrylonitrile membranes with pruritus and who had also high concentrations of interleukin-II during haemodialysis the described elevation of interleukin-II concentration was interrupted after changing to Cuprophane membrane.

Further studies on toxicity, compatibility of blood and tissue are necessary as far as contact of blood with foreign surfaces is concerned.

Thrombogenicity is an essential factor of blood compatibility.

Platelet activation, platelet turn-over, endothelial alterations, fibrinolysis activation and drop in leucocyte number, activation of granulocytes and complement (12) are reproducible parameters of our interaction model and allow to compare different membranes used in haemodialysis.

REFERENCES

1 Arnaout MA,Hakim RM,Todd RF,Dana N,Colten HR (1985) N. Engl. J. Med. 312:415
2 Baldamus CA,Koch KM,Schoeppe W (1983) Contribution to Nephrology 36:1
3 Beveniste J,Camussi J,Polonsky J (1977) Monogr. Allergy 12:138
4 Bingel M,Lonnemann G,Shaldon S,Koch KM,Dinarello CA (1986) Nephron 43:161
5 Born GVR (1962) Nature 927
6 Camussi G (1986) Kidney Int. 29:469
7 Chestermann CN (1986) Clinics in Hematology 15:575
8 Cheung AK,Henderson LW (1986) Am. J. Nephrol. 6:81
9 Damle NK,Doyle LV,Bender JR,Bradley EC (1987) J. Immunol. 38:1779
10 Dawes J,Pumphrey CW,McLaren KM,Prowse CV,Pepper DS (1982) Thromb.Res. 27:65

11 Fareed J,Walenga JM,Hoppenstaedt D,Huan X,Rocanelli A (1988)
   Haemostasis 18 (suppl. 3):3
12 Hakim RM,Fearon DT,Lazarus JM,Perzanowski CS (1984) Kidney Int.
   26:194
13 Heinrich D,Thilo Körner DGS (1986) Hämostaseologie 6: 57
14 Hörl WH,Hjalmar B,Steinhauer HB,Schollmeyer P (1985) Kidney
   Int. 28:791
15 Jungi FW,Spycher MO,Nydegger UE,Barandun S (1986) Blood 67:629
16 Kawaguchi M (1987) ISAO PRESS 308:42
17 Simon P,Ang KS,Cam G (1987) Nephron 45:172
18 Smith MC,Danviriyasup K (1982) Am. J. Med. 73:669

SELECTED DISCUSSION REMARKS

Jerusalem: The acute rejection of kidney allografts is largely
        dependent on the activation of the complement cascade. Can
        we prevent the acute rejection by treatment with the cobra
        venom before transplantation? Because cobra venom activates
        almost the rest of all the molecules of the complement
        cascade.

Seyfert: This is a very important comment and its on the same
        line. We have to concentrate our activities on these
        special topics like activation of granulocytes, complement
        inhibitors of coagulation, protein S/protein C interaction
        and cytokines in order to create an interaction model.

Figure 1:
Mean values of the ADP induced platelet aggregation
max. amplitude in dialysis patients do not differ
from that in normals.
NP 0 Normals, PAT = dialysis patients
ADP I, II, III, IV = different concentrations of ADP

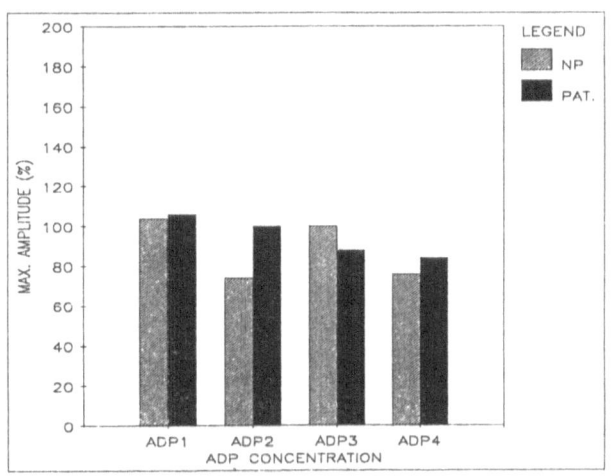

Figure 2:
Course of leucocytes during hemodialysis. Drop within the
first 30 min after beginning of dialysis (T 1) and
independly of the antithrombotic substances used.
T 0 pre hemodialysis (HD), T 1 10 min after begin of HD
T 2 30 min after begin of HD, T 3 150 min after begin of HD,
T 4 HD end, T 5 30 min after end of HD

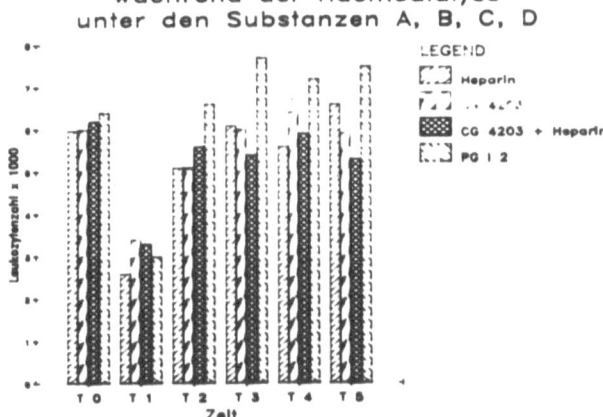

Figure 3:
Leucocytes in % of initial value (T 0) in correlation to Delta
max. platelet amplitude (in % initial value) at different
times (T 1, T 2, T 3, T 4) during heparin hemodialysis

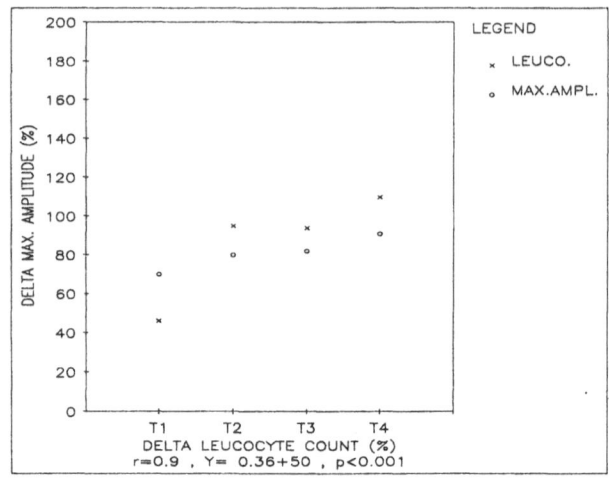

Figure 4:
Leucocytes in % of initial value (T 0) in correlation to Delta
maximal platelet amplitude (in % initial value) at different
times (T 1, T 2, T 3, T 4) during CG 4203 hemodialysis

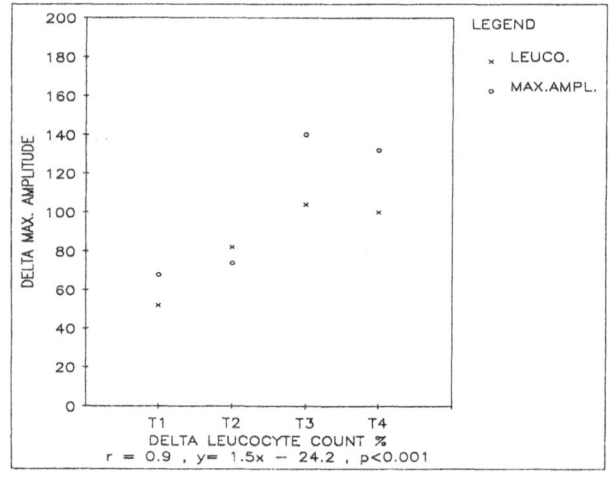

Figure 5:
Leucocytes in % of initial value (T 0) in correlation to Delta
max. platelet amplitude (in % initial value) at different
times (T 1, T 2, T 3, T 4) during heparin-CG 4203 hemodialysis

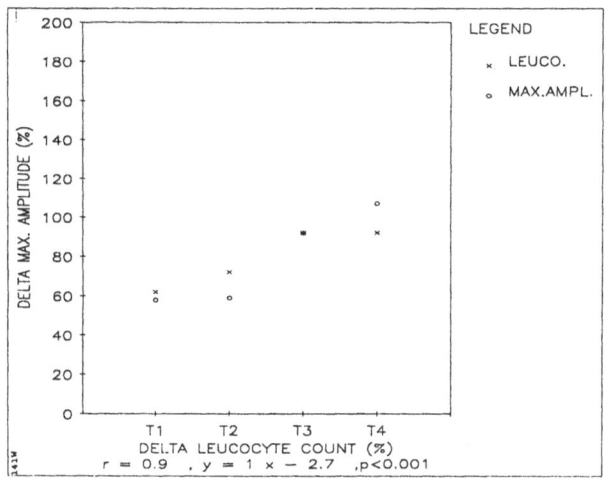

Figure 6:
Leucocytes in % of initial value (T 0) in correlation to Delta
maximal platelet amplitude (in % initial value) at different
times (T 1, T 2, T 3, T 4) during PG I 2 hemodialysis.

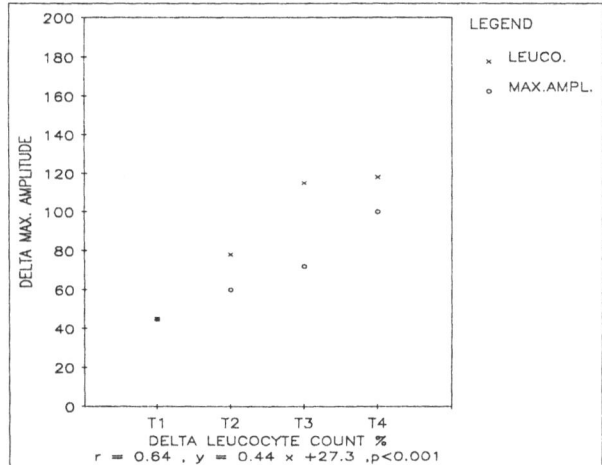

Figure 7:
Platelet aggregation studies with ADP 4 during hemodialysis with
heparin, CG 4203, heparin + Cg 4203, PGI 2.
Within the first 30 min of hemodialysis a significant drop
in ADP induced platelet aggregation was observed.

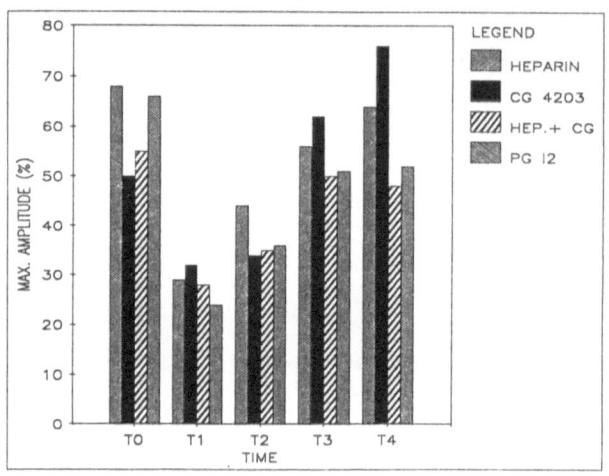

Figure 8:
Differences in antifactor Xa activity during therapy with
conventional vs. low molecular weight heparin.

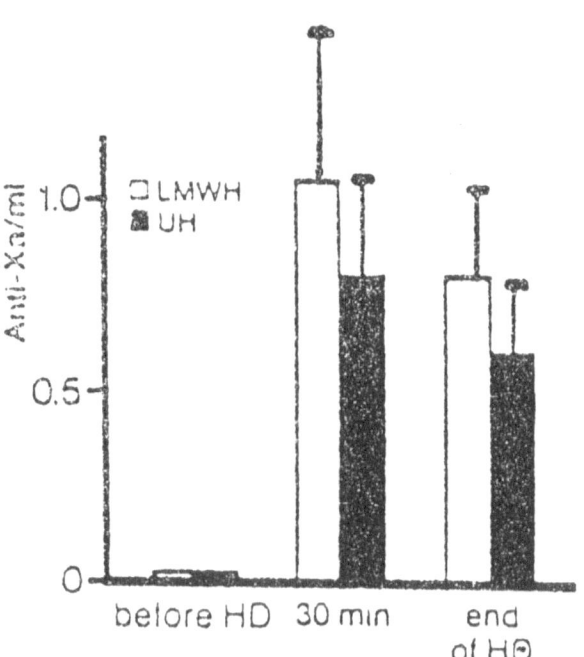

54

Figure 9:
Morphological images of filters in heparin hemodialysis

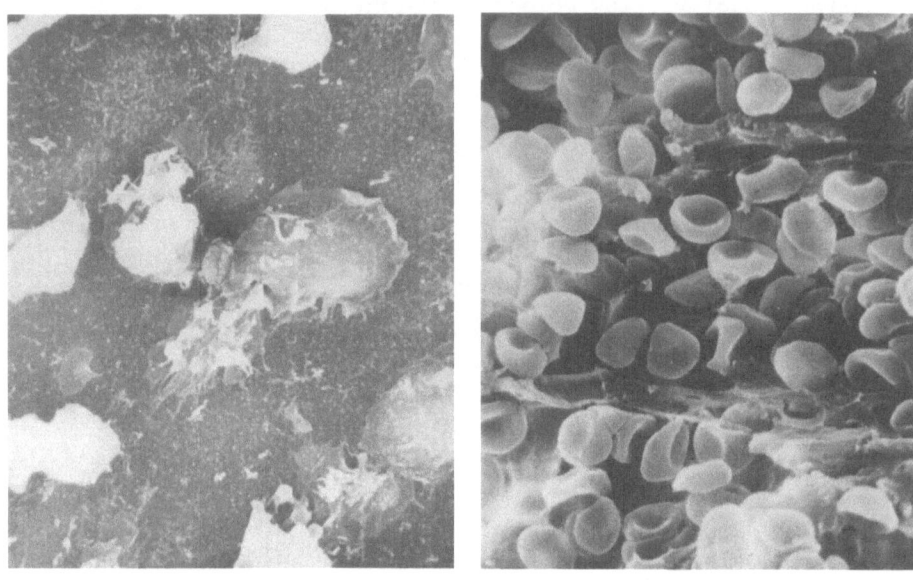

(6)

Figure 10:
Morphological images of filters in PG 12 hemodialysis

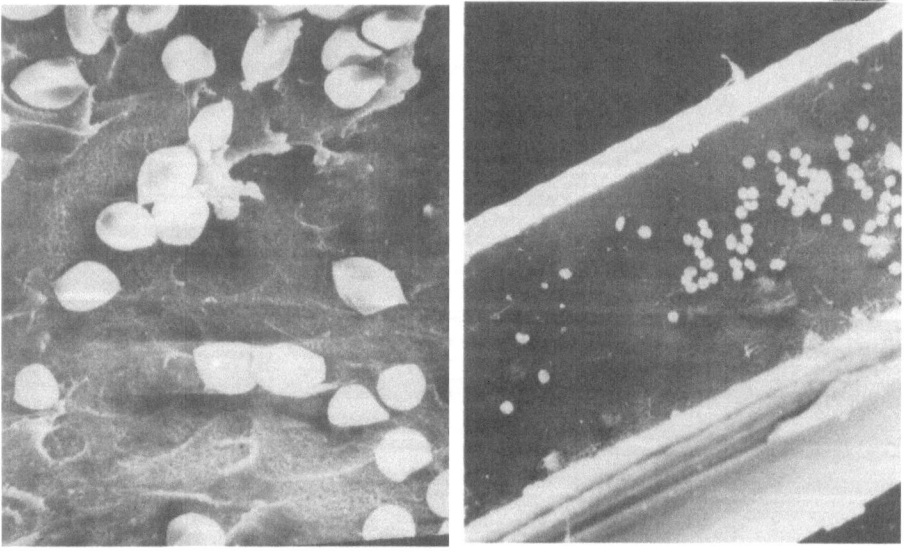

Figure 11:
Morphological images of filters in CG 4203 hemodialysis

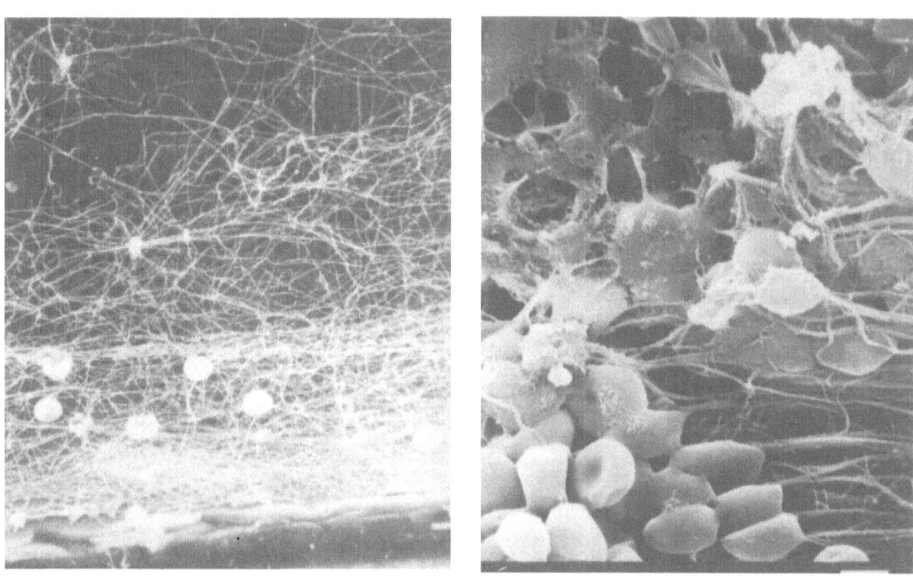

Figure 12:
Leucocytes during Cuprophane vs. AN 69 hemodialysis - no
significant drop during AN 69 hemodialysis.
T 1 = pre HD, T 2 = 25 min after begin of HD, T 3 = 150 min
after begin of HD, T 4 = HD end

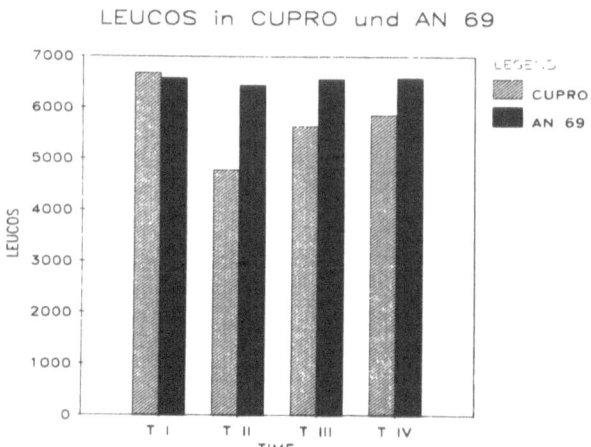

Figure 13:
Factor VIII:C levels during Cuprophane vs. AN 69 hemodialysis

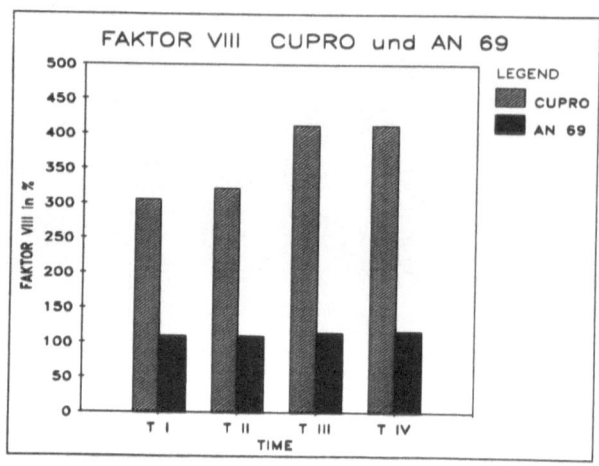

Figure 14:
F - Dimer levels during Cuprophane vs. AN 69 hemodialysis

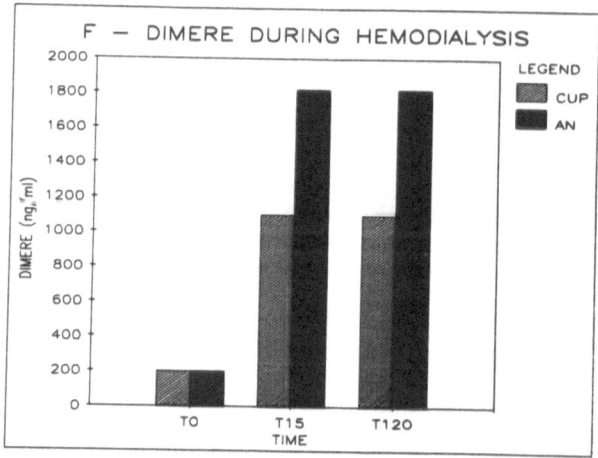

Figure 15:
Our ex vivo model confirms a heparin induced PF 4
(platelet factor 4) exliberation depending on the used heparin
concentrations.
HEP I = 0.5 U/ml, HEP II = 1 U/ml, HEP III = 1.5 U/ml heparin
concentration

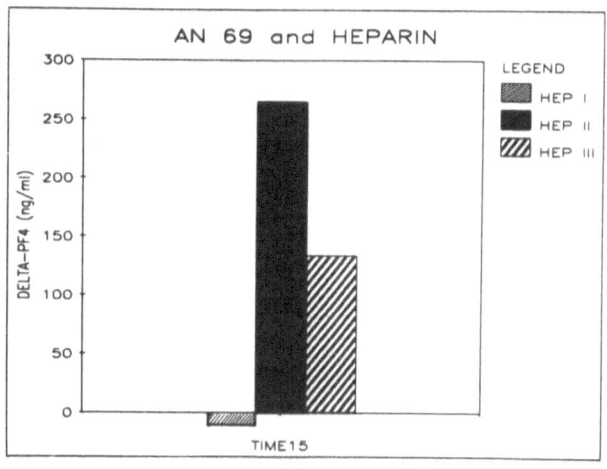

Figure 16:
Morphological images of AN 69 filters (scanning electron
microscopic pictures)

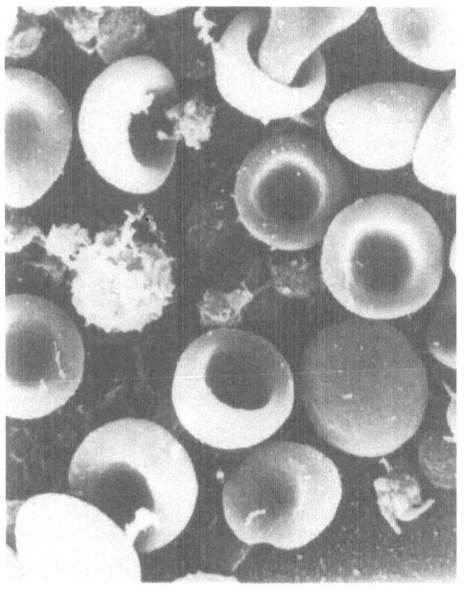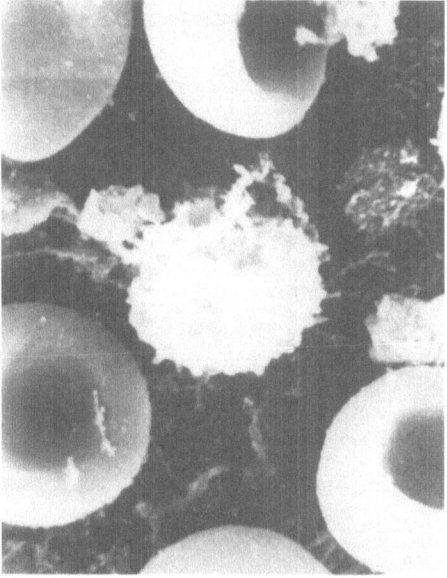

58

Figure 17:
Significant increase in viscosity and fibrinogen levels
in patients with endstage kidney failure

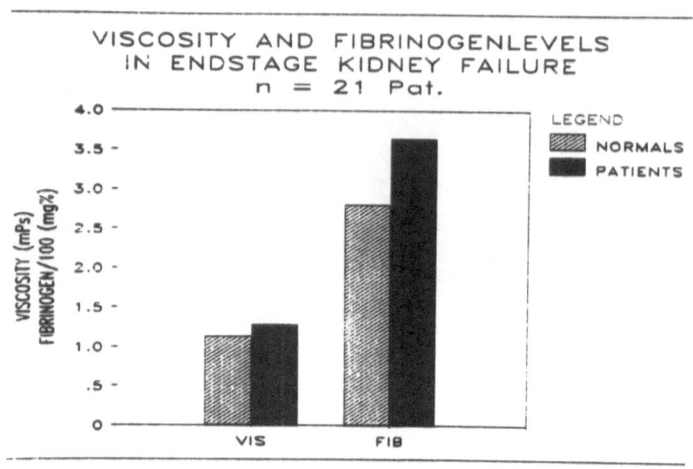

Figure 18:
Course of hemorheological parameters during hemodialysis
(n = 10 patients) - AN 69 filters. No discrimination between
Cuprophane and AN 69 filters by these special hemorheological
parameters

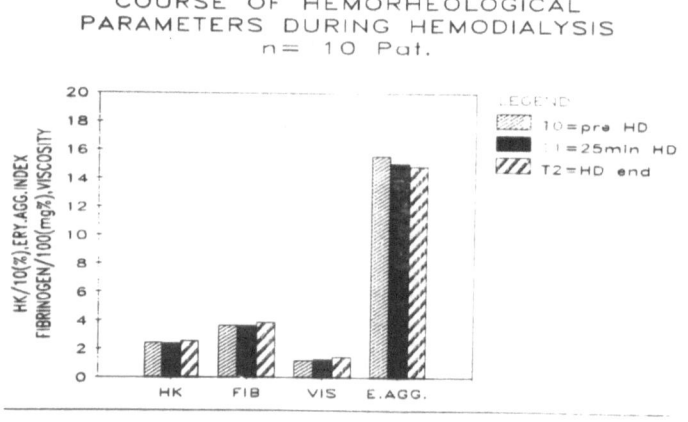

Figure 19:
Short schematic follow-up of blood – foreign surface
interaction (1 = foreign surface, 2 = platelets, 3 = leucocytes,
4 = erythrocytes, 5 = protein adsorption, 6 = fibrin)

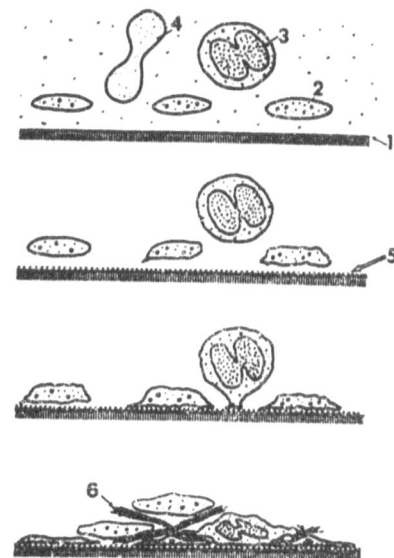

Figure 20:
Interactions of blood with artificial surfaces

INTERACTIONS OF BLOOD WITH ARTIFICIAL SURFACES

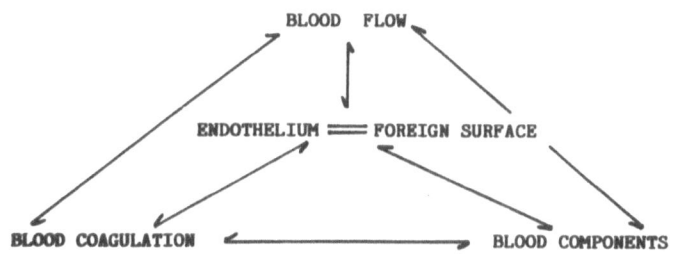

MACROPOROUS TEXTILE AND MICROPOROUS NONWOVEN VASCULAR PROSTHESES: HISTOLOGICAL ASPECTS OF CELLULAR INGROWTH INTO THE STRUCTURE.

Jerusalem, C.R. and Hess, F.

Dept. of Cell Biology, K.-University of Nijmegen,
(P.O.B. 9101) 6500 HB Nijmegen (The Netherlands).

S U M M A R Y.
The trellis concept of healing is suspended by the **preclotting of the porous prosthetic fabric.**

The absorption of the precoagulated blood and an occasional peri-prosthetic hematoma does not take place as rapidly as is usually surmized. Therefore, the presence of unorganized blood coagula can inhibit the stabilization of a fibrin layer which forms an early blood flow interface and can result in thrombotic complications.

Subsequently, the development of a cellular neointima which spreads exclusively from vascular tissues at the anastomotic site is impeded .

In contrast, the **coated prosthesis (UNI-GRAFT {R} DV)** can be used immediately without prior preclotting, thus also in patients being heparinized or suffering from coagulopathy.

The coating effectively prevents the imbibition of the prosthetic wall with blood as well as primary and secondary bleeding into the prosthetic bed. The coating smoothes the inner surface and imme-diately forms a highly hemocompatible (noncollagenous) contact in-terface to the blood flow and promotes the formation of a dense and coherent primary fibrin layer as the "guide-rail" for a com-plete cellular neointima.

On the gelatine coating, the autologous fibrin film remains cohe-rent and stable for long periods, probably indefinitly, when af-ter clinical implantation the cellular healing of the graft re-mains poor or is absent.

In the absence of endothelium, a stable fibrin flow surface, the nature's own means of covering a denuded region of the vascular wall, is superior to both the teflon/air flow surface of **ePTFE**, and the glutaraldehyde-fixed collageneous surface of the umbilical vein biograft.

The microporous flow surface of the **experimental ffPUR** prosthe-sis allows the firm attachment of a very thin fibrin film. Even under haemorrheologic disadvantageous conditions (loop-shaped conduit), this fibrin film forms both the athrombogenic interface to the streaming blood for many months and together the guide-rail with excellent anchoring possibilities for neointimal cells.

According to the present study, the patient's (diseased) vessels in which a prosthesis was implanted, the flow surface consisted of fibrin rather than of endothelium.

In contrast to preclotted blood, the coating can be easily infil-

H. Planck M. Dauner M. Renardy (Eds.)
Medical Textiles for Implantation
© Springer-Verlag Berlin Heidelberg 1990

trated by cells which contribute to the healing of the graft.

As a result of improved cellular immigration and penetration the **UNI-GRAFT {R} DV** prosthesis is better attached to the surrounding tissue and the fibrin layer is better stabilized, thus allowing a more rapid spread of endothelium and neointimal smooth muscle cells.

I N T R O D U C T I O N.
Arterial grafting began in 1906/1908 when CARREL and GUTHRIE demonstrated that homologous and heterologous veins and arteries could serve as arterial substitutes experimentally and clinically. Nevertheless, 40 years elapsed until methods of preservation had been developed and stored arterial allografts could be used satisfactorily in humans (GROSS et al. 1948).

However, with increasing number of implanted preserved arterial allografts, complications such as allograft aneurysms, infections (probably also rejection phenomena), and inconstant durability, were reported with increasing frequency (SZILAGY et al. 1957).

The era of vascular prosthesiology proper started in 1952/1954 by the demonstration of VOORHEES et al. on the use of a synthetic Vinyon "N" cloth fashioned into a tubular configuration, as an arterial substitute.

The selection of Nylon as the fibre of choice was obviously a mistake (EDWARDS 1978), since HARRISON (1958) reported that Nylon loses 85% of its tensile strength already 3 months after experimental implantation. In contrast, Teflon and particularly Dacron, as inert fibres seemed more likely to last in the body for many years. The durability of the implant shoud be superior to the life expectance of the recipient (GUIDOIN et al. 1988).

In 1955 the crimped configuration of the tubes was introduced (EDWARDS and TAPP). This crimping could largely reduce the kinking and bending of the strait tubes on the one hand, but on the other hand, the rugged flow surface is more thrombogenic and is more prone to bacteremic colonization, as demonstrated with the human umbilical vein graft (JULIEN et al. 1989).

Although more than 30 years of further intense research on vaslar prostheses have passed, several main questions are still debated, that are:

1. The requirement of an endothelial lining within the scope of "complete healing" of the vascular graft.

2. The optimal texture of the prosthetic wall.

3. The porosity of the vascular wall. Tightly woven (filamentous), expanded polytetrafluorethylene (ePTFE), and fine-fibrillar polyurethane (ffPUR) grafts are microporous, permeable for water but impermeable for cells. The braided or knitted filamentous fabrics can be manufactured with large, medium sized or small pores.

4. The advantages of an external and/or internal velour surface.

5. The implication of compliance mismatch at the anastomosis between rigid graft and flexible artery (EDWARDS 1978)

The complete or full-wall healing (BERGER et al. (1972), which
SAUVAGE et al. (1980) defined as incorporation of the entire graft
within a fibrous tissue matrix whose flow surface is covered with
endothelium, is the desirable objective.

According to these authors, only the complete healing of the prosthetic flow surface, with the formation of an intact, prostacy-
clin-secreting endothelial flow surface, can provide long-term
freedom from thrombotic occlusions, since the preclotted large calibre Dacron vascular prosthesis has a flow surface with a low degree of blood compatibility (MERHI et al. 1988).

SAUVAGE et al. (1980) stressed that a filamentous, porous wall of
the vascular substitute serves as a trellis to which cells can
attach and upon which they can migrate inward, facilitating the
healing process.

To consider are three sources of tissue for healing a synthetic
prosthesis.

 Firstly, the tissues of the vascular stumps at both ends of the
implant.

 Secondly, the perigraft tissue surrounding the implant. It is
suggested that cells of the perigraft tissue differentiate into
endothelium after transmural migration through the pores of the
graft fabric.

FLOREY et al. (1961) and recently KOGEL et al. (1989) suggested
that small vessels, down to capillary size, can in certain circumstances pass through the wall and provide the source of lining
endothelium and postulated that the the size of the pores should
as large to allow the ingrowth of capillaries.

 Thirdly, seeding on the flow surface of blood-born cells (so-
called "fall-out healing").

An unavoidable consequence of the "trellis concept" (SAUVAGE et
al. 1980) or "WESOLOWSKI concept" (WESOLOWSKI et al. 1961, 1964,
1968) of improved healing is the essential preclotting to seal the
pores. The size of these pores is poorly defined because the "porosity factor" means permeability to water, not really considering
that the same amount of water which passes through a few large pores can penetrate through many small pores.

SAUVAGE et al. (1980) stressed that the "porosity of the fabric
would ideally be such that cellular ingrowth would be encouraged
while complete and effective preclotting would also be assured.
The balance between these two factors is a delicate one". Since
increasing the porosity to promote healing "may simultaneously
make preclotting more difficult and bleeding more likely".

Particularly inserted in the thoracic aorta, knitted tubes, even after careful preclotting, were dangerous in patients receiving heparin sodium (EDWARDS et al. 1959).

In prothesiology, a breakthrough has come with the coating and impregnation with gels of the knitted graft fabric.

The industrially sealing of these vascular prostheses has three advantages:
Firstly, the pores are maintained which allow predictable and healthy ingrowth of perigraft tissue.
Secondly, the sealing (coating) renders precoagulation super-fluous.
Thirdly, the coating smooths the inner surface and represents a contact interface to the blood flow of distinctly higher haemo-compatibility than preclotted blood.

Several coated (impregnated) knitted Dacron prostheses are cur-rently available. As a coating substance respectively albumin, collagen or gelatine are used. The coating has to be stabilized, either by fixation with formaldehyde or by cross-linkage for ex-ample with isocyanate.

Previous studies of BENSLIMANE et al. (1986, 1987,1988) have de-monstrated that albuminated polyester arterial prostheses induce mild inflammatory reactions in rats and in dogs.

The collagen coating remains somewhat thrombogenic, although less than it is the precoagulated blood.

Formaldehyde is not entirely cross-linked with the substrate and can be liberated after implantation. This may be the reason that several recipients of these types of prostheses can suffer from general malaise for several days.

The presently optimal combination appears the isocyanate cross-linked gelatine-coated **UNI-GRAFT** {R} **DV** [B. Braun Melsungen AG (GFR)] vascular prosthesis.

For bypassing smaller arteries (e.g. femorofemoral or femoropopli-teal) ePTFE fabrics have replaced glutaraldehyde-processed umbili-cal vein allografts. Although ePTFE grafts are almost athromboge-nic (BIBBY et al. 1987, ASHLEY et al. 1989) reported patency rates vary considerably (SAUVAGE et al. 1980).

The aim of the present experimental study was to obtain reference data on the healing of conventionally preclotted and of industri-ally coated vascular grafts, and to compare the results with the course of events in microporous vascular grafts, as well as in various types of vascular substitutes noticed after clinical im-plantation.

M A T E R I A L   A N D   M E T H O D S.

Three commercially available coated textile vascular prostheses were submitted to physical, toxicologic and morphologic examination. Two fabrics were excluded from systematic studies.

In one fabric the coating (formaldehyde stabilised gelatine) was incomplete, as demonstrated by increased leakage (2500 ml H_2O/ h) and insufficient pressure holding, resulting in hemorrhagic infiltration of the prosthetic wall and the wound bed after pilot implantation.

The coating of this prosthesis and that of another one (formaldehyde stabilized collagen) were cytotoxic as shown by temporary inhibition by more than 50% of in vitro growth of T.pyriformis according to GRÄF's (1985) Ciliate Test of Erlangen), probably because of evaporation of formaldehyde.

Subsequently, the optimally sealed (leakage of 175 ml H_2O only) and obviously non-toxic (proliferation of T.pyriformis by +117%/24 h) vascular graft (**UNI-GRAFT {R} DV**), coated with isocyanate cross-linked gelatine, was selected for systematic studies and implanted in the infrarenal aorta of 37 dogs (group A).

For comparison, another 36 dogs received a conventionally preclotted vascular prosthesis of the same texture and length (group B).

Respectively 2 animals of each group were sacrificed after 1,2,4,6 and 8 weeks, 10-11 (A,B) animals each after 3 and 6 months, and 4-5 (A,B) dogs after 9 months.

Included in this study are experiences gained with more than 700 microprotheses. It concerns a ffPUR fabric (BRAUN et al. 1988), and ePTFE specimens, both with a porosity between 10 and 60), with an inner diameter of 1.5 mm, and lengths between 1.0 cm and 10 cm (10 cm loop-shaped) implanted in the rat aorta, and with an inner diameter of 3.0 mm and lengths of 4 cm and 20 cm (20 cm loop-shaped) implanted in the dog femoral and carotid artery (HESS et al. 1984, 1986, 1987, 1988; JERUSALEM et al. 1987). The observation periods ranged from 1 hour to 24 months p.i..

Furthermore, 54 clinical biopsies of textile vascular prostheses (40 uncoated, 1 coated, 13 ePTFE), are studied histopathologically. These specimens were explanted because of various complications, between 5 hours and 22 years after implantation.

The recipient's response to the implant was studied histologically and the development of neointima determined morphometrically, in ffPUR and ePTFE grafts also by means of scanning electronmicroscopy. Selected specimens were processed for transmission electron microscopic examination.

For histology specimens were embedded in JB$_4$-plastic, 2 μm thick sections cut with a glass knife and stained with Giemsa.

RESULTS.

CLINICAL COURSE.
Among the recipients of a Dacron aortic graft, one dog of group A died 24 hours after surgery, and 2 dogs of group B respectively

after 12 and 24 hours. In the other animals the postoperative
course was uneventful. There was no primary mortality among recipients of ffPUR and ePTFE vascular substitutes.

HISTOLOGY.
Texture and feature of vascular prosthesis before implantation.

Histologic sections of the coated prosthesis (UNI-GRAFT {R} DV)
reveal an continuous coating present at both sides which smooths
the surfaces by incorporation also of the projecting velour fibres. The pores (Ø up to 0.08 mm), too, are entirely stuffed with
the coating, and the individual fibres of the twisted threads are
embedded within the sealing agent.

In ePTFE specimens between the expanded trabeculae the distance
is +/- 10 μm, and between the fibrils bracing the bars, less than
2 μm. In the ffPUR microprosthesis the size of pores varies be-
ween 5 μm (inner surface) and 15 μm (outer region).

Please remember that human red blood cells have a mean diameter of
7.1 μm, while fibroblasts need a space at least of 3 μm.

Haemocompatibility of the flow surface of vascular prostheses
and early events after vascularization.

O n e   t o   2 4   h o u r s   p.i..
In all vascular prosthetic grafts, coated or uncoated, a fibrin
layer (with various numbers of erythrocytes interspersed) is layed
down on the inner surface during the first 5 to 24 hours after the
vascularization. This fibrin deposit also fills up the valleys of
crimped fabrics and its precipitation ceased, when the turbulent
blood flow has normalized to a slip stream.

In the coated UNI-GRAFT {R} DV prosthesis implanted in dogs, the
fibrin layer is continuous, smooth, thin (0.03 mm) and almost free
of red blood cells.

In the precoagulated Dacron prosthesis implanted in dogs, the
fibrin deposit is irregular, leaves uncovered internal velour
threads, and can locally have incorporated so many erythrocytes
that the secondary deposit is indistinguishable from the primary
(pre-) coagulum.

After 5 to 24 hours, the inner surface of clinically implanted,
precoagulated Dacron prosthesis largely resembles those implanted in dogs. Several flat mural thrombi can bridge the crests of
the crimped fabric.

In ePTFE grafts (rat, dog) the intertrabecular spaces are filled
with cellular debris. A coherent fibrin film is only present in
shorter prostheses (1 cm, rat).

In both, rats and dogs the inner surface of ffPUR prostheses is
covered with a very thin fibrin coat (3-10 μm) alredy after 1 hour
p.i., which masks the texture of the fabric.

Cellular reactions to the vascular prosthetic graft.

One to 2 weeks p.i..
In the coated UNI-GRAFT {R} DV prosthesis (dog), undifferentiated cells, probably deriving from the pericyte population of the capillaries, migrate into the coating and subsequently differentiate into myofibroblasts (MF).

Simultaneously the coating disappears, suggesting that the coating serves as a food for immigrating cells. Bulk phagocytozing macrophages usually do not appear. A few monocytes join early to form foreign body giant cells (FBGC). However, giant cells are usually scarce.

Already 1 week after implantation the outer coating is removed by cellular activity and the MF have started to migrate transmurally through the pores of the prosthetic fabric. After 2 weeks several MF have got to the inner coating.

At the inner surface, the thin fibrin layer remains stable and does not increase in thickness.

From the vascular tissues at both anastomoses, endothelial cells have started to proliferate and migrate upon either the gelatine coating or the fibrin film.

In the precoagulated Dacron prosthesis (dogs), only few MF are present in the prosthetic bed, while practically all fibrils which are not incorporated into the (pre-) coagulum, were clasped by FBGC.

Despite careful preclotting small periprosthetic haematomas had developed locally.

In a clinically implanted, preclotted Dacron prosthesis, explanted after 14 days, no cellular reactions of any kind were noticed.

While a fibrin film is absent in several longer experimental **ePTFE** prostheses, short implants (and particularly those with higher porosity (60) exhibit a continuous fibrin coat. In an irregular pattern, the micropores of the prosthetic wall are stuffed with debris of circulating blood cells.
In contrast, all experimental **ffPUR** grafts are provided with a thin (3 - 10 μm) and coherent fibrin film. MF and fibroblasts migrate into the outer pores of the prosthesis.

In both the **ePTFE** and the **ffPUR** vascular substitute endothelial cells start to migrate from the anastomoses towards the middle of the prosthesis, at the end of the first week p.i..

Four to 8 weeks p. i..
Coated Dacron (UNI-GRAFT {R} DV prostheses (dog). Between the 4th and the 8th week after implantation the immigrating MF progressively absorb the entire coating, and the cells not only occupy all interstices but immigrate, too, into the fibrin layer, which was layed down immediately after vascularization.

Endothelial cells migrate further from the vascular tissues at
both anastomoses to form the cellular flow surface. The speed of
spread of endothelium is approximalety 0.25-0.35mm/day in distal
and 0.1-0.2mm/day in proximal direction.
In a second wave smooth muscle cells (SMC) infiltrate the coating
and the fibrin film beneath the endothelium.

Preclotted Dacron prostheses (dog). Four of the 6 dogs showed
more or less extended flat retroprosthetic haematomas, indicating
that fibrinolytic activity may lasten for longer periods or even
increase during the course of time.

Cellular reactions were poor and restricted to those areas where
haemorrhagic infiltration of the prosthetic bed was absent.

At these places, MF infiltrated the clot, and locally the cells
had reached the inner surface after 6 to 8 weeks .

The absorption of the (pre-) clotted blood and of retroprosthetic
coagula is very slow.

The inner surface of the preclotted implant is irregular, exhibits
craters and occasionally small flat mural thrombi.
These irregularities obviously impede the spread of the endothelium. In the case of fibrin retraction, outgrown endothelial cells
are buried under secondary fibrin deposits, and suffocate.

Clinically implanted preclotted Dacron prosthesis. Only one
specimen was available, explanted because of perigraft infection
after 8 weeks. Only neutrophils were present in the wall but neither MF nor fibroblasts.

Short (1 cm) **ePTFE** prostheses, implanted in the rat aorta, can
be provided over the entire length with a neointima, consisting of
endothelium and underlaying SMC, after 6 weeks. In longer implants (4 cm, dog, carotid or femoral artery), a short cuff of
neointima (3 - 4 mm) is only present immediately adjacent to the
anastomosis. In the middle of the prosthesis a fibrin coat may be
present or absent.

In all short **ffPUR** prostheses (1 cm, rat aorta) a complete neointima covers the entire internal surface after 4 - 6 weeks. In
longer grafts (4 - 20 cm, dog) the neointimal cuff at both sides
measures 4 - 5 mm. In the other regions a fibrin film of constant
thickness (3 - 10 μm) is firmly attached to the prosthetic fabric
and represents the flow surface.

T h r e e   t o   9   m o n t h s   p. i..
Coated UNI-GRAFT {R} DV prostheses (dog). Three months after implantation the coating is entirely replaced by MF and by subsequently differenciated fibroblasts. Because these cells also have
infiltrated the fibrin to form an inner capsula, they largely stabilize those portions of the flow surface which are not yet
replaced by neointima and by endothelium.

With increasing implantation periods the size of that areas

increases, where the "inner capsula" (mainly MF) joins the neo-intima, consisting of endothelium and SMC. The uniting starts at both sides adjacent to the anastomoses and progresses towards the middle of the prosthesis.

At times where in the middle of the prosthesis only an inner cap-sula had developed but not yet a neointima, island of pseudoendo-thelium may be present.

Preclotted Dacron prostheses (dog).
The longer the implantation period, the more it becomes evident that retroprosthetic haematomas disturb the development of a neointima.

In almost al cases, immediately beneath a localized mural thrombus at the flow surface, an unabsorbed coagulum was present in the prosthetic bed.

Morphometric studies clearly revealed that the coated **UNI-GRAFT {R} DV** prosthesis heales faster and more completely than con-ventionally preclotted fabrics.
After 3 months, 34.6% (mean of n=11) of the inner surface of coa-ted grafts was covered with endothelium as compared to 24.7% (n=11) after preclotting.
For 6 and 9 months p.i. the figures were 72.4% (n=10), and 38.3% (n=11), and 92.8% (n=4), and 68.8% (n=5) respectively.

Coated Dacron prosthesis (human). Of our collection, the only coated prosthesis was an axillofemoral bypass explanted because of seroma and kinking (cause/effect?) 8 months p.i..
The coating was entirely absent, probably it was digested enzy-matically, and replaced by fibrin but not by cells. The inner sur-face was free of thrombotic deposits.

Preclotted Dacron prostheses (human). Up to an implantation time of 9 months the histologic picture varied considerably.

Healing events were entirely absent in 2 cases. The Dacron frame-work, embedded within the precoagulated blood presented the aspect of a sequestrum. It was embedded between a flat mural thrombus at the flow surface and a retroprosthetic haematoma.

In contrast, at a Dacron-ePTFE anastomosis abundant, tissue of the inner capsula, deriving from the Dacron portion, had let to a fi-brous stenosis of the adjacent ePTFE segment.

In 3 **clinically implanted ePTFE prostheses** (femoral artery) a fibrous hyperplasia has led to a stenosis of the anastomosis be-tween the 6th and the 8th months p.i.. The fibrous pannus was short, not attached to the prosthetic wall, and not covered by endothelium. In 2 specimens a recent thrombus had developed at the stepped declivity of the ring-shaped pannus.

From histologic findings can be suggested, that in **experimentally implanted ePTFE prostheses** the neointima advances towards the middle of the implant for several weeks, but than may become

shorter with increasing implantation periods. At the bare segments, a fibrin film may be present or absent.
Calcification of the prosthetic wall is a common phenomenon of
experimentally implanted ePTFE grafts.

In contrast, in the **experimental ffPUR prosthesis** the neointima advances continuously and with the same rate of growth as
noticed in the coated UNI-GRAFT {R} DV prosthesis, that are approximalety 0.25-0.35mm/day in distal and 0.1-0.2mm/day in proximal direction. Therefore, 4 cm long grafts are completely lined
with endothelium 3 months p.i..
The middle portion of longer grafts (10 - 20 cm) remains covered
by the very thin fibrin film.

L o n g - t e r m r e s u l t s .
This chapter deals with histologic findings of **32 clinically implanted, preclotted vascular Dacron prostheses of various fabrics,** which were removed totally or portions of it, 3 to 11
years post surgery, **and one woven prosthesis** with an implantation
time of 22 years.

In 17 preclotted specimens and in the woven fabric cellular
events, as required to define a healing process, were absent. This
concerns not only the lack of neointima, but also the inability to
absorb the precoagulated blood. According to phase-contrast microscopy, ghosts of firmly pressed erythrocytes were demonstrable
even 4-8 years after surgery.

Several specimens had a bare surface with the Dacron fibres as an
interface, in others a fibrin layer served as a flow surface.
Mural thrombi of various size were present in almost **every specimen.**

The fibrin layer, however, was instable, because it **was not an-**
chored by cells of an inner capsula. In one case, a **large flap of**
a thick fibrin deposit had exfoliated, and had closed **abruptly**
the two branches of an Y-prosthesis like a valve .

In 16 of the 21 explants missing cellular reactions, a sufficient
large portion of the natural vessel (aorta, iliac artery) **was ex-**
cised together with the anastomosis, allowing the diagnosis of pathologic changes.

All vessels were severely damaged, mostly converted into an avital
(cell-free) tube of connective tissue. Several specimens exhibited large atheromatous plaques, and others small aneurismal dilatations.

In all these cases the flow surface consisted of a more or **less**
coherent fibrin layer only. Intimal and particularly **endothelial**
cells were totally absent.

In 14 preclotted Dacron prostheses (implantation time 3 to 11
years), phenomena of healing of various intensity were **detectable.**
Most of these patients were younger than those who showed no
response to the vascular graft.

In most implants the perigraft tissue, mainly MF, had penetrated
and simultaneously removed the (pre-) clotted blood, totally or
locally. In several grafts, the fibrin layer was largely stabilized by immigration of MF and fibroblasts.

Sporadically islands of pseudoendothelium were present.

In grafts sutured to a vessel with an intact intima, at the anastomosis a thick but short neointimal pannus had developed. This
pannus, however, was covered with endothelium only in the 2 cases
where the endothelium of the natural intima, too, was preserved.

In all other specimens the flow surface of the natural vessel was
made up by a fibrin film.

In 8 clinically implanted ePTFE prostheses, mainly explanted
because of prosthetic and anastomotic aneurisms, healing events
were entirely absent.

No long-term results with experimentally implanted ePTFE grafts.

Long (10 - 20 cm) ffPUR prostheses implanted in rats and dogs,
showed a stable neointima and or a thin fibrin film. It appears
that with the increase in length of the graft the rate of proliferation of endothelial cells somewhat decreases. In 10 cm long
prostheses (loop-shaped extensions of the rat aorta) the total endothelialization requires approximately one year, in the 20 cm
long loop-shaped extensions of the dog femoral artery, about 2
years.
In no case a once developed neointima increased in thickness,
particularly fibrous hyperplasia at the anastomosis was entirely
absent.

D I S C U S S I O N.

Although the possibility of endothelization of vascular prostheses
remains the desirable objective of graft fabric design, from the
present study it is deducible that vascular prostheses of both,
small and large diameters can function very well without endothelium for many years, provided they are covered with a coherent and
stable fibrin layer.

In most of the patient's diseased vessels which are replaced by a
vascular substitute, fibrin, too, forms the interface to the
streaming blood.

It appears that the majority of thrombotic complications is not
due to the lack of endothelium but arises from an instable fibrin
film, and/or decrease in blood flow velocity (SAUVAGE et al 1980).

In the Dacron prosthesis, the stabilisation of this fibrin deposit
is promoted by an internal velour surface, but stronger by transmural ingrowth of cells, which requires a porous wall. The adequate coating of the prosthetic fabric facilitates cellular immigration, and simultaneously prevents from the adverse accumulation
of coagulated blood in both the prosthetic wall and bed.

Immediately after vascularization of a vascular prosthesis the inner surface is smoothed by a fibrin layer. Histologically, the gelatine coat of the **UNI-GRAFT {R} DV** fabric appears largely haemocompatible, since only few erythrocytes are incorporated within the fibrin film. This implies that already after a few hours not the gelatine, but the likewise athrombogenic autologous fibrin forms the interface to the streaming blood.

In contrast, much more red blood cells are incorporated into the fibrin deposit at the inner surface of precoagulated grafts, suggesting a distinctly higher thrombogenicity of the flow surface.

By means of an ¹¹¹-Indium assay ASHLEY et al. (1989) confirmed this observation: To the **preclotted flow surface** adhere approximately 30 (!) times more platelets than to the gelatine coating of the **UNI-GRAFT {R} DV** prosthesis.

In contrast to the concept of SAUVAGE et al. (1980) that "the properly preclotted graft is in fact no longer foreign body, having become an autogenous protein (fibrin) conduit", the clot in which the Dacron framework is buried, is real a clot, consisting of all components of the streaming blood.

However, this clot is badly absorbed, particularly when a retroprosthetic haematoma has developed. In about one third of clinially implanted grafts the (pre-) clotted blood can be demonstrated after 5-8 years.

The combination of coagulated blood in the prosthetic wall and retroprosthetic haematoma appears to maintain thrombogenetic activity for long periods.

Furthermore, not all fibres of the prosthetic fabric are "buried" within the clot. Isolated fibres (velour) and the border of the tightly twisted bundles of fibres are occupied by FBGC. The FBGC appaer to block up the immigration of MF and fibroblasts and thus the healing of the implant for long periods.

In contrast, in the coated prosthesis, the sealing substance is penetrated first by MF, fibroblasts and occasionally by smooth muscle cells, the main cellular elements of the healing process, while FBGC develop (in limited numbers) only later, when the several fibres are liberated from the coating by immigrating cells.

However, there are indications that endothelium is never derived from transmurally ingrown perigraft connective tissue.

Recent studies of our laboratory on experimentally and clinically implanted vascular grafts revealed that endothelium exlusively derives from vascular tissues at the anastomotic site, thus only from the natural vessel (HESS et al. 1984, 1986, 1987, 1988; JERUSALEM et al. 1987).

Furthermore, in rats and dogs, the period necessary for endothelialization in its overall length of a microporous (impenetrable to cells) vascular prosthesis, is essentially the same as for a porous one.

The same holds true for the so-called "Fall-out healing". Like the transmurally immigrated cells of the connective tissue series, the circulating blood cells are too specialized to be able to differentiate to an endothelial cell.

Islands which show a pattern reminiscent of that of vascular endothelium but less regular, has been called "pseudoendothelium". This pattern is due to the presence of cells normal to healing tissue, that are collections of mononuclear macrophages interspersed with fibroblasts and other mesenchymal cells (PUGATCH 1964). At the flow surface of the prosthesis, the mononuclear cells are probably blood-born monocytes.

In man, complete healing including endothelialization of the flow surface, even with prostheses of greater porosity, has not been observed. The reason is not a compliance mismatch at the anastomisis between rigid graft and flexible artery (EDWARDS 1978), since the diseased vessel is almost as rigid as the graft, but due to to the largely devitalized recipient's arteries.

It shoud be stressed that the lack or restricted events of healing of vascular grafts are not characteristic of every clinical implant. Implanted in younger patients (traumatic surgery), the cellular reactions may be as intense as in experimental animals.

In these cases, too, the porous prosthesis is advantageous, because the proliferating neointimal cells can securely grip the fabric surface. In **tightly woven vascular Dacron or in the ePTFE prosthesis**, the interstices are so small as to render the wall microporous, making proclotting unnecessary, but they are also too small to allow cells to use the graft as a support.

Therefore, at the aporous flow surface, the proliferating cells can detach and tear off. The defect will be substituted by a mural thrombus which subsequently becomes organized. The organization tissue, in turn, can detach, too, and the repetition of the course: mural thrombosis and organization leads finally to a fibrous occlusion of the anastomosis.

In contrast, in the **ffPUR** vascular prosthesis, the micropores at the internal (flow-) surface allow the firm attachment of a very thin fibrin film and the permanent anchoring of neointimal cells.

R E F E R E N C E S.
>ASHLEY,S., BROOKS,SG., LATIF,AB.,GEHANI,AA., RAJAH,SM., KESTER, RC.: The influence of collagen and gelatine coating on the thrombogenicity of Dacron vascular grafts. Abstr. subm. XIX World Congr. Int. Soc. Cardiovasc. Surg. Toronto 1989.
>BENSLIMANE,S., GUIDOIN,R., ROY,PE., FRIEDE,J., HÉBERT,J., DOMURADO,D., SIGOT-LUIZARD,MF.: Degradability of cross-linked albumin as an arterial polyester prosthesis coating <u>in vitro</u> and <u>in vivo</u> rat studies. Biomaterials, 7, 268-272,1986.

>BENSLIMANE,S., GUIDOIN,R., MARCEAU,D., KING,M., MERHI,Y., RAO, TJ., MARTIN,L., LAFRENIERE-GAGNON,D., GOSSELIN,C.: Albumin-coated polyester arterial prosthesis: is xenogenic albumin safe? Biomat. Art. Cells Org. 15, 435-481,1987.

>BENSLIMANE,S., GUIDOIN,R., MERHI.Y., KING,M., DOMURADO,D., SIGOT-LUIZARD,MF.: An _in vivo_ evaluation of polyester arterial grafts coated with albumin: the role and importance of cross-linking agents. Eur.Surg.Res. 20, 66-67,1988.

>BERGER,K., SAUVAGE,L.R., RAO,A.M., WOOD,S.J.: Healing of arterial prostheses in man. Ann.Surg. 178, 18- 27,1972.

>BIBBY,SR., CROW,MJ., SHEEHAN,SJ., KESTER,RC.: Should pre-clotted Dacron grafts still be used? Abstr. p. 29. Ann. meeting Vasc. Surg. Soc. Great Brit. Ireland. 26/27 nov. 1987.

>BLAKEMORE,AH., VOORHEES,AB.jr.: Use of tubes constructed from vinyon "N" cloth in bridging arterial defects: Experimental and clinical. Ann.Surg. 140, 324-334,1954.

>BRAUN,B., GRANDE,P., LEHNHARDT, F.-J., JERUSALEM,C. HESS,F.: Herstellung und tierexperimentelle Untersuchung einer kleinlumigen mikroporösen Polyurethangefässprothese.
Vasa, Suppl. 22, 1-38,1988.

>CARREL,A.: Results of the transplantation of blood vessels, organs and limbs, abstracted. JAM 51, 1662-1667,1908.

>CARREL,A., GUTHRIE,CC.: Uniterminal and biterminal venous transplantations. Surg.Gynicol. Obst. 2, 266-286,1906.

>EDWARDS,WS.: Progress in synthetic graft development: An improved crimped graft of teflon. Arch.Surgery. 45, 298-309,1959.

>EDWARDS,WS.: Arterial grafts. Arch.Surg. 113, 1225-1233,1978.

>EDWARDS,WS., TAPP,JS.: Chemically treated nylon tubes as arterial grafts. Surgery, 38, 61-70,1955.

>GRÄF,W.: Der Erlanger Ciliatentest. Ein in vitro Verfahren zur Ermittlung von Zellverträglichkeit und Zytotoxizität. GIT Fachz. Lab. 6, 601-614,1985.

>GROSS,RE., HURWITT,ES., BILL,AH JR.: Preliminary observations on the use of human arterial grafts in the treatment of certain cardiovascular defects. N. Engl. J. Med. 293, 578-579,1948.

>GUIDOIN,R., COUTURE,J., ASSAYED,f., GOSSELIN, C.: New frontiers of vascular grafting. Int. Surg. 73, 241-249,1988.

>HARRISON,JH.: Synthetic materials as vascular prostheses: II. A comparative study of Nylon, Dacron, Orlon, Ivalon sponge and Teflon in large blood vessels with tensile strength studies. Am.J. Surg. 59, 16-14,1958.

>HESS,F., BRAUN,B., JERUSALEM,C., VAN DET,R., STEEGHS,S., SKOTNICKI,S., GRANDE,P.: Endothelialization of polyurethane vascular prostheses implanted in the dog carotid and femoral artery. J.Cardiovasc.Surg. 29, 458-463,1988).

>HESS,F., JERUSALEM,C., BRAUN, B., GRANDE,P.: Significance of the inner surface structure of small calibre prosthetic blood vessels in relation to the development, presence and fate of a neointima. A morphological evaluation. J.Biomed. Mater.Res. 18, 745-755, 1984.

>HESS,F., JERUSALEM,C., BRAUN,B., GRANDE,P.: Patency and neointima development in 10 cm long microvascular polyurethane prostheses implanted in the rat aorta. Thorac. Cardiovasc.Surgeon 32, 283-287, 1984.

>HESS,F., JERUSALEM,C., BRAUN, B., GRANDE,P.: The inner prosthetic surface structure and re-endothelialization: an experimimental

study using two types of microvascular prostheses for aortic implantation Microsurgery, 7, 29-37,1986.

>HESS,F., STEEGHS,S., JERUSALEM, C., WIJN,P., SKOTNICKI,S.: Determination of the patency of vascular prostheses implanted in the rat aorta by means of ultrasonic blood flow measurement. Microsurgery 8, 5-10, 1987.

>HESS,F., STEEGHS,S., JERUSALEM,C., BRAUN,B., GRANDE,P.: Implantation of 20 cm long polyurethane vascular prostheses in the femoral artery of the dog. priliminary results. Thorac. Cardiovasc. Surgeon. 36, 348-350,1988.

>JERUSALEM,C., HESS,F., WERNER,H.: The formation of a neointima in textile prostheses implanted in the aorta of rats and dogs. Cell Tissue Res. 248, 505-510,1987.

>KOGEL,H., VOLLMAR,JF., CYBA-AYUNBAY,S.: Ingrowth of microvessels into vascular grafts, a prerequisite for complete endothelial lining / animal experiments. Thorac. Cardiovasc. Surgeon,
 37, Suppl.I, 82,1989.

>MERHI,Y., GUIDOIN,R., FOREST,J-C.: Fate of poyester arterial prostheses implanted as thoraco-abdomonal by-pass in dogs: Haematology, pathology and biiochemistry. Clin.Invest.Med. 11, 403-416,1988.

>JULIEN,S., GILL,S., GUIDOIN,R., GUZMAN,R., CHARARA, J., ROY,P-E., MAROIS,G., BATT,M., ROY,P., SERISé,J-M., MAROIS,D.: Biologic and structural evaluation of 80 surgically excised human umbilical vein grafts. CJS 32, 101-1017,1989.

>PUGATCH,EMJ.: The growth of endothelium and pseudoendothelium on the healing surface of rabbit ear chambers. Proc.Roy.Soc. B,
 160, 412-422,1964).

>SAUVAGE,L.R., BERGER,K., WOOD, S.J., RITTENHAUSEN,E.A., DAVIS, C.C., SMITH,I.C., HALL,D.G., MANSFIELD,P.B.: Grafts for the 80's. Monograph from the Bob Hope International Heart Research Institute, Seatle, Washington 1980.

>SZILAGYI,DE., McDONALD,RT., SMITH,RF.: Biologic fate of human arterial homografts. Aarch. Surg. 75, 506-529,1957.

>VOORHEES,A.B., JARETZKI,A., BLANKENASE, A.H.: The use of tubes constructed from vinyon-N cloth in briding arterial defects. Ann. Surg. 135, 332-336,1952.

>WESOLOWSKI,SA., FRIES,CC., KARLSON,KE., DEBARKEY,MC., SAWYER,PN.: Porosity: primary determinant of ultimate fate of synthetic vascular grafts. Surgery 50, 91-96,1964.

>WESOLOWSKI,SA., FRIES,CC., HENNINGAR,G., FOX,LM., SAWYER,PN., SAUVAGE,LR.: Factors contributing to long-term failures in human vascular prosthethic grafts. J.Cardiovasc.surg. 5, 544-576,1964.

>WESOLOWSKI,SA., FRIES,CC., MARTINEZ,A., McMAHON,JD.: Arterial prosthetic materials. Annals NY Acad.Sc. 146, 325-344,1968.

SELECTED DISCUSSION REMARKS:

Gogolewski: We heard before that we need rough surfaces just to have an anchorage of the cells growing on the surface. But usually the surface covered with gelatine is smooth. Does the formation of the new intima on gelatine coated vascular prostheses have to do with the degradation and bioresorption of the gelatine?

Jerusalem: The gelatine is a carbohydrate and has principally a low thrombogenicity and then it is smoothing the surface. If you analyze the fibrin layer laid down on the preclotted prostheses for instance, it contains much more erythrocytes and also thrombocytes. But they are absent on the gelatine coated prostheses. That's still an experience, we have no explanations.

Gogolewski: Which was the longest distance on which your prostheses were endothelialized?

Jerusalem: At the experimental prostheses the longest section was 60 cm.

Seyfert: In the coagulation labs we know that isocyanate can cause formation of soluble fibrin monomers and these monomers can lead to disseminated intravascular coagulation. Is it possible to differentiate between gelatine and isocyanate effects?

Jerusalem: There is no differentiation. It is only that the deposite on the gelatine coating is apparently a pure fibrin. The effects of split products and others we can only evaluate by exclusion tests.

CYTOTOXICITY TESTS IN THE BIOLOGICAL EVALUATION OF MEDICAL DEVICES

Wolfgang G. K. Müller-Lierheim

DR. MÜLLER-LIERHEIM AG, Biological Laboratories
Behringstraße 6, D-8033 Planegg

Abstract: Since the early sixties mammalian and human cell cultures have been used as sensitive tool in the biological evaluation of medical and dental materials and devices. The three main
areas of interest for the application of cell culture tests are:
- cytotoxicity tests
- mutagenicity tests, and
- cell models for studying the interaction between device surface
  and surrounding tissue.

Cytotoxicity tests are frequently used in the development, preclinical testing and quality assurance and are now subject to international standards. Based on the experience derived from testing
more than 1000 medical and dental devices over a period of nine
years some general remarks about cytotoxicity testing and examples
of cytotoxicity test protocols are given.

Introduction: As European legislation is under elaboration in the
medical devices sector there is an increasing demand for biological safety testing.

Biological test methods for medical and dental materials and devices are subject to many different national and international standards. There is a growing need for uniform standard test methods.
This demand is reflected in a recent workshop on medical devices
organized by the Joint European Standards Organization CEN/CENELEC
in December 1988 in Brussels, and the forthcoming establishment of
a new technical committee ISO/TC 194 Biological Testing of Medical
and Dental Materials and Devices by the International Standards
Organization in July 1989 in Pforzheim.

H. Planck M. Dauner M. Renardy (Eds.)
Medical Textiles for Implantation
© Springer-Verlag Berlin Heidelberg 1990

Biological safety tests are performed on several levels
- clinical test on humans,
- preclinical toxicity and usage tests on animals, and
- in vitro tests on isolated mammalian and human cells and
  tissues.

In vitro tests can answer a range of different questions in biocompatibility assessment such as
- cytotoxicity tests for the acute toxic potential,
- mutagenicity tests for the screening of carcinogenic activity,
- cell models for immunogenicity, fibrogenicity and teratogenicity and
- cell models for studying the interaction of the device surface
  with surrounding tissues, e.g. the adherence, growth, and
  differentiation of human endothelial cells on the surface of
  coated vascular grafts.

The following paragraphs will focus on cytotoxicity tests for the
evaluation of the acute toxic potential of medical and dental materials and devices.

General remarks: Cytotoxicity tests are advantageously applied in
three different phases during the product life cycle

1. in research and development, e.g. for the development of
   polymerisation and extraction procedures, surface coatings,
   cleaning procedures or evaluation of changes due to different
   sterilization processes,

2. in the preclinical testing for acute toxic effects, and

3. in the quality assurance of raw material and finished product
   batches using simplified test protocols and short standard
   reports.

Criteria for design and evaluation of cytotoxicity tests have to
take into account
- sample preparation,
- cell type,
- biological endpoint, and
- interpretation of test results.

Acute toxic effects of medical devices can usually be attributed
to leachables. Therefore, and for reasons of good reproducibility
of test results cytotoxicity tests should preferably be performed
on extracts or eluates. We generally get reliable test results
with the use of cell culture medium (including 10 % fetal calf serum) and the solvent Dimethylsulfoxide (DMSO). Whenever possible
we use per ml of solvent 6 cm^2 surface area of the medical device
in the case of cell culture medium and 20 cm^2 in the case of DMSO.
Elution takes place over a period of 24 hours at 37 $^{\circ}$C. Before use
the DMSO eluate is diluted 1:100 with cell culture medium.

If possible we look for a dose-response relationship in our cytotoxicity experiments. For this purpose we test a logarithmic dilution series of the eluates in cell culture medium (e.g. 100 %,
30 %, 10 %, 3 %...).

There has been much discussion, whether primary cell cultures or
established cell lines should preferably be used in cytotoxicity
testing. Primary cell cultures show a high degree of differentiation. But cells tend to dedifferentiate after only a few passages,
and this gives rise to changes in their metabolism. For this reason, tests on primary cell cultures tend to be less reproducible
than those on permanent cell lines, despite the higher efforts involved in the application of primary cell cultures.

Permanent cell lines have a lower degree of differentiation than
primary cells. But these cell lines have the benefit of a well defined cell physiology and metabolism which gives rise to high reproducibility of cytotoxicity test results. The experimental
efforts are lower with permanent cell lines as compared to primary
cells. Due to the high reproducibility of cytotoxicity test
results with permanent cell lines, these methods are suitable for
standardization.

Our experience shows that

- the use of permanent cell lines (e.g. L 929 mouse fibroblasts)
  is adequate for the cytotoxicity testing of medical devices, and
  that

- the quality of test results depends mostly on test design and
  experience in the interpretation of test results.

The choice of the biological endpoint regarded in a cytotoxicity
experiment is another criterium to be considered. Toxic substances
interact on the cellular level with:

- cell membranes,
- mitochondriae (cell metabolism),
- protein synthesis,
- DNA (DNA synthesis and cell division).

We found that toxic effects arising from biomaterials can be detected in all established cytotoxicity test protocols no matter
which biological endpoint they primarily consider.

Most important in the biological evaluation of medical devices by
the means of cytotoxicity tests is the interpretation of test results.

As well-designed cytotoxicity tests are highly sensitive, positive
cytotoxic responses do not necessarily mean the rejection of a
biomaterial. Test results cannot be considered as absolute values.
Results on new materials and devices must be compared with devices
that have proved good in clinical application and can, therefore,
be used as test standards in each experiment.

Last but not least the interpretation of test results by an expe-
rienced cell biologist with a background in toxicology and materials sciences is essential.

<u>Examples of cytotoxicity test protocols:</u> The grandfather of all
standardized cytotoxicity test protocols is the AGAR OVERLAY TEST
dating back to 1965 (1). A monolayer of cells in a Petri dish are
washed, covered by an agar layer and stained by Neutral Red vital
dye. Pieces of the material to be tested are placed on the agar
surface, and incubated for 24 hours. If toxic leachables diffuse
into the agar, cells in the surrounding of the tested material may
eventually decolorize and be destroyed. The Agar Overlay Test is
evaluated semiquantitatively with respect to the diameter of the
decolorized area (zone index) and the percentage of dead cells
within this area (lysis index). Figure 1 shows cells in the vicinity of a nontoxic material (a and b) and a cytotoxic piece of
plastic (c and d). The Agar Overlay Test is subject to many national and international standards.

A well established quantitative cytotoxicity test method is the
CELL GROWTH INHIBITION TEST on extracts or eluates. With the use
of different solvents (e.g. cell culture medium and DMSO, see
above) this test is the best reproducible and most sensitive cytotoxicity test with the highest predictive value (2). Different extract concentrations are added to cell cultures and cell growth
rate is quantified at the beginning and after 72 hours in comparison with control cultures. Test results can be quantified for example by counting cell numbers or determining the protein content
by the Lowry method (3). Figure 2 shows the growth of L 929 mouse
fibroblasts in presence of different extract concentrations from a
cytotoxic medical device; the extract dilutions are 1:10'000 in
figure 2a, 1:1'000 in 2b and 1:100 in 2c. Note that a high percentage of rounded cells also indicates a cytopathic effect.

In cases where the protein determination by the Lowry method leads
to false results due to an interaction between extract and reagents, Neutral Red dye can be used instead of the Lowry reagents
and cell density be quantified by optical absorption measurement.

Instead of cell growth other parameters of physiological activity
of the cells may be used, e.g. DNA synthesis or metabolic activity. In the ^{3}H THYMIDIN TEST DNA synthesis is quantified by
measuring the incorporation of tritium labelled thymidin into to

cell DNA. A very sensitive method in the evaluation of cytotoxicity tests is the use of a Tetrazolium dye (MTT). This dye is originally yellow and turns dark blue when metabolized in the mitochondriae of cells (see figure 3). Thus metabolic activity can be quantified by simple optical absorption measurement in the MTT TEST (4).

Conclusions: Cytotoxicity tests with standardized test protocols are sensitive methods of high predictive value in the assessment of the acute toxic potential of medical and dental meterials and devices. Such test are used in
- research and development,
- preclinical testing, and
- quality assurance (batch testing).

In growing demand for uniform test protocols for all medical and dental materials and devices is assumed to lead to new international standards in the near future.

Literature:

(1)   W. L. Guess, S.A. Rosenbluth, B. Schmidt, and J. Autian, "Agar Diffusion Method for Toxicity Screening of Plastics on Cultured Cell Monolayers", J. Pharm. Sci., 54, 1545 - 1547 (1965)

(2)   Elwood O. Dillingham, "Primary Acute Toxicity Screen for Biomaterials: Rationale, In Vitro/In Vivo Relationship and Interlaboratory Performance", in: Cell-Culture Test Methods, ASTM Special Technical Publication 810, pp. 51-70, S.A. Brown ed., Philadelphia 1983.

(3)   O.H. Lowry, N.J. Rosebrough, A.L.Farr, and R.J.Randall, "Protein Measurement with Folin Phenol Reagent", J. Biol. Chem. 193, 265-275 (1951)

(4)   S. Roßberger, "Standardverfahren für die Zytotoxizitätsprüfung: Der MTT-Test", Z. Zahnärztl. Implantol. IV, 251-253 (1988)

Fig. 1: Agar Overlay Test

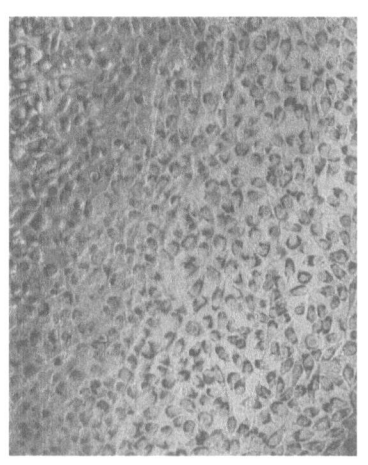 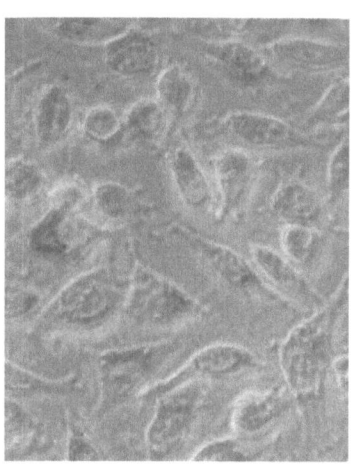

a   b

Cells in the vicinity of a nontoxic polymer

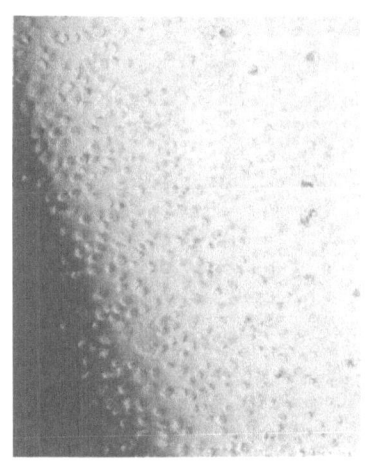 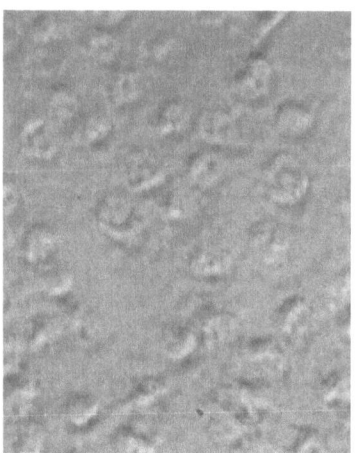

c   d

Cells in the vicinity of a cytotxic polymer

Figure 2: Growth of L 925 mouse fibroblasts in the presence of different extract concentratior
from cytotoxic medical device in the Cell Growth Inhibition Test

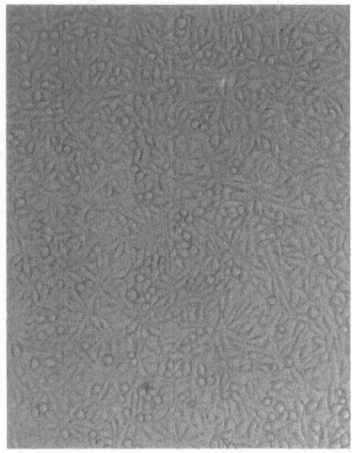

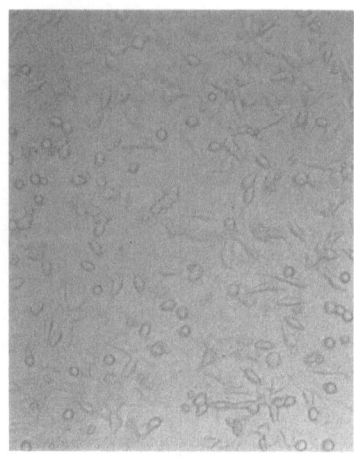

Fig. 2a: dilution 1:10000

Fig. 2b: dilution 1:1000

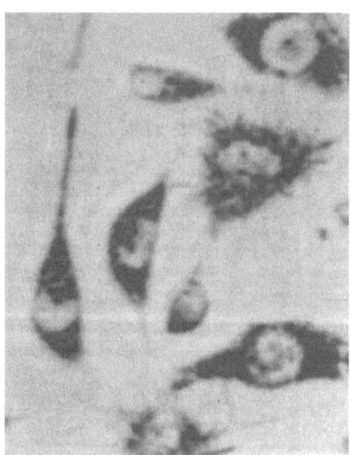

Fig. 2c: dilution 1:100

Fig. 3: Metabolization of Tetrazolium dye in mitochondriae
in the MTT-Test

FACT DATABASE MEDIPLAST
Polymers In Medicine

H. Planck, G. Umutyan, S. Konle, A. Kunberger

Institut für Textil- und Verfahrenstechnik, D-7306 Denkendorf

SUMMARY

A textual-numerical database about polymers and polymeric products used in the medical area has been established by the Biomedical Engineering Division of our Institute. Up to now the database contains informations about 290 trade names and 230 different models of medical products. The database language is German and English. MEDIPLAST is available either as inhouse or as online version.

INTRODUCTION

For developing medical products it is absolutely necessary to get an overview about experiences which already are available before starting the project.

Only this way unnecessary attempts and failures can be avoided and suited materials can be chosen.

With MEDIPLAST now a database is available which closes a gap in the information supply concerning polymer medical technique, development of materials for implantation as well as medical accessories.

LITERATURE SOURCES

Literature from 1980 to the present is evaluated.

Facts are taken from:

- Company informations
- Journals
- Books
- Theses
- Conference reports.

H. Planck M. Dauner M. Renardy (Eds.)
Medical Textiles for Implantation
© Springer-Verlag Berlin Heidelberg 1990

STRUCTURE

The database comprises at the present time 240 categories. The categories describe test methods, material properties or product identifications.
One category consists of one or several fields.

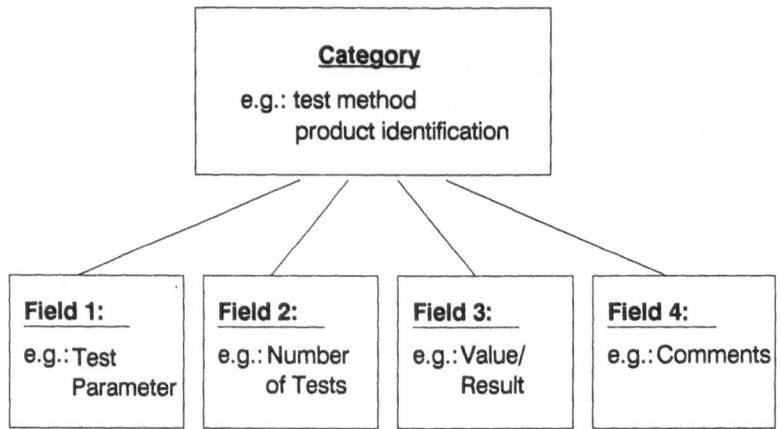

Fig. 1: Definition of a MEDIPLAST category

ITEMS OF MEDIPLAST

The following items are covered by MEDIPLAST:

- Source information
- Material information
- Polymer characterization
- Manufacturing data
- Biocompatibility tests
- Biodegradation tests
- Flow criteria
- Medical criteria
- Mechanical tests.

BIOCOMPATIBILITY TESTS (IN VIVO, IN VITRO, EX VIVO)

As parameters for biocompatibility the following items are considered:

- Toxicity
- Thrombogenicity
- Hemolysis
- Teratogenicity
- Mutagenicity
- Cancerogenicity
- Infection behaviour
- Allergy.

For each biocompatibility test exists a special test description.
As an example for a test used for evaluation of biocompatibility the MEDIPLAST category for the corneal implantation test with the appropriate test description is shown.

ARRANGEMENT OF A RECORD

In MEDIPLAST a record is defined as the sum of facts given for one polymer or one model of a medical product taken from one reference.

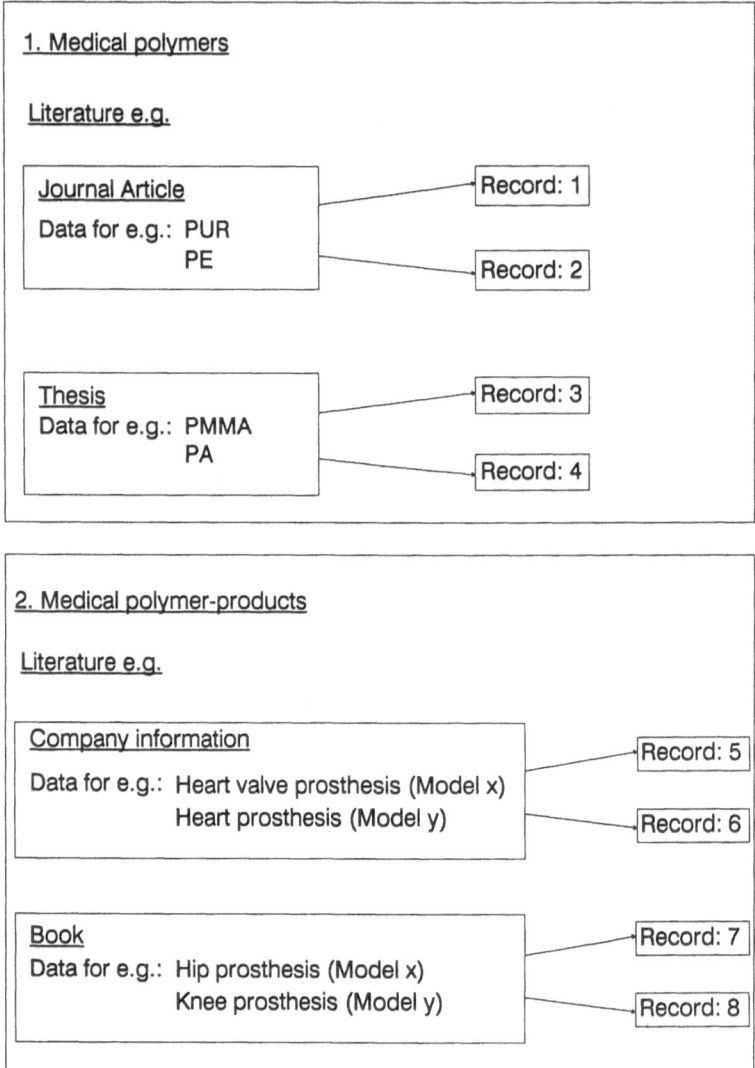

Fig. 2: Arrangement of a record in MEDIPLAST

ITV FDB MEDIPLAST (Englisch) M0470   Laufnr:M_____ Wdh.:__

3370 Corneal Implantation Test

Species: _____

Implantation Time: _____
Implant: _____

Reference Value not Pathologic: _____

Reference Value Pathologic: _____

Value/Result: _____

Comments: _____

Fig. 3: MEDIPLAST mask 470 for corneal implantation test.

Mask 0470

The test material in form of powder or pellets is implanted into an artificially produced cornea pocket of test animals (e.g. rabbits). In fixed time periods the tissue compatibility of the test material is examined macroscopically or microscopically. Macroscopic signs of cornea inflammation are reddening and clouding. Parameters of histologic examination are: formation of oedemas, cellular infiltration and neovascularization.

Literature: Gimbrone, M.A., Cotran, R.S., Leapman, S.B., Folkman, J.:
                   Tumor Growth and Neovascularization: an Experimental Model using the
                   Rabbit Cornea.
                   Journal of the National Cancer Institute 42: 413-427 (1974)

Fig. 4: MEDIPLAST test description for mask 470.

CONTENTS OF MEDIPLAST

The file size is 3600 records. At the present time the database includes 290 polymer trade names and 230 different models of medical products. Up to now facts to the following groups of polymers and medical products are available (listed in alphabetical order):

Table 1: Polymers and medical products in MEDIPLAST

Polymers:	Products:
Cellulose derivatives	Artificial kidney
Polyacetal	Bone cement
Polyamide	Blood circulation assisting system
Polyamino acids	Blood filter
Polycarbonate	Catheter
Polyethylene	Drug delivery system
Polyglycolic acid	Heart prosthesis
Polymethyl methacrylate	Heart valve prosthesis
Polylactic acid	Hip prosthesis
Polyoxymethylene	Knee prosthesis
Polypropylene	Ligament prosthesis
Polystyrene	Membrane
Polysulfone	Oesophagus prosthesis
Polyterephthalate	Patella prosthesis
Polytetrafluoroethylene	Prosthetic ring for valvuloplasty
Polyurethane	Protective bandage
Polyvinyl alcohol	Skin replacement
Polyvinyl chloride	Surgical mesh
Polyvinyl fluoride	Suture
Silicone rubber	Tube
	Vascular prosthesis

INHOUSE (PC) VERSION

The inhouse version is based on dBASE III (Ashton Tate, Inc.) and Clipper (Nantucket, Inc.).

It is menue-directed and structured in:
- mask establishing program
- input program
- retrieval program
- output program
- additional programs (e.g. descriptions for biocompatibility).

ON-LINE VERSION

The online version was set up in cooperation with FIZ Chemie, Berlin and is implemented at the host INKA.DAT, Karlsruhe, at the present time as a test version with 1500 data units.

This project was supported by the German Ministry for Research and Technology.

THE CENTRAL REGISTER FOR SIDE EFFECTS OF BIOMATERIALS

U.T. Seyfert, G. Bohnert, G. Pindur, *R. Bambauer, *G. Jutzler,
H. Kiesewetter, E. Wenzel

Department of Clinical Haemostaseology and Transfusion Medicine,
*Department of Nephrology, University of Saarland,
D-6650 Homburg/Saar

ABSTRACT

Attempts to realize on a national as well as international level a
Central Registry for documenting clinically relevant side effects
of biomaterials will be reported in detail. Implementation of the
registry involves the review of present literature and standardization procedures. An bioincompatibility report/questionnaire
has been developed. The aims of registering and evaluating this
report will be demonstrated and discussed in detail.
Keywords: Biocompatibility - Central Register - Bioincompatibility
report

INTRODUCTION

In 1984 the Department of Biomathematics in conjunction with the
Department of Hemostaseology and Transfusion Medicine at the
University of Saarland were accorded the task of establishing on
an international level a Central Register for documenting
clinically relevant side effects resulting from the use of
biomaterials. In 1989 the ISAO (International Society of
Artificial Organs) support the activities of the registry.
Thus we need to investigate clinical aspects of biocompatibility
of foreign materials according to the efforts undertaken in the
field of pharmaceutical side effects on a national and to some
extent international level, e.g. the reports of the "Deutsche
Arzneimittelkommission der Deutschen Ärzteschaft" in the last 50
years.

H. Planck M. Dauner M. Renardy (Eds.)
Medical Textiles for Implantation
© Springer-Verlag Berlin Heidelberg 1990

ORGANIZATION OF THE REGISTRY

The actual organization of the registry is listed up in the flow diagramm (figure 1,2). There must be an intensified exchange of knowledge and information between different bodies (figure 3). The registry of biomaterial incompatibilities will attempt to document clinically significant side effects of biomaterials that have occurred during randomized, prospective clinical studies, experimental studies, retrospective studies and as case reports. A bioincompatibility report has been developed according to the report on pharmaceutical side effects of the Deutsche Arzneimittelkommission der Deutschen Ärzteschaft. This incident report contains several informations and data, that will be protected under the information laws and will be stored in coded form (figure 4). This form should serve to facilitate easy documentation of suspicious or obvious cases of biomaterial incompatibility. This questionaire shall be obtainable from physician associations, DAB/GTH, ICTH, ISAO etc.

Thus our computer system (IBM) and our computer language (DBaseIII) will allow convenient access to the Central Data Bank for input and withdrawal of specific information. Bulletins with reports and editorials will or can be published on a half yearly basis.

Main criteria for registering and evaluation can be summarized:

1., objective evaluation by standardization of laboratory and diagnostic methods
2., appointment of reference laboratories and submitting of proposed and tentative standards
3., differentiation between toxicity, blood - and tissue incompatibility
4., relationship of biomaterial to physico - chemical basis of the biomaterial device
5., interaction of flow characteristics and bioincompatibility of specific devices
6., characterization of clinical aspects of biomaterial incompatibility
7., statistical evaluation of frequency, clinical intensity and duration of side effects

8., incidence of side effects with and without use of a specific biomaterial

9., interaction with medication and specific pharmaceutical agents

10., convenient access to information in the data bank

Conclusion

Immediate and ultimate goal of the Central Registry is prompt recognition of unexpected side effects and evaluating of biomaterials. This can also be achieved by intensified national and international collaboration.

Request for reprints

Dr. U.T. Seyfert
Abt. für Klin. Hämostaseologie
und Transfusionsmedizin
Haus 75
Uniklinikum
D - 6650 Homburg Saar
FRG

Figure 1: Organization of the Registry I

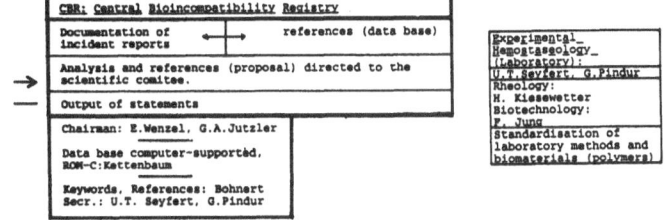

Figure 2: Organization of the Registry II

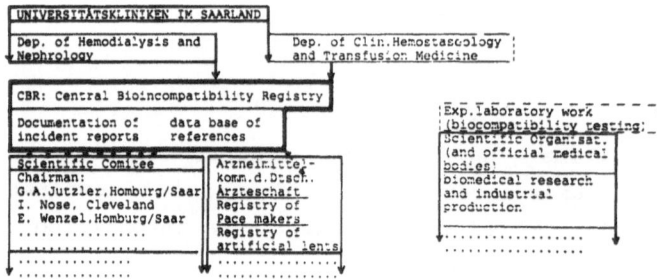

Figure 3: Exchange of knowledge and information between different
bodies

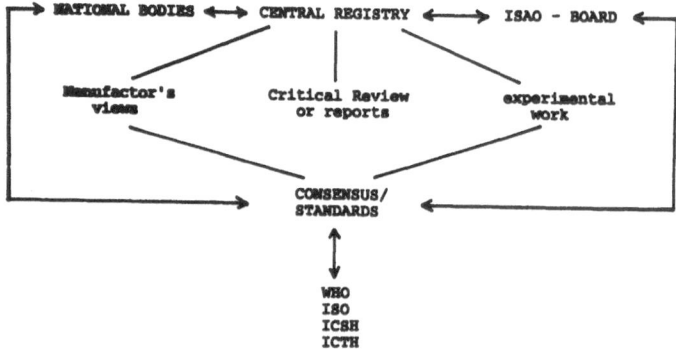

Central- Registry - Organization

Figure 4: Biomaterial incompatibility incident report

1., Patients data
2., Description of clinical aspects
3., Type of biomaterial used (duration, reason for use, etc.)
4., List of medications and dosages
5., Underlying disease process
6., Laboratory parameters related to the adverse effects
7., Therapy and course of adverse effects

INFLUENCE OF PROCESSING PARAMETERS FOR ARTIFICIAL LIGAMENTS

Dauner, M., Planck, H., Syré, I., Dittel, K.-K.[*]

Institut für Textil- und Verfahrenstechnik, Forschungsbereich Biomedizintechnik,
D-7306 Denkendorf
* Marienhospital, Chirurgische Klinik 7000 Stuttgart 1

ABSTRACT

In principle, the structure of ligaments and tendons, composed of a multiple of fibrils and fiber bundles, can be compared with polymeric textiles. Following disapprovel of nontextile structures for ligament replacement reasons are given for the superiority of braid technology over other conventional textile processing. The influence of the processing parameters onto the mechanical and biological properties of a braid are discussed more intensively.

INTRODUCTION

The performance of an implant will be determined mainly by the properties of the material, by an adequate design and by the proper implantation procedure.

Producing an implant the processing parameters and the design have to be adjusted to the material and the specific requirements of the destinctive application. For artificial ligaments the requirements are shortly listed in Fig. 1.

I. Biological	II. Biomechanical
- Biocompatibility	- Physiological progressive stress-strain behaviour
- Long term stability	- Low creeping
- Supporting tissue proliferation	- High shear strength
- As few non-viable material as possible	- High bending resistance
	- Low bending stiffness

Fig. 1: Requirements on a Cruciate Ligament Prosthesis

H. Planck M. Dauner M. Renardy (Eds.)
Medical Textiles for Implantation
© Springer-Verlag Berlin Heidelberg 1990

By changing the processing parameters one can influence within limits the performance of the device relating tissue ingrowth and the mechanics in general as they are: physiological stress-strain behaviour, high shear strength and bending resistance, low bending stiffness and low creeping.

An often stated requirement for an implant is the so called isoelasticity, means the stress-strain behaviour should be equivalent to that of the natural ligament. Yet only in combination with the theoretical isometric implantation an isoelastic prosthesis can work ideally under physiological conditions. Up to now a non isometric site has to be regarded as the normal case. As very clearly shown by Müller /1/, but by others before him also, any divergence from the theoretical ideal site will lead to overloading and overstretching and thereby to destruction of the implant /Fig. 2/. A strong prosthesis yet will restrict the knee flexion at first, but can cause severe degeneration of the knee joint as well.

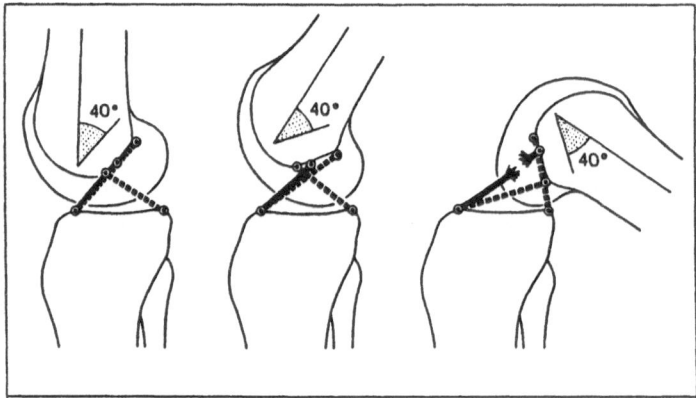

Fig. 2: Non-Isometric Implantation leading to overload/-stretching and thereby to Destruction of the Implant and/or the Bone (from Müller /1/).

Thus, besides the general requirements of biocompatibility, functionality and implantibility, we have to demand that the prosthesis should have a strength not beyond that of natural ligament; and the prosthesis should have, as the engineer calls it, a safety limit, which helps to avoid a damaging of natural structures. The safety limit ideally leads not to rupture of the implant, but will allow an overstretching of the prosthesis by creeping of the material. Differences from the optimum should be balanced by means of a compensation member.

Textile structure for ligament replacement

One of the earliest prostheses on market was built up by a rod of a bulk polymer. Fig. 3 shows the implantation site, which is quasi-isometric as recommended by the inventors /2/.

Now in general ligament prostheses will be bent at the insertion points while knee flexion.

Its a base of mechanics at a bent rod you have a neutral phase from which into one direction tension, into the other direction compression increases with the distance from the neutral phase (Fig. 4). The greater the diameter of the rod, the higher the tension and compression will be at the outer layer. Therefore kneeflexion produced high alternating bending load at this prosthesis, a stress which will be borne hardly by any material. Thus failure of the prosthesis did occur, and as far as we know, it will not be implanted any more. That means, to minimize the bending load one has to reduce the diameter.

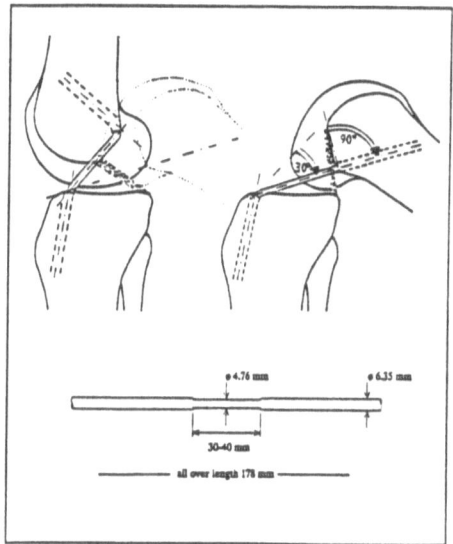
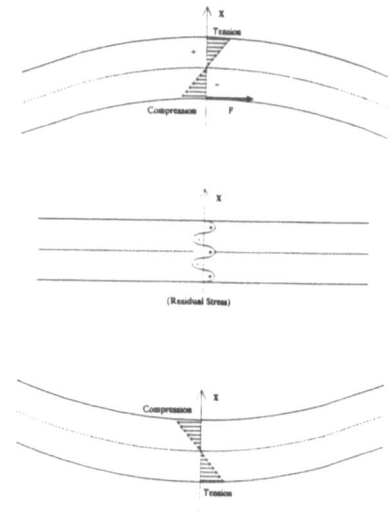

Fig. 3: Implantation Site of the RICHARDS- Fig. 4: Tension/Compression Inversion
         Polyflex UHMWPE-Prosthesis   Following Alternating Bending Load

Looking to nature of ligament as shown very clearly by Kastelic /3/, Prockop /4/ and others, it can be seen that in electron microscopy ranges peptide chains form helices which may be comparable to a twisted yarn (Fig. 5). That is the configuration of a collagen molecule. Like staple-fibers a multiple of those are building up the microfibrilles and fibrils. A large number of fibrils are bundled to the fascicle, which is sheathened by the fascicular membrane. The arrangement of the fascicles finally build up the ligament.

As an approach the structure of a ligament can be compared with twisted and bundled multifilaments consisting of stable fibers. The characteristic progressive stress-strain behaviour may be caused in a crimped structure of the fascicles.

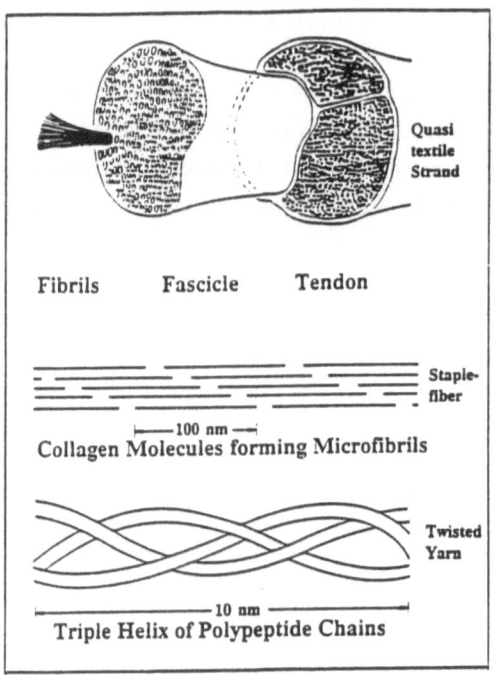

Fig. 5: From Peptide Chains to Tendon (Following Kastelic et al /3/ and Prockop et al /4/)

Requirements onto a textile structure

Now the use of a textile structure for ligament prostheses is obvious, and one has to look how to fulfil the requirements of the implant by a textile processing.

As any polymeric material, parallel aligned multifilaments normally have a linear Young's modulus at first, which decreases at higher strain. The progressive stress-strain behaviour can be produced in a yarn by crimping. But this processing will impair thermically the mechanical and the long term properties of the material. Thus the required behaviour should be achieved by a conventional textile manufacturing.

Low bending stress and the flexibility are a consequence of the low diameter of the fibers and will facilitate the implantation.

Tissue proliferation onto a surface is mainly dependent on the surface structure of the material used. But if this is given, the textile configuration enables tissue ingrowth and thereby a complete incorporation of the implant. One can discuss if this should be aspired, but below arguments will be given from a mechanical point of view for the advantage of the ingrowth of tissue.

Conventional textile technologies

From the conventional textile structures, the nonwovens are generally unsuitable because of their poor strength.

Also the weft knitting can not be taken for load bearing device, because stretching of a **weft knit** is possible at first without essential force (see Cpt. 1, Planck). When the loops are extended, any further elongation will be performed by the material only and thus the Young's modulus of the weft knit increases promptly to that of the material. On the contrary the natural ligament will bear load at a small extension already but allows considerable additional elongation at high forces.
The behaviour of the **warp knit** is comparable to that of the weft knit, though not extremely so (see Cpt. 1, Planck).
Both of these two technologies have in common, that they are unsuitable for the processing of high modulus fibers as carbon or aramid fibers. Because of their poor bending strength, break at the top of the loops would be unavoidable.
As far as we know, weft knitting is not used in ligament prostheses, but warp knitting technology is employed for getting enhanced tissue ingrowth.

With **woven fabric** there is no essential extension beyond the material elongation (see Cpt. 1, Planck). Though high modulus fibers can be manufactured well by this technology, the lack of extensibility of the fabric seems to exclude its usability. Yet one woven prosthesis should be mentioned: Contzen /5/ takes a dense fabric made of high strength polyester fibers. The extensibility and the progressive stress strain behaviour is reached to some degree by twisting the fabric on the longitudinal axis.

Finally the stress strain behaviour of a **braided structure** is roughly comparable to that of the natural ligament (Fig. 6). The difficulty is to combine material and processing parameters to fulfil all the requirements.

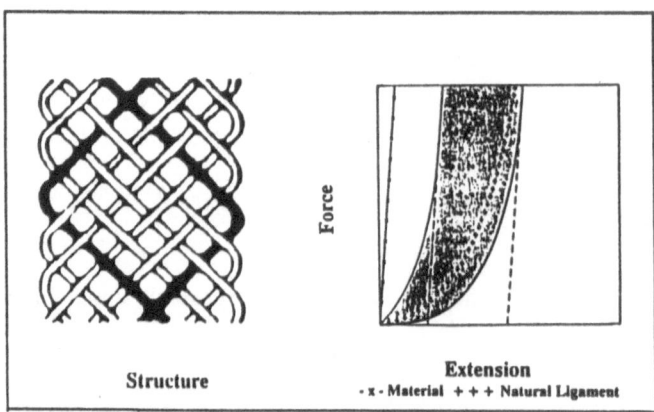

Structure   **Extension**
                 - x - Material + + + Natural Ligament

Fig. 6: Structure and Stress Strain Curve of a Braid

With braids the key to extensibility and to the progressive stress-strain behaviour is the crossing angle of the yarns, the so called braid angle, which can be defined as twice the angle between the fibers and the longitudinal axis of the braid.

While stretching the prosthesis the angle will be reduced, the fiber bundles pressed and the fibers themselves will be deformed at the crossing points. The free space between the fibers and fiberbundles decreases nearly to zero. Then further elongation will be performed by the material only.

Processing parameter

Starting at the sum of requirements the materials have to be chosen. Braiding technology generally allows the usage of high strength fibers as polyethyleneterephthalate fibers, for example Dacron (Du Pont de Nemours) or Trevira (Hoechst/Frankfurt); and as well the usage of high modulus fibers as carbon fibers for example.

While high modulus fibers have a poor bending strength, high strength fibers on the other hand are tending to creep.

Using two or more different materials, one has to consider that the less wear resistant material will be abraded gradually producing particles which may cause severe body reactions.

The shape can be tubular or tape like. To protect the fibers at the bending edges of the insertion points, sometimes the prostheses were coated or sheathened. Thereby tissue ingrowth will be excluded. With tubular braids the lumen can contain a non textile core.

Within limits the titer of single filament can be chosen. The titer per carrier has to be adjusted to the number of carriers for getting a total titer which brings up sufficient strength.

Before spooling the yarn onto the carrier you can ply, twist or just braid it. While twisting improves the processability of the yarn and in consequence the strength of the braid, between only plyed fibers, it means parallel aligned fibers, tissue has or can make space to ingrow. Relating extensibility of the braid there is no difference of these processes. Trials to enhance the extensibility of the prosthesis by using a prebraided yarn didn't bring the wanted result.

Stationary threads lying parallel to the longitudinal axis of the prosthesis are enclosed by the crossing fibers. The material must have an elongation at break at least comparable to the extensibility of the braid. That requires the usage of elastomeric fibers. Nevertheless by condensing the braiding structure stationary threads will reduce the extensibility of the braid yet improve the reversibility of extension. The same effect core threads can have, but with sufficient elongation at break, they will reduce the extensibility of the braid less.

Pick counts per length, that's the number of crossing points of the fibers per a defined length, generally per french inch, that is 27.8 mm. The pick counts are very closely connected to the braid angle, as will be discussed below.

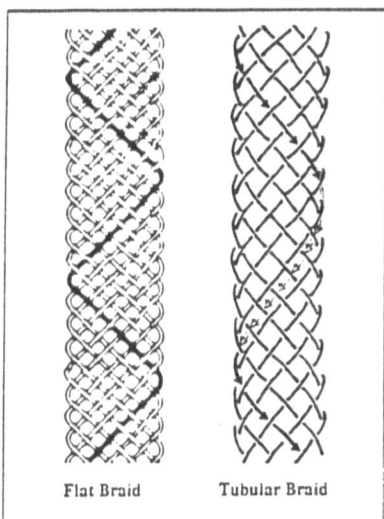

Fig. 7: Course of the Yarn in a Braid

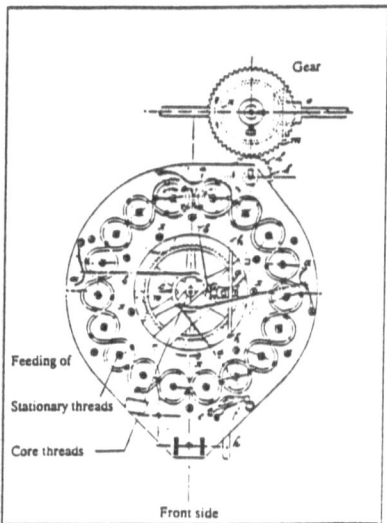

Fig. 8: Top View onto a Braiding Machine

Relating the shape of a braid it's obvious, that a tubular braid offers the option to enclose a core, which following implantation can consist also out of ingrown tissue. But it can be seen from Fig. 7 there is also a mechanical reason to produce a tubular braid especially for high modulus fibers. At the edges of a flat braid the yarn has to be turned relatively sharp. The turning angle is equal to 180 degree minus the braid angle, as can be easily checked. At tubular braid the yarn is wound helically without strong bending.

Looking at the top of a braiding machine (Fig. 8) one can imagine the course of the carriers, half turn to right and the other to left.

Stationary threads will be fed through the axis of the impeller wheels, core threads through the center of the machine.

Influence of Filament Titer

Choosing the titer of single filament of the yarn, the following has to be considered, being strongly valid yet only for one destinct material (Fig. 9).

As already mentioned the bending stiffness and the tension is increasing with the filament-diameter or titer affecting the bending behaviour negatively. The titer related tensile strength decreases when the titer increases because of the lower cristallinity of the fibers.

Yet the absolute breaking strength of single fiber will increase with the titer. Therefore break of filaments will occur sooner at a lower single filament titer.

Tissue needs space for ingrowth. It is obvious that the interstice between two fibers or two fiber bundles increases with the respective diameter. As we assume there may be an optimum of free space for inducing tissue ingrowth (Fig. 9 + 10).

Single fiber Titer (dtex/denier)	High	Low
Bending behavior	— < +
Tensile strength of multifilament	— < +
Break of single fiber	+ > —
Tissue ingrowth (between single fibers)	+ > —

Titer per carrier Number of carrier	High	Low
Tensibility of braid [f (braid angle)]	— < +
Tensile strength	+ > —
Tear strength	+ > —
Tissue ingrowth (between fiber bundles)	+ > —

Fig. 9: Properties of Yarn depending on Titer of Single Fiber

Fig. 10: Properties of Braid depending on Yarn Titer per Carrier

Tensile strength and tear strength of a braid will increase when the titer per carrier increases and accordingly a smaller number of carrier is used, because then the shear forces will decrease. That's important for high modulus fibers especially. Consequently on the other hand the tensibility of the braid will be lower. That's a disadvantage when using high modulus fibers.

Braid Angle

A key position in the stress strain behaviour is the often mentioned braid angle. It cannot be chosen independently but it is rather a result of the other braiding parameters (Fig. 11).

When increasing the filament titer per carrier or the number of carriers the braid angle will increase too. Yet in general enlarging the one you have to reduce the other parameter to get a destinct total titer. To define the resulting braid angle now is more complicated.

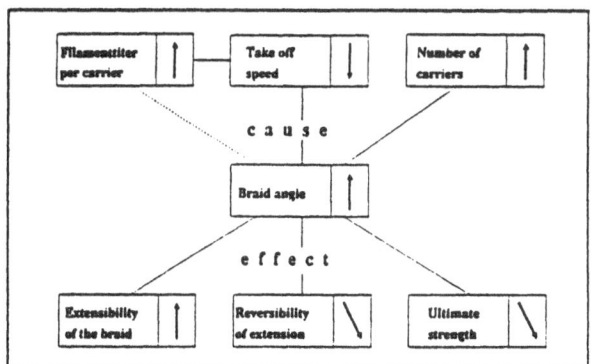

Fig. 11: Cause and Effect of Increasing Braid Angle

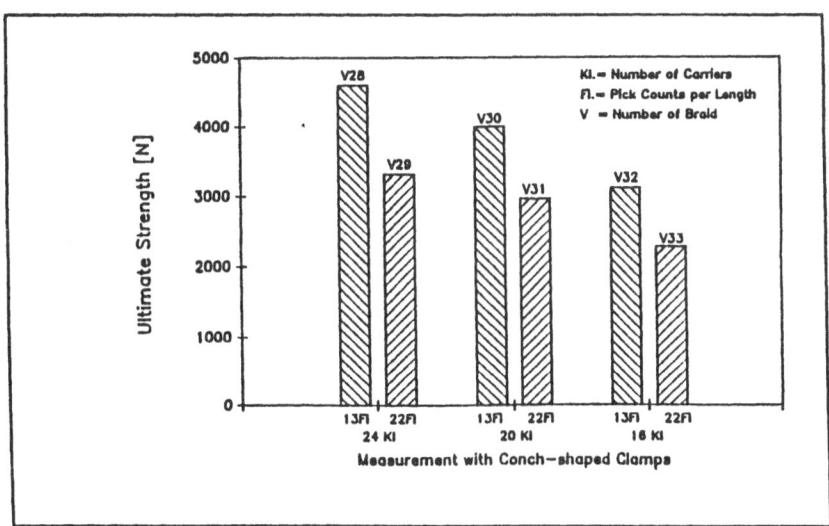

Fig. 12: Ultimate Strength as a Function of the Pick Counts per Length, Braids of KEVLAR 49

A direct possibility to enlarge the braid angle is to choose a smaller take off speed. Hereby the pick counts per length will increase proportionally. Yet higher shear forces on the filaments will take effect, resulting in a lower ultimate strength of the braid. This is shown in Fig. 12 for three braids of the high modulus aramid fiber. A nearly proportional reduction of the strength for the different braid constructions follows the increasing of the pick counts.

A second disadvantage may be the increase of irreversibility of extension, especially when the braid angle is extremely enlarged.

Yet the great advantage of the braiding technology is the possibility to adjust the extensibility of the prosthesis by changing the braid angle even with high modulus fibers. To maintain a sufficient elastic behaviour of the implant further efforts may be necessary.

Reversibility of Extension

Though not a processing problem, but essential for the design of a ligament prosthesis is the influence of the preload during implantation onto the stress-strain behaviour of the braid (Fig. 13). As considerable parts of elongation happen at a range of low forces, a preload will take away these parts limiting the function of the prosthesis. Elongation at low forces is partially irreversible (Fig. 13, curve 4). That can be solved with a high elastic element, for example a core material. To keep the extensibility of the braid a material with a low hardness is required. Elastomeric fibers can adapt to the reduction of the braid under load best.

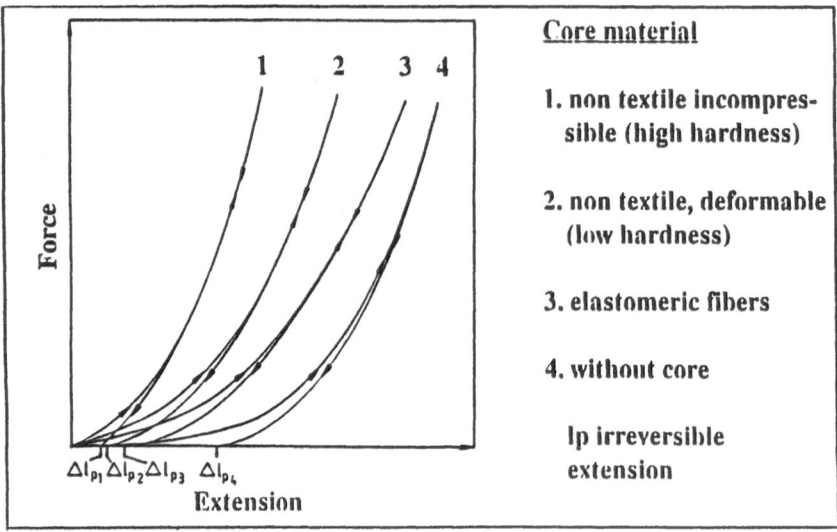

Fig. 13: Dependence of Extensibility and Reversibility of the Core of a Tubular Braid

This is the point to argue the advantage of the ingrowth of tissue, which will act as an elastomeric core. Beside the complete sheathing and thereby protection of the filaments, there are reasons to assume that ingrown tissue can be within limits a corrective for the insufficiency of an implanted ligament prosthesis.

Looking at curve 4 one may try to eliminate the irreversible elongation by bringing up a preload, putting up with a reduction of the over all extensibility.

Yet under the premises that the prosthesis is implanted without a high preload and that the surface chemistry of the material is appropriate, ingrown connective tissue may hold the prosthesis straight and thereby eliminates the irreversible part of the extension nearly, maintaining the over all extensibility. A stress strain curve like curve 3 may result.

This means, ingrown tissue will reduce the overall extensibility by a small degree, yet enhances considerable the elastic, reversible extensibility. This may be the key of the good performance of prostheses which are relatively stiff at first, but seem to be more elastically stretchable after a one year period of implantation.

Our - the engineers and the surgeons - problem is that we cannot manage reproducable that tissue ingrowth.

References:
1 Müller, W.: Das Knie, Springer-Verlag (1982)
2 Patent US 503990-12331
3 Kastelic, J. et al: The multicomposite structure of tendon,
          Connective Tissue Research, 6 (1978), 11-23
4 Prockop et al: Collagen diseases and the biosynthesis of collagen
          Hospital Practice, Dec. (1977), 61-68
5 Contzen: Alloplastischer Bandersatz, Unfallchirurgie 11 (1985), 242-246

SELECTED DISCUSSION REMARKS:

Claes: What is about the primary fixation of the ligament? We know that the fixation is also an elastic system. And the deformation of the fixation is added to the deformation of the prosthesis itself.

Dauner: The elastic behaviour of the fixation may be neglectible at the forces you have under physiological conditions. It may be essential perhaps at forces greater than 400 or 500 N.

Freudiger: You have mentioned the possibility to use a core. Do you actually use one in the braid?

Dauner: We didn't use a core. If you have a core of a polymeric material, you have the problem that the friction between the two materials will produce abrasion particles from the material with the lower wear resistance. For example if you would use a carbon fiber braid with a silicone core, you will have silicone rubber particles in the joint. But we have seen with the aramid fiber, and it's the same with the carbon fiber prosthesis, that tissue can ingrow into the lumen of a tubular braid and that could work just like an elastomeric core.

TREATMENT OF LIGAMENTOUS INSTABILITIES OF THE KNEE-JOINT UNDER THE SPECIAL
ASPECT OF PROSTHETIC LIGAMENT RECONSTRUCTION

*Dittel, K.-K., **Dauner, M., **Planck, H., **Syré, I.

*Department of Surgery, Marienhospital Stuttgart
**The Institute of Textile Research and Chemical Engineering, Denkendorf

The origin of surgery dates back a few thousand years, the development of
surgery on joints however only about one hundred years. Over a period of 50
years a multiple number of operative procedures have been described to achieve
stabilisation of the so-called "ACL-deficient knee" (anterior-cruciate-
ligament-deficient knee). The effective development of the surgery on the knee
joint up to today's standards does not reach back more than 25 years. Especially
the introduction of anaesthesia and the consequent development of the vital
precondition for a minimal contamination with germs gave the surgery of
ligamentous injuries of the knee-joint important impulses.

The present state of surgery of this region would be unthinkable without
scientific and experimental research. Experimentation on animals - which is
often criticized today - has had a rather considerable share in the increase of
basic knowledge of this special part of surgery.

Historical view:

The development of knee-joint surgery concerning injuries of the ligaments can
be classified in 4 stages of treatment. The stage of therapeutically
conservative nihilism was succeeded in the twenties by the stage of surgical
treatment by primary suture of ruptured ligaments which was marked by the first
experiments with anatomical reconstruction. Intra- as well as extraarticular

H. Planck M. Dauner M. Renardy (Eds.)
Medical Textiles for Implantation
© Springer-Verlag Berlin Heidelberg 1990

stabilizing methods were developed using local autologous material. The procedures of autologous syndesmoplasties which was inaugurated in the early fifties led to a considerable improvement of surgical results, also in regard to primarily ignored injuries and their resulting secondary instabilities. For about 20 years now alloplastic materials are being tested in the course of experimentations on animals as well as in the clinical sector since the results in the application of autologous materials have not always satisfied the longterm expectations.

Up to the turn of the century ligamentous knee injuries were hardly noticed, as there were neither the necessary diagnostic measures taken, nor was there an acceptable therapy available. In 1901 Hints was the first one to report on traumatically caused ruptures of the collateral ligaments of the knee indicating the possibility of resulting instabilities of the knee-joint. Goetjes as well as Jones and Smith reported in 1913 about injuries of the cruciate ligaments and

Figure 1:

Prognosis of the ACL deficient knee

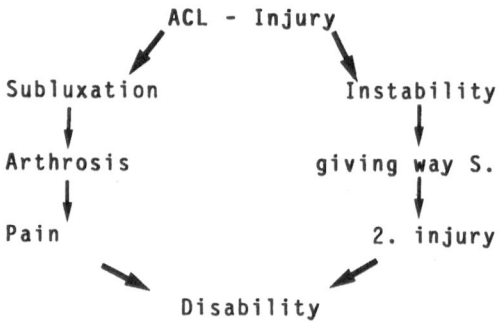

they were the first to propagate the possibility of surgical reconstruction. In 1917 Hey-Groves underlined the general opinion concerning the conservative therapy of the ligamentous injuries to the knee-joint at that time in a dramatic way:

"A rigid plaster or leather-tube carried for one year followed by a tubular appartus represents the generally accepted method!"

Therapy:

The conservative treatment of capsular and ligamentous lesions located at the knee-joint was the almost exclusively propagated and practiced therapy until the end of the forties. The German speaking countries were essentially influenced in this fashion by the Vienna school of Böhler. Collateral ligament lesions as well as ruptures of the ACL, were immobilized over a period of 3 months in a plaster dressing after the first results of surgical treatment had been strikingly worse and had shown longer terms of treatment than by using a conservative procedure.

In the past decades opposing viewpoints have accompanied the syndrom of the ruptured ACL in regard to the cause of accident, symptomatology, clinical diagnostics and possible methods of treatment. Especially diagnostic as well as therapeutic necessities were discussed very enthusiastically and with a lot of discrepancy. The consequences of primarily unnoticed fresh ruptures of the ACL were considered from "seriously handicapping" to "completely unimportant" (Arnold 1979).

In 1938 Palmer was one of the first to plead for a primary surgical treatment of lesions, because he had seen better results after operative procedure in his own cases. The decisive turning point in traumatology of the kneejoint within the anglo-american countries was brought about by O'Donoghue in 1950. He introduced the term of the so-called "unhappy triad" and claimed as a goal to any treatment

114

Operative procedure after ACL-injury

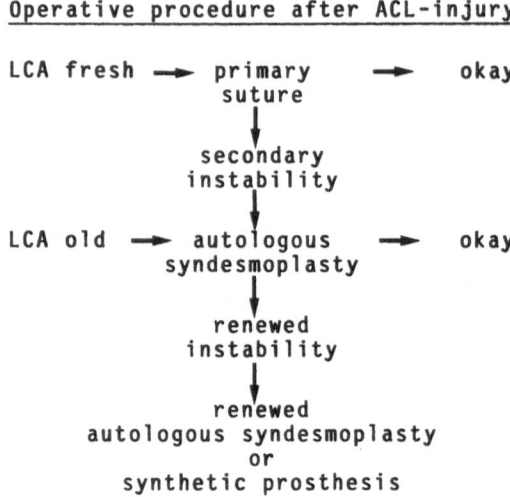

of ligamentous lesions a consequent anatomical reconstruction. He recognized the essential significance of the ACL and its insufficiency as a main cause for reduced or terminated capability in sports.

Because acceptable therapeutical results with primary conservative treatment could be expected in only 1/4 of all cases a mental change of direction towards the formerly scorned method of surgical treatment of ligamentous combinated injuries took place.

The resulting experiences from the surgical procedures led to the development of mumerous methods of intra- and extraarticular stabilization in the sense of primary anatomical reconstruction or the transposition of tendons which insert in the periarticular area. Initially intraarticular techniques based on autologous tissue transfer were mostly used. An excursion to the development of extraarticular procedures took place, but when bad longterm results became apparent a return to intraarticular reconstruction of the ACL took place again.

In reconstructive surgery to the knee the application of autologous materials has proved to be successfull. This is true for primary anatomical reconstruction as well as for autologous syndesmoplasties. But the application of own tissues for augmentation as well as for a complete substitute, for which we use primarily pediculate or free tissue parts from the ligament of the patella, the tendon of the quadriceps muscle or from the fascia lata, can show obvious disadvantages next to the undisputable advantages. The decision for a trans- position or for the removal of a strip of fascia includes, besides the primary traumatic damage and accompanying damage of intact structures, an extension of the operation not only in terms of time. The risk of an additional weakening of stable structures is thus obvious. Finally an immobilisation of the operatively treated knee is necessary to guarantee the process of healing. The disadvantages of a long term immobilisation in a plaster dressing must not be underestimated.

The success of any surgical stabilizing measure can't be judged solitarily under the aspect of biologic adaption of the transplant, but it depends essentially on the surgical technique and the solidity of the used autologous tissue as well as the anchorage in the bone tunnels. Unsatisfactory late results have inforced the search for alternate possibilities of treatment. Another reason has to be seen in the fact that secondary instabilities are to be expected in about one third of the patients after surgical treatment. From the beginning of the seventies various synthetic materials were tested under experimental research on animal as an alloplastic substitute for the ACL. The main advantage of immediate possible mobilisation, which is granted through the primary stability is pitted against the possible disadvantages in regard to the biomechanical characteristics of the used filament material including the problems of biocompatibility. Through the possibility of an unlimited early functional treatment the disadvantage of an immobilisation in a plaster dressing is cancelled. But it should not be

Table 1:   <u>Disadvantages by immobilisation</u>

- 5 to 6 weeks lack of active motion
- risk of thrombosis and embolism
- functional loss of neighbouring joints

- consequences to soft tissue and bone (dystrophy, cartilage destruction)
- disability after removal
- need of secondary surgery

Table 2:   <u>Operative principles for reconstruction</u>

<u>of the anterior cruciate ligament</u>

1. Isometric position by the one shot tunnel method
2. Increase of stability and ingrowth by tight connection between tunnel and prosthesis
3. Stable fixation at femur and tibia

4. Tension of the substitute should be in range of the normal knee
5. A notch plasty can avoid ligament rupture
6. Reconstruction of additional lesion can reduce rotation

forgotten, that when a biocompatible synthetic substitute is implanted in a knee joint, the practiced surgical treatment has to stand comparison to the results of "conventional" methods of treatment.

Problems will arise if a knee-joint becomes again unstable in connection with a failed reconstruction using autologous material. For these problematic cases the alloplastic ligamentous substitution can be a real alternative. Besides the big advantage of immediate postoperative stability a further advantage can be seen in connection with the unlimited availability and the protection of intact soft tissue used for a syndesmoplasty. Under these circumstances all synthetic implants have to be jugded critically since they are not biological materials.

Experimental research:

In this experimental paper a newly conceived braided ligamentous prosthesis produced out of kevlar 29 is presented. The filament material consists of a

Figure 3:   **Operative Series (2/86 - 11/86)**

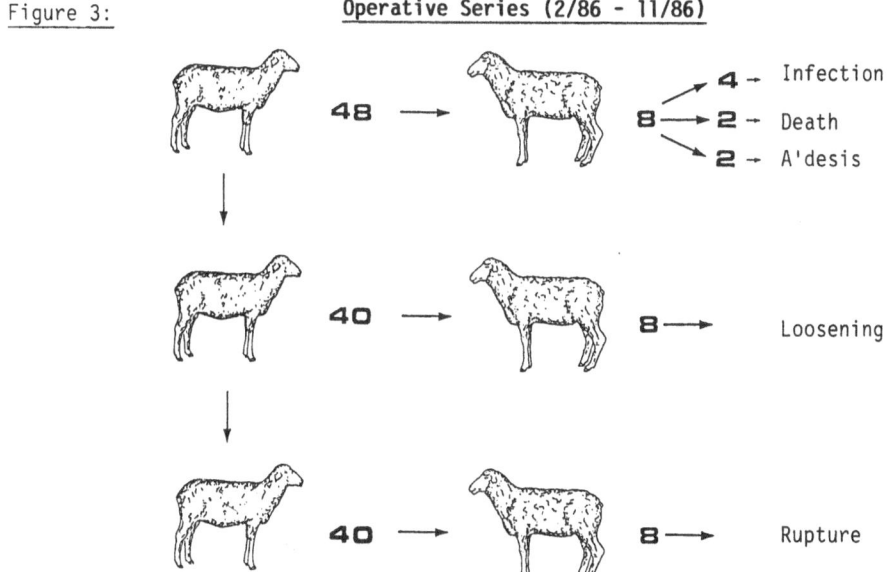

fiber thickness of 12 μm with a tensile strength of 3,6 kN/mm² for the single fiber and a breaking elongation of 4 %. The number of single filaments in the prosthesis structure is n = 18.000. The tensile strength of the prosthesis is 3.500 N and the breaking elongation 9 %. The extension of the ligament is 6 x 1 x 300 mm. The prosthesis has the shape of a tubular implant that is produced out of prefabricated kevlar cords with a 60° angular network.

For the research series, we chose sheep as experimental animals because their knee-joints have similar dimensions to the human ones. Between February and November 1986 48 female Merinoland sheep aged between 4 and 7 years were used to implant the prosthesis as an alloplastic substitute for the anterior cruciate ligament after prior resection of the ACL in the right knee-joint. The implantation was done isometrically using the 4,5 mm drill of the AO. The extraarticular fixation was done with jagged ring plates and small fragment screws. There was no postoperative immobilisation of the joints. They were defined as stable for stress and treated functionally early. 8 animals were lost for the scheduled postoperative examinations. In intervals of 3 months in the course of one year a quarter of the sheep was sacrificed and the knee-joints were examined macroscopically, microscopically and under biomechanic aspects.

The macroscopic findings showed an arthrosis in 5 animals, the degree of degeneration corresponding to the findings at the time of implantation. Between the 6th and 9th month there was a destinct integration of the filament material through connective tissue growth and complete surrounding of the prosthesis. Intraarticular infections could not be found. Furtheron there could be found no abrasion particles in any joint.

In 12 joints respectively was existing a loosening of the prosthesis or an interruption of the continuity. So far no intraarticular tumor growth could be found.

Figure 4:   **Tear-out forces of the kevlar prosthesis**

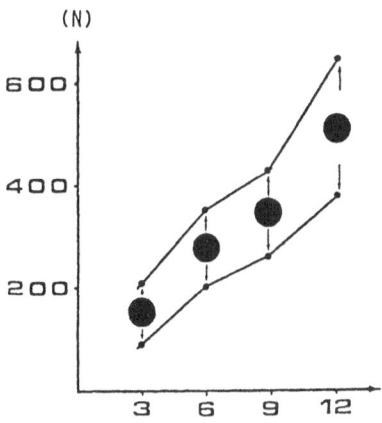

(N)

Time of implantation (months)

There was a decrease in the motion range on the operated knee-joints depending on the implantation time. An increase of periarticular scar formation with a reactive tendency for contraction after the medial arthrotomy as well as the progressive connective tissue and callous integration of the prosthesis within the bony canals were seen as an explanation for these findings.

Standardized tests for the stability of the joints showed a decrease of the anterior drawer symptom depending on the duration of implantation, contrary to the increase of limited mobility. This was explained by the increase of intra- as well as extraarticular fixation.

The tearing out forces showed an increase on the average of 67 % between the third and sixth month with another boost of 30 % until the end of 9 months and with regard to the overall observation period another increase of 48 %. The increase of the average tearing strength from 115 N (3 months) to 515 N (12 months) corresponds to a mean increase of 232 % over the total observation period. At the end of 12 months the tearing strength had been defined to be at a minimum rate of 355 N and at a maximum rate of 695 N.

The loosenings were explained by osteoporotic bone catabolism and pressure peaks caused by the material around the intraarticular bone canal openings. The intraarticular ruptures along the kevlar prosthesis could be explained by newly formed osteophytes in the intercondylar area in the sense of a secondary impingement after the performance of a primary notchplasty.

The histological examinations showed an encapsulation reaction against the kevlar prosthesis without a complete osseus transformation of the capsular layers around the prosthesis and thereby no consolidating osseus integration. When comparing the test results from 3 up to 12 months an increasing osteogenesis along the bone canals was found however without an integrating bone growth into the textile structure of the kevlar prosthesis so that it could be described as an "extraterritorialization" of the prosthesis. At no given time there were indications for an integrating bony ingrowth. The additionally performed examinations for foreign body reactions in the parenchymatous organs gave no evidence of abrasion particles.

Conclusions:

In conclusion can be said, that there were histological signs of an active biodegradation to be seen in the kevlar material that became stronger the earlier the organism succeeded to widen the wallstructures of the prosthesis. Since it could be found that an increase of the biodegradation reactions is connected to the frequency of dissociation of the single fibers, the close-meshed tubular network appeared to be a promising concept for the implant. A broad-meshed network would seem to be more suitable for an osseous integration of the prosthesis. However it could be expected a broader reaction and working area with regard to the biodegradation.

The presented Aramid-fiber prosthesis is suited to grant the necessary stability for a physiological motion range in the sheep even after an implantation of one year with an anatomically correct implantation and with a stable fixation of the ends of the ligaments. Secondary neoplasms by osteophytes, especially in the intercondylar area, including secondary arthrotic developments and decomposition under the ring plates as well as around the intraarticular openings to the bone canals had to be considered as complicating factors. The results of the animal experiments however cannot be transfered to clinical conditions uncritically under the aspects of the successful implantation in 28 animals (70 % of the usable findings) even though the technical proceedings appear to be practicable.

Table 3:   **Indication for a synthetic substitute of the ACL**

➤ Stabilization after prior failed reconstruction

➤ Primary measure in chronical instability

➤ Contraindication for immobilisation
   • therapy resistant atrophy of the quadriceps
   • risk of activating a gonarthrosis
   • relevant secondary follow up damages

➤ Missing cooperation concerning rehabilitation

➤ Special situation of the patient
   • imminent loss of work with long disablement
   • professional sportsman
   • self-employment

➤ Temporarily limited measure
   (secondary retreat to autologous syndesmoplasty)

SELECTED DISCUSSION REMARKS:

Gogolewski: Do you have any opinion about the different results presented by Prof. Claes and you, both with aramid fibers? Could you find any kind of reason as far as the chemical structure of the polymers is concerned?

Dittel: We have used the same type of prosthesis. The stabilization Prof. Claes has done in Ulm was much stronger than the stabilization we did. We used small cancellous bone screws with toothed discs, so that the prosthesis had the chance to move and the load peaks weren't as high as in the case of the very strong fixation performed in Ulm with the ligament fixation plates.

Claes: It is a difference of percentage of failure, not in the principle. We both have seen ruptures. The question is, if there is any biodegradation and if it is a chemical or an enzymatic degradation. But an answer may be given in the next paper.

HISTOLOGY OF ARAMIDE CORDS (KEVLAR {R}) USED AS A CRUCIATE KNEE LIGAMENT SUBSTITUTE IN THE SHEEP.

C.R. Jerusalem[1],   M. Dauner[2],   H. Planck[2],   K.-K. Dittel[3]

1) Dept. of Cell Biology, K.-University of Nijmegen (NL),
2) Institut f. Textil- und Verfahrenstechnik Denkendorf (D),
3) Marienhospital Stuttgart, Stuttgart (D).

SUMMARY.

A higher porosity of the graft fabric could certainly improve the healing, that is the incorporation of the entire graft in a connective tissue body and subsequently in an osseous matrix, of aramide cords (KEVLAR {R}), used as a cruciate knee ligament substitute in the sheep.

Accelerated healing could probably also reduce biodegradation of superficial aramide fibers, through the early replacement of macrophages by connective tissue and osteoprogenitor cells.

INTRODUCTON.

Various methodes are recommended for the replacement of the cruciate ligament. Free or pedicle tendon autografting is frequently used, since large sheathless tendons have little tendency to form adhesions and can be securely anchored to soft tissue and bones because their suture holding is good (JÄGER und WIRTH 1977).

The disadvantages of autografting are the exhaustion of depot if repeated operations are required, and additional cosmetic disfigurement.

Preserved connective tissue substitutes have certain advantages. In particular dura mater is superior to other biological materials (STERNEMANN and VOORHOEVE 1973) because is has low extensibility and good suture holding properties (JÄGER AND WIRTH 1977).

Trials using synthetic material in the preformation of a gliding bed of sheathed tendons and the implantation of alloplastic tendons were already undertaken in the beginning of this century (BIESALSKI 1910). The demonstration that a nylon thread, inserted as permanent replacement for a flexor tendon, guaranteed full performance for 15 years, is of recent date (SARKIN 1973).

Experiences with synthetic vascular prostheses, however, revealed that the selection of Nylon as the fibre of choice was obviously a mistake (EDWARDS 1978), since HARRISON (1958) reported that Nylon loses 85% of its tensile strength already 3 months after experimental implantation. This is in contrast to the postulate that the durability of the implant shoud be superior to the life expectance of the recipient (GUIDOIN et al. 1988).

The synthetic aromatic polyamide (aramide, KEVLAR {R}) appeared to meet the two main requirements of connective tissue substitution namely, the life-expectancy of the implant is superior to that of the recipient, and secondly, that it maintains its extrordinarily high durability and tensile strength permanently.

H. Planck M. Dauner M. Renardy (Eds.)
Medical Textiles for Implantation
© Springer-Verlag Berlin Heidelberg 1990

To explore the in-vivo stability of KEVLAR {R} material, we have examined by histologic and scanning electron microscopic studies, specimens subcutaneously implanted in the rat, and KEVLAR {R} cords, used as a cruciate ligament substitute in the sheep.

MATERIAL AND METHODS.

In a first examination series, the braided aramide cords (200 den) were implanted subcutaneously in 42 rats. Groups of 6 animals were sacrificed after 1, 3, 6, 9, 12, 18 and 24 months, and of each group the implants of 3 rats were examined histologically for their tissue compatibility. The rest were submitted to physico-mechanical examination.

In a second study, the KEVLAR {R} cords were used as a substitute of the anterior cruciate ligament of the knee in 51 Merinoland sheeps. Groups of 4,5, and 4 implants were studied histologically, respectively 6,9, and 12 months post surgery. The rest, too, were submitted to physico-mechanical examination.

For the replacement of the anterior cruciate ligament, the aramide cord was passed through a channel, which was drilled obliquely through the condyle of the femur and the head of the tibia, and the ends of the substitute were fixed with screws.

The preparation of histologic section raised some difficulties. Routine histological techniques proved to be entirely unsuitable and comparable with attempts to cut rubber bands (elastics), immersed in a plumpudding, with a blunt kitchen knife.

Therefore, specimens were soaked in the monomer of JB4 plastic for 3 to 4 weeks before polymerization, and the hardening pronounced at 60^0 C. Two μm thick sections were cut with a (freshly broken) glass knife, and stained with Giemsa.

The KEVLAR {R} cords implanted as a cruciate ligament substitute were roughly prepared with an iron-saw, and small pieces, including the cord and adjacent bone tissue processed with a motor-driven fret-saw.

After rinsing, the bone was simultaneously decalcified and degreased in a mixture of 7.5% HNO_3 and 70% ethanol (pay attention, danger of explosion of the mixture!), and subsequently embedded in JB4 plastic.

Randomly selected specimens of KEVLAR {R} implants were macerated in 20% KOH, and prepared for scanning electron microscopy.

This report deals with the morphological changes only.

RESULTS.
Texture of the implant.
The fabric consisted of approximately 30 fascicles (Ø +/- 340 μm), each composed of +/- 500 fibrils (Ø +/- 18 μm), coiled to form a tube. The fascicles were approximately 18 μm apart, while the fibrils were closely packed together.

Subcutaneous implantation.
One month after implantation the fascicles were surrounded by densely packed foreign body giant cells (FBGC) which advanced thin cytoplasmic extensions between the outer 2 to 4 rows of fibrils. The interfascicular space and the centre of the tubular fabric were occupied by mobile cells of the loose connective tissue and by several myofibroblasts. Collagenous fibers were absent. Several mplants were entirely isolated from the host tissue by a dense membrane and were free of cells.

Between 2 and 22 months p.i.. the histologic features did not differ essentially from those after 1 month. During excision for histologic preparation, several implants were squeezed out of their capsula, even after an implantation period of 22 months.

However, in several implants the FBGC had succeded in separating the fibrils of several fascicles. The formation of connective tissue remaind scarce and the implants were mechanically not well-fixed to the surrounding tissue.

Several FBGC were found with stored, and obviously biodegraded aramide material.

Cruciate liga ents substitutes.
M a c r o s c o p i c a l l y,
cross-sectioned specimens had a cockade-like appearance: The surrounding bone (grey-white), an external capsula (white), the implant (yellow), and an internal casula or cylinder (white).

H i s t o g i c a l l y,
the external capsula consisted of three layers.

Six months after implantation, as an outer rim of the external capsula, a fibrous callus had developed which calcified after longer implantation periods (9 to 12 months).

After 12 months signs of remodeling were noticed because locally the original fibrous callus was turned to lamellar bone. Several osteons, too, had developed.

The middle portion of the external capsula consisted of a dense fibrous tissue, provided with numerous capillaries. The longer the implantation period the more this tissue showed the characteristics of a fibrous callus, in which histiocytes and some osteoprogenitor cells prevailed.

The inner rim of the external capsula was rich in cells. There were local accumulations of lymphocytes, numerous mononuclear macrophages and adjacent to the implant, abundant FBGC.

Six months after implantation, the FBGC were mainly found to cover the outer circumference of the implant only. After 9 and 12 months, the FBGC had locally also squeezed between the fibrils of the densely twisted fascicles and dispersed occasionally their fibrils.

After 9 and 12 months, although septa of connective tissue, deriving from the middle portion of the external capsula, had grown between several fascicles, the total incorporation of the implant in a dense connective tissue body and/or in newly formed bone substance was relatively poor.

However, the interdigitation of the perigraft tissue with the irreglular profile of the KEVLAR {R} cord surface could serve as an additional anchoring mechanism.

In several specimens an inner capsula was absent 6 months after implantation but present in all implants after 12 months.
The composition of the tissue of the inner capsula was essentially the same as that of the inner rim of the external capsula.

Petechial haemorrhages whithin the inner rim of the external capsula, and flat but more extended bleedings located between the implant and the external capsula were present in 2 of 4 animals after 6 months, in 3 of 5 after 9 months and in 2 of 4 after 12 months.

Surprisingly, indications of biodegration of the aramide material were much more obvious than after subcutaneous implantation.

Large numbers of macrophages, crowded with degraded aramide material, were arranged in circular rows not only directly adjacent to the implant, but also within all layers of the external capsula, even in autochthoneous Havers channels, and within the synovial folds and villi.

The substance stored in macrophages had a microfibrillar or needle-like shape, it was strongly birefringent like the fibers of the original implant (about 10 times more than the bone substance), but in contrast to the original implant, occasionally the incorporated material stained with the methylene blue component of the Giemsa compound.

The largest number of macrophages were present in specimens 9 months after implantation (3 of 5), while after 6 and 12 months only one specimen each (of 4) exhibited abundant macrophages.

However, a small number of cells with ingested birefringent material coud be detected around every implant.

All specimens selected for scanning electron microscopy could be exposed without cutting, though after 9 and 12 months under application of increased tractive forces and/or leverage.

The results of scanning electron microscopy, however, were less conclusive, because the effect of maceration after fixation in formaldehyde for several months is unsatisfactory. For this reason, probably disintegrated aramide fibrils were hardly to distinguish from remaining collagenous and/or elastic fibers.

Nevertheless, broken and occasionally splitted aramide fibrils, and fibrils sticking out from the fascicles were present in every specimen, while changes of the otherwise smooth surface of the fibrils, suggestive of biodegradation, were rare.

DISCUSSION.

As postulated for grafting of synthetic vascular substitutes, the full-wall or complete healing, which is definded as incorparation of the entire graft within a fibrous connective tissue matrix is the ultimate goal of all prosthesis design (BERGER et al. 1972, SAVAGE et al. 1980).

A "conditio sine qua non" (indispensible condition) for predictable and healthy ingrowth of perigraft tissue is the porosity of the graft fabric (WESOLOWSKI concept, 1964).

The size of these essential pores is poorly defined, because "porosity factor" means permeability to water, but according to own studies, a minimum pore size of 60 - 100 μm would be favorable.

The braided KEVLAR {R} cord used in this study appears microporous. Cells can obviuosly immigrate only between the densely twisted fascicles. Therefore, the incorporation of the entire substitute in a connective tissue body, to say nothing of incorporation in an osseous matrix, comparable to Sharpey's fibers, is unsatisfactory, and similar to the poor healing of tighly woven Dacron vascular subsitutes (JERUSALEM and HESS 1989).

A roughly knitted cord of KEVLAR {R} fibers would heal better, if provided with pores which remain open even under exposure to that high tensile forces usually acting on the cruciate ligament.

As recognized in histologic sections, the substance which is stored in macrophages most probably is biodegraded aramide material.

Concerning differential diagnosis microshivers of bone, deriving from drilling of the channel coud be discussed. However, the phagocytozed material is much more birefringent than bone, it is stored in mononuclear macrophages and foreign-body giant cells, and not in osteoclasts (by the way, osteoclasts never phagocytise).

The staining with methylene blue of the incorporated material could indicate that the polyamide chain was broken open (by an adaptively synthetized polyamidase?), ollowing the dye to react.

Therefore, to avoid biodegradation, masking the aramid fibrils and changes of the texture of the implant to fascilitate anchorage of cells and intercellular substances, have to be discussed.

REFERENCES.

>BERGER,K., SAUVAGE,L.R., RAO,A.M., WOOD,S.J.: Healing of arterial prostheses in man. Ann.Surg. **178**, 18- 27,1972.
>BIESALSKI, K.: Über Sehnenscheidenauswechslung. Dtsch. med. Wschr. **36**, 1615-1618 (1910).
>EDWARDS,WS.: Arterial grafts. Arch.Surg. **113**, 1225-1233,1978.
>GUIDOIN,R., COUTURE,J., ASSAYED,f., GOSSELIN, C.: New frontiers of vascular grafting. Int. Surg. **73**, 241-249,1988.

>HARRISON,JH.: Synthetic materials as vascular prostheses: II. A comparative study of Nylon, Dacron, Orlon, Ivalon sponge and Teflon in large blood vessels with tensile strength studies. Am.J. Surg. **59**, 16-14,1958.

>JÄGER, M., WIRTH, C.J.: Handb. allg. Path. VI/8 'Tranplantation'. Transplantation of connective tissue (tendon, cutis, fascia and dura) pp. 359-401, Springer-Verlag Berlin Heidelberg New York (1977).

>JERUSALEM,C., HESS,F.: Macroporous textile and microporous nonwoven vascular prostheses. Histological aspects of cellular ingrowth into the structure. This volume.

>SARKIN, T.L.: A fifteen-year follow-up of the Nylon replacement of cut flexor tendons of the finger. In: Orthop. Surg. Traum. A'dam: Exc. Medica 1973.

>SAUVAGE,L.R., BERGER,K., WOOD, S.J., RITTENHAUSEN,E.A., DAVIS, C.C., SMITH,I.C., HALL,D.G., MANSFIELD,P.B.: Grafts for the 80's. Monograph from the Bob Hope International Heart Research Institute, Seatle, Washington 1980.

>STERNEMANN, H.O., VOORHOEVE, A.: Die Kreuzbandplastiken des Kniegelenks. Arch. Orthop. Unfall-Chir. **74**, 329-335 (1973).

>WESOLOWSKI,SA., FRIES,CC., KARLSON,KE., DEBARKEY,MC., SAWYER,PN.: Porosity: primary determinant of ultimate fate of synthetic vascular grafts. Surgery **50**, 91-96,1964.

>WESOLOWSKI,SA., FRIES,CC., HENNINGAR,G., FOX,LM., SAWYER,PN., SAUVAGE,LR.: Factors contributing to long-term failures in human vascular prosthethic grafts. J.Cardiovasc.surg. **5**, 544-576,1964.

Keywords.
Polyamide, aramide, implantation, cruciate ligament, biodegradation

Author's address: see pg 61

SELECTED DISCUSSION REMARKS:

Claes: What may be the degradation mechanism?

Jerusalem: We have the impression that after subcutaneous implantation the degradation of this material is less significant than in a homostatic position, where the cord is really exposed to mechanical forces.

Claes: I cannot believe that it's the tensile load on the ligament, because this aramid fiber is used for tows because it is extremely strong to tensions. I think it must be the enzymatic activity or something, that is specific for our body, that is destroying this material.

Chu: In our suture research both our group at Cornell University and Prof. Williams in Liverpool, UK, we had to demonstrate that biodegradable fibers under stress conditions were degraded faster than under normal conditions. And using biodegradable fibers in ligament devices this factor has to be considered under considerable stress.

MECHANICAL PROPERTIES
OF VARIOUS LIGAMENT PROSTHESES

L. DÜRSELEN and L. CLAES

Sektion für Unfallchirurgische Forschung und Biomechanik, Universität Ulm
Helmholtzstr. 14, 7900 Ulm/Donau, Federal Republic of Germany

Abstract: Ligamentous injuries can lead to chronical joint instabilities which can be reduced by implantation of ligament prostheses. A couple of such devices available on the market were compared concerning tensile and bending behaviour. Stiffness, static and dynamic creep rate as well as tensile strength were determined in a materials testing machine. To test the bending fatigue strength a special loading apparatus was designed. The different ligament replacements partially showed enormous differences in properties. Polymer materials showed more creep than others, whereas e.g. carbon fiber exhibited lowest shear strength. For application as ligament prosthesis all materials being tested seem to have advantages and disadvantages. So it could be concluded that from a mechanical point of view an ideal replacement device does not yet exist. But at least they provide improvement of patients life quality for a limited time.

INTRODUCTION

Untreated ligament lesions or failed reconstructions often lead to an unstable joint. This is most fatal in the knee joint because it is highly loaded and strongly needs ligamentous guidance due to the lack of covering muscles and the incongruency of its joint surfaces. Of course there are knee spanning muscles, which certainly have stabilizing function. But in comparison with e.g. the hip joint it becomes obvious that the strong joint covering abductor muscle group and the ball shaped joint itsself contribute much more to joint stability than the knee spanning muscles and the incongruency of the knee joint surfaces do.

There are only few methods available to stabilize a knee joint. Joint braces applied to the knee can provide stabilization to external varus and valgus

H. Planck M. Dauner M. Renardy (Eds.)
Medical Textiles for Implantation
© Springer-Verlag Berlin Heidelberg 1990

moments. They can also prevent the joint from hyperextension or -flexion. But an anterior or posterior drawer cannot be avoided by such braces. Several operative treatments have been developed using autografts to replace the destroyed ligament structures. Many surgeons e.g. use a patellar tendon graft to reconstruct the anterior cruciate ligament. Most problems using this technique are due to lengthening of the tendon graft. Another possibility of treating an unstable knee joint is the use of allografts. Here it can be distinguished between biological tissue and synthetically manufactured materials. Biological tissue means e.g. preserved animal or human transplants. This paper focusses on artificial ligament prostheses. The main difficulty of design of such synthetic devices is to simulate the properties of natural ligaments as well as its anchorage to the bones. Natural ligaments consist of different fiber bundles stretched in different motion and loading situations of the knee joint and being attached to a large insertion area. Artificial devices mostly show a uniform structure and must be inserted at one special point punctiformly. Therefore an approximate isometric implantation is required. But this is very difficult to achieve. To provide a whole life time function natural ligament continuously remodel by means of its vascularization. Artificial devices of course do not possess this biological effect. For this reason it must be protected against mechanical wear, fatigue and biological degradation. Therefore mechanical tests were performed considering the specific stress situation of the ligaments in the knee joint.

MATERIALS AND METHODS

Six different ligament replacements already being implanted as ligament prostheses were investigated (see table 1). The Kennedy LAD and the resorbable PDS are only recommended for use as augmentation of an autologous transplant.

Table 1

List of the samples being tested

name	company	material	structure
LaFil	Braun	carbon fiber	braided tube
Dacron	Stryker	polyester	woven
Gore-Tex	Gore	polytetrafluoroethylene	braided
Leeds-Keio	OEC	polyester	woven tube
PDS	Ethicon	polydioxanone, resorbable	woven
Kennedy LAD	3M	polypropylene	woven

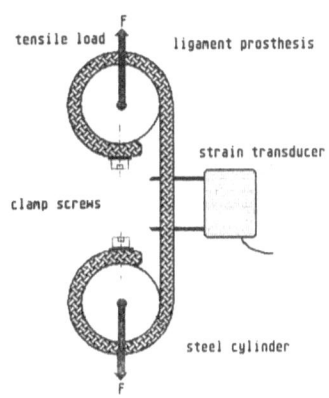

Fig.1. Clamping set-up

The most important characteristical mechanical parameters regarding the application of materials as ligament prosthesis are stiffness, static and dynamic creep rate, tensile strenght and bending fatigue strength. The tensile tests were carried out in a materials testing machine (Zwick 1454: static tests, Schenck PSA: dynamic tests). Here the strain was measured using an Instron strain transducer. The transducer was pierced through the woven or braided fiber structure of the tests ligaments (see Fig.1). The applied strain rate was constantly kept at 10 % of the initial length per minute. The samples were clamped by winding them around a steel cylinder (diameter = 20 mm, see Fig.1). All ligaments were tested under both dry and wet conditions.

To measure the stiffness of the ligament prostheses tensile tests in the range from 0-200 Newton were performed. The static creep behaviour was investigated by applying a constant tensile load of 200 Newton and recording the increasing length by time. To simulate a physiological tensile characteristic e.g. during walking in a cruciate ligament an additional dynamic creep test in a hydraulic materials testing machine was carried out. Here a sinusoidal load between 30 and 200 Newton was applied and the maximum strain at 200 Newton was recorded by time. A rupture test revealed the ultimate tensile strength of the materials.

The bending fatigue strength appears to be a very important factor due to the fact that clinical and experimental findings revealed that ruptures of prostheses most often occur at the bony insertion points, where bending and friction wears the material. In order to determine the bending fatigue strength a special apparatus was designed. Under an adjustable tensile load the ligament prostheses were bent cyclically around a polished steel cylinder (diameter = 8 mm, see Fig.2). This cylinder was fixed in a way that it could not turn around its axis. Thus not only bending stress

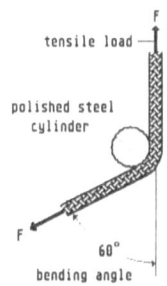

Fig. 2
Bending fatigue test

but friction between the ligament and the cylinder surface occured. The bending cycles until failure were counted. During the test the samples were kept wet.

RESULTS

Stiffness

The measured tensile load-strain curves (Fig.3) illuminated the tensile behaviour of the ligament prostheses in a load range, which corresponds to the loads occuring in normal knee joint ligaments.

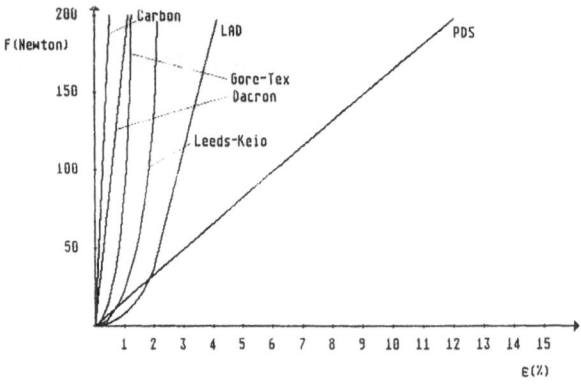

Fig. 3. Force-strain diagram of tensile tests in the range from 0 - 200 Newton

It is known that during normal walking cruciate ligament forces of only 30-60 Newton occur (ref. 1). Only external loads as there are varus or valgus and rotational moments and of course trauma situations rise the ligament forces to higher values. Figure 3 shows the carbon fiber ligament to be the stiffest one followed by Dacron (Stryker), Gore-Tex and Leeds-Keio. The augmentation devices Kennedy LAD and resorbable PDS turned out to be weaker than the others. Between dry and wet testing conditions only small differences could be observed.

Static Creep

The carbon fiber ligament replacement showed lowest static creep. After two hours under a constant load (200 N) it lengthened less than 0.1 % . The Leeds-Keio creeped about 0.5 % and the Dacron (Stryker), Gore-Tex and the dry PDS around 1.0 %. Wet testing of the resorbable PDS led to a rupture of the sample already after one hour load. The Kennedy LAD exhibited 2 % creep after two hours and even went on creeping. Except the PDS ligament again no marked differences between wet and dry testing was recognized.

Dynamic Creep

The sinusoidal load test revealed similar results (fig.4). Dacron (Stryker) and Gore-Tex asymptotically reached a creep length of about 1 %. This final elongation was observed after 300,000 load cycles at the Dacron ligament whereas Gore-Tex needed 600,000 cycles to reach its final length. Carbon fiber showed more creep than in the static experiment but still less than 0.5 %. The Leeds-Keio prosthesis creeped least. PDS already ruptured after less than 1,000,000 load cycles and the Kennedy LAD creeped most as already seen in the static test and even did not reach an asymptotic value after 5×10^6 load cycles.

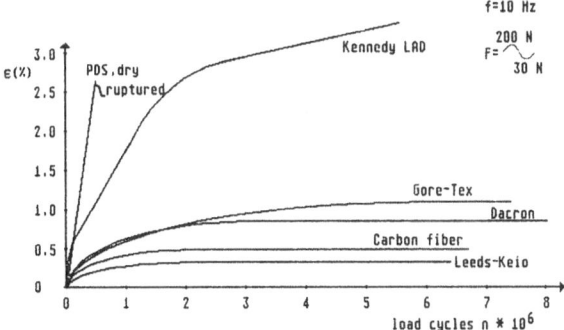

Fig.4. Diagram showing dynamic creep behaviour under a pulsating tensile load

Rupture Test

Large differences were also found out for the tensile strength. Table 2 shows the Gore-Tex ligament to be the strongest and the resorbable PDS to be the weakest one.

TABLE 2

name	rupture force (Newton)
LaFil	2300
Dacron(Stryker)	1300
Gore-Tex	>5000
Leeds-Keio	800
Kennedy LAD	1300
PDS	250

Bending Fatigue Test

This test partially showed enormous differences in bending behaviour of the prostheses materials (see Table 3). Under 40 N tensile load the Stryker ligament showed the best resistance against cyclic bending. But it failed much earlier after increasing the tensile load. Better bending strength at increased tensile load was provided by Gore-Tex. Lowest endurance at 120 N tensile load exhibited the carbon fiber ligament, whereas the Kennedy LAD showed the worst result of all under 40 N load.

Table 3

name	tensile load		n	
	40 N	120N		
LaFil	437,369	14,077	4	
Dacron(Stryker)	2,702,578	151,475	1	number of bending cycles
Gore-Tex	2,100,000	1.400,000	1	until failure
Leeds-Keio	1,400,000	200,000	1	bending radius 4 mm
Kennedy LAD	169,100	77,186	4	

DISCUSSION

Comparing the tensile behaviour of the ligament prostheses with the force-strain characteristic of a natural knee ligament (see Fig. 3) it becomes obvious that most of the materials are much stiffer than the biological tissue which it should replace. This seems to be one of several critical factors in ligament replacement because a real isometric implantation cannot be achieved. Thus implanting a too stiff replacement device in a knee joint under a certain tensile preload may cause restriction of the knee mobility either in flexion or extension depending on the position in which the device was pretensioned and fixed. Only the resorbable PDS and to some extent the Kennedy LAD seem to have a similar stiffness as a natural ligament. But the nonlinear characteristic increase of stiffness with increasing strain could not be observed. It was seen in animal experiments that e.g. when using carbon fiber ligaments such nonlinear behaviour could be achieved by the ingrowth of connective tissue in between the fibers. Thus the extensibility of the structure came much closer to that of a natural ligament (ref. 1). But just from the in-vitro point of view the prostheses appear too stiff. It would be most desirable to have a highly extensible structure in the intraarticular part and a stiffer portion in the bony drilltunnel in order to avoid micromovements supporting better bony ingrowth.

Regarding the tensile strength of the materials it must be said that it is not the most important parameter. Rupture forces of 800 N and more (see Table 2) only occur in trauma situations. So the measured strengths are sufficient except the PDS (250 N). But at least this device may provide a little support for an autologous transplant. Tensile strength too high in combination with high anchorage strength even may not be desirable, because in case of trauma a prosthesis rupture would be better than damage of the bony insertion site.

Creep is another important mechanical criteria for a ligament replacement device. A pretensioned prosthesis elongates by time. In long term this may lead to loosening and thus to a new instability of the knee joint. One must consider that creep does not only occur in the intraarticular portion but also inside the bony channels because prostheses often are only surrounded by connective tissue instead of bone. So the entire length of a ligament replacement can measure 100 mm and more. So lengthening of the prosthesis by creep of 2% may lead to an increase of the anterior drawer of about 3 mm. The static as well as the dynamic creep test revealed that all prostheses show creep below 2 % exept the two augmentation devices (PDS, Kennedy LAD).

Regarding the bending fatigue results we could not find a good correlation between in-vitro and in-vivo behaviour. E.g. the Stryker ligament showed a good bending fatigue in-vitro, but in an animal experiment with sheep it was ruptured more often than others. One reason for that is probably the varying amount of connective tissue ingrowth. It can provide a protection of the material against mechanical wear.

However, for all prostheses a careful surgical procedure is required which should provide rounding of all bony edges and creating large bending radii to minimize the bending stress.

REFERENCE

1 L. Claes, L. Dürselen, Biomechanical requirements on ligament prostheses, in: C. de Putter, G.L. Lange, K. de Groot and A.J.C. Lee (Ed.), Implant Materials in Biofunction, Advances in Biomaterials, Volume 8, Elsevier, Amsterdam, 1988, pp. 17-23

DILATABILITY AND STRETCHING CHARACTERISTICS OF POLYESTER
ARTERIAL PROSTHESES. EVALUATION OF THE ELASTIC BEHAVIOUR

E. Debille, R. Guidoin, J. Charara, D. Torché, M. Bernier-Cardou,
D. Marceau, D. Boyer, C. Chaput, L. Dadgar and A. Cardou

Biomaterials Institute, Room F1-304, St.François d'Assise Hospital,
Quebec City, Qc, G1L 3L5, Canada.

SUMMARY

The static elastic behaviour of polyester arterial prostheses of medium and large diameter has been characterized theoretically using a two-dimensional incremental model. The circumferential deformation in the isotonic mode under static pressure from 0 to 340 mm Hg and the longitudinal deformation under longitudinal tension from 0 to 456 gf has been measured for 36 polyester prostheses. These prostheses represented all classes of textile polyester arterial grafts commercially available, i.e. woven, weft knitted, warp knitted, impregnated and externally supported grafts.

Woven grafts presented a large range of dilatability with circumferential stiffness parameter D_θ ranging from 10×10^5 to more than 250×10^5 dynes/cm at a longitudinal tension of 456 gf. The stiffer grafts were those with low water permeability and their elastic behaviour depended widely on longitudinal tension.

The weft knitted grafts behaved in a similar manner with more dilatability ($D_\theta = 22.0 \times 10^5 \pm 2.3 \times 10^5$ dynes/cm at T = 114 gf) and they were not influenced by longitudinal tension. The behaviour of the warp knitted grafts was similar to that of weft knitted ones but the former was less likely to dilate as internal pressure increased. The velour graft's circumferencial deformation was variable.

All the above grafts had high longitudinal distensibility due to their crimp structure, while knitted grafts stretched more ($D_z = 25.9 \times 10^4 \pm 23.8 \times 10^4$ dynes/cm).

Impregnation of the grafts, with bioerodible compounds, which led to composite structures, maintained the polyester fibers together and the stiffer material was less elastic.

Finally, externally supported grafts had an excellent dimensional stability with D_θ ranging from 68.9×10^5 to more than 250×10^5 dynes/cm and D_z from 115.6×10^4 to 648.6×10^4 dynes/cm at a longitudinal load of 456 gf.

Running title: The elastic behaviour of polyester grafts.

Key-words: Vascular grafts, polyester, mechanical properties, deformation.

H. Planck M. Dauner M. Renardy (Eds.)
Medical Textiles for Implantation
© Springer-Verlag Berlin Heidelberg 1990

INTRODUCTION

Polyester arterial prostheses are widely used as blood conduits for replacement of medium and large diameter arteries, i.e. 8 mm and above (1). They can be manufactured as straight tubes or as bifurcations. Without a doubt, the polyester presents outstanding endurance and adequate biostability. Thus, the choice of this material is astute and not disputed. All other synthetic materials except for expanded polytetrafluoroethylene (Teflon®) are now out of use. However, due to the many number of failures, manufacturing of a valuable commercial and medically sound prosthesis is still controversial. The required characteristics for developing new improved grafts are as follows:

1) Good handling characteristics, i.e., easy of handling and suturing, without any fraying at the ends regardless of the cutting direction. The graft should be soft enough to allow continuous suturing. Preclotting must be easy and efficient so as to prevent blood from oozing through the wall after the blood flow has been restored.

2) Good healing characteristics, i.e., rapid tissue ingrowth through the textile material is of great importance. The flow surface of the grafts should be smooth, glistening and, ideally, endothelialized so as to lower the thrombotic threshold velocity.

3) Good mechanical stability, i.e. the grafts should neither dilature nor stretch. The effective diameter should be constant and close to the nominal diameter.

4) Good chemical stability, i.e. neither degradation of the filaments nor humidity and lipid uptake should occur.

Much of the information concerning the characteristics of the polyester Dacron® grafts were obtained from several clinical and experimental reports and served to develop new designs. These designs, from the structural point of view, include two major categories of woven and knitted (warp and weft knitted). Also, recently models, such as velour, prostheses impregnated with a bioerodible coating in order to avoid the inconvenience of preclotting and externally supported in order to prevent kinking and compression at areas of flexion were developed.

Despite the new developments in the design of the synthetic grafts, the need for a more effective arterial substitute remains to be desired (2,3). Among the many factors that may influence the long-term patency of the vascular graft, the effect of the mechanical properties is of particular interest, and the importance of the design of the vascular graft is generally recognized. These properties are important since the rigidity of the graft wall could influence the pulse propagation and the stress applied on the host artery to which the blood conduit is anastomosed. In addition, stress at anastomoses due to the differences of distension between host vessel and graft (suture line stress) could be associated with the rupture of the suture line, leading to aneurysmal formation. Other graft-related causes of anastomotic aneurysm could either be yarn slippage or yarn breakage. Furthermore, the graft is subjected to a static pressure as well as to the repeated stress of pulsation, leading to some degree of permanent dilation which could be aggravated by the yarn fatigue.

Our laboratory has developed few test procedures to predict the overall performance of new polyester prostheses (4-6) and also to characterize the mechanical and structural properties of grafts so as to provide the surgeon with guidelines likely to help him in his decision making (7-9).

The present study was carried out to examine the dimensional stability of medium diameter polyester (Dacron[R]) arterial prostheses representative of all classes of textile polyester arterial grafts commercially available. It permits the prediction of the in vivo deformation and also helps to avoid the possible related failures.

1. THEORETICAL ANALYSIS

A three-dimensional stress-strain relationship was required to analyse the mechanical and particularly the elastic behaviour of larger diameter arterial prostheses with interlaces between strains in different directions. The graft was modelized by a cylindrical tube submitted to inner pressure and longitudinal tension, and the stress distribution was expressed in cylindrical polar coordinates r, θ and z (radial, circumferential and longitudinal directions respectively) with the axis located at the centre of the tube. The wall thickness could be considered thin enough to assume negligible normal stress in the radial direction σ_{rr} compared to the normal stress in the circumferential direction $\sigma_{\theta\theta}$, and the stresses $\sigma_{\theta\theta}$ and σ_{zz} were uniform throughout the wall thickness, with a resulting error evaluated by Fung ranging from 3% to 11% (10). Therefore, the problem was reduced to a two-dimensional analysis and the stresses applied were as follows (10,11):

$$\sigma_{\theta\theta} = \frac{P_i r_i - P_o r_o}{h} \tag{1}$$

$$\sigma_{zz} = \frac{F + P_i \pi r_i^2 - P_o \pi r_o^2}{\pi(r_o^2 - r_i^2)} \tag{2}$$

where P is the pressure applied on the graft, F the longitudinal tethering force applied to the end of the prosthesis and h the wall thickness of the graft. The subscripts i and o denote the inner and outer surfaces. Equation (1) was derived from Laplace's law with proper modifications due to the thickness of the wall

(12). However, for large diameter grafts, the wall thickness was generally less than 10% of the radius and the internal radius could therefore be considered equivalent to that of the external (13). If the measurements were taken at the relative pressure, the term P_o-P_1 would be equivalent to the experimental pressure, P. Using these assumptions, the equations were reduced to:

$$\sigma_{\theta\theta} = \frac{Pr_o}{h} \tag{3}$$

$$\sigma_{zz} = \frac{F + P\pi r_o^2}{2\pi r_o h} \tag{4}$$

If there was no end plate on the graft, the pressure P, in equation (4), did not occur and the equation was reduced to:

$$\sigma_{zz} = \frac{F}{2\pi r_o h} \tag{5}$$

The yarn structure and the complex fabric design of the Dacron graft provided a non linear stress-strain behaviour. However, since any curve could be approximated as a sequence of linear segments, using small controlled variations of pressure and stress, the material could be considered linearly elastic within a small operating region of the stress-strain curve (around specific pressure values) (14,15). This incremental analysis using small variations made the material Hookean within a small range of the stress-strain curve. These assumptions, which allowed the linear stress-strain theory to describe a non linear stress-strain behaviour and therefore the circumferential and longidutinal deformations became;

$$\epsilon_\theta = \frac{dr_o}{R_o} = \frac{\sigma_{\theta\theta}}{E_\theta} - \gamma_{\theta z} \frac{\sigma_{zz}}{E_z} \tag{6}$$

$$\epsilon_z = \frac{dl}{L} = \frac{\sigma_{zz}}{E_z} - \gamma_{z\theta} \frac{\sigma_{\theta\theta}}{E_\theta} \tag{7}$$

where L was the length of the tube, E_θ and E_z were Young's moduli in circumferential and longitudinal directions, γ was the Poisson ratio, and R_o and L were the reference dimensions of the radius and the length within the considered region of curves.

By knowing the inner pressure and the longitudinal stretching forces, and measuring the external deformation, $\sigma_{\theta\theta}$, σ_{zz}, ϵ_θ and ϵ_z could be calculated. As four parameters were unknown in these equations (E_θ, E_z, $\gamma_{\theta z}$ and $\gamma_{z\theta}$), two separate experiments were required to simplify these equations and to determine

the Poisson ratios. Two tests were performed, one describing the dilatability of
the prosthesis and the other its stretching behaviour.

1.1 Dilatability

The circumferential deformation under internal pressure which has been
described in earlier studies was observed in natural vessels where the
radial motion was highly predominant compared to the longitudinal one (16).
Several theoretical expressions relating pressure to radius were derived
from different elasticity models; some of them, having taken into account
the wall thickness, have resulted in complicated relationships (17). However, numerous studies have expressed the mechanical behavior of the grafts in
term of compliance, which is a simple and convenient parameter. The compliance, C, was mainly expressed as the percentage change in the external
diameter, D, per unit change in pressure assuming a constant vessel length
and neglecting changes in the thickness of the wall:

$$C = \frac{dD}{DdP} * 100 \tag{8}$$

Compliance was frequently used to discuss arterial and graft mechanics and
its influence in the patency of arterial graft (11,18-23). Compliance can be
defined as the inverse slope of the line between two points taken on the
pressure-circumferential strain curve. Natural blood vessels exhibit viscoelastic behaviour were described by a non linear stress-strain curve.
Therefore, the compliance value was dependent on two chosen points (i.e. the
pressure range considered). Although the stress-strain behaviour of the
synthetic arterial graft was closely linear (24,25), compliance failed to
describe the physiologic situation. As a matter of fact, using a compliance
assumed that there was a constant graft or vessel length which was true for
the in vitro testing where the conduit ends were both immobilized. On the
other hand, implanted grafts were submitted to a longitudinal tethering
force from the host artery and the constant length assumption was not
correct. Therefore, the compliance was not used here, and measurements of
the graft circumferential dilatations were performed under constant longitudinal stress such that the length remained constant (11). The consideration of a constant length was such that no longitudinal force was imposed
(i.e. F = 0). Therefore equation (5) was reduced to $\sigma_{zz} = 0$ and in
substituting it into equation (6) we obtained:

$$\epsilon_\theta = \frac{dr_o}{R_o} = \frac{\sigma_{\theta\theta}}{E_\theta} \tag{9}$$

Then, substituting equation (3) into this result gave:

$$\epsilon_\theta = \frac{Pr_o}{hE_\theta} \tag{10}$$

4.2. Stretching

In the absence of tethering, vascular grafts seemed to be very sensitive to elongation, especially the crimped polyester prostheses. The particular crimped geometry of the conduits were introduced to improve their patency, as well as their bending without kinking at areas of flexion (26).

As mentioned above, the complex structure of the polyester graft made it anisotropic, viscoelastic and non linear, corresponding to an hysteresis stress-strain curve. By contrast to a circumferential deformation, this hysteresis was more pronounced in the longitudinal direction. Since we had used the incremental analysis that allowed the use of linear stress-strain behaviour within a small range of the longitudinal stress-strain curve, in the absence of inner pressure equations (5) and (7) became;

$$dF = 2\pi r_o h E_z \frac{dl}{L} \tag{11}$$

where dF and dl represented the small variations of F and L respectively. Young's modulus E_z varied with dF and dl and was determined from the quasi-steady stress-elongation curve where the graft samples was stretched at a very slow fixed rate (27). The resulting E_z was not exactly the moduli defined for incremental analysis but was a good approximation (14).

In order to obtain the quasi-steady state curve, the material was submitted to several cyclic loading and unloading forces. The stable state of the curve was the steady state. Therefore, this preconditioned graft was considered as a pseudo-elastic material (10). However, in this study, the data obtained during the loading portion of the hysteresis curve were used for further calculations.

Our objective was to choose simple parameters to analyse the mechanical properties in circumferential and longitudinal directions. Since the polyester prostheses were very stiff compared to natural vessels, the radius and the thickness of the pressurized vessels stayed close to the reference radius and thickness (i.e. h = H and $r_o = R_o$). Using these assumptions, equations (9) and (10) produced:

$$E_\theta = \frac{PR_o^2}{Hdr_o} \tag{12}$$

and equation (11);

$$E_z = \frac{dF}{dl} \cdot \frac{L}{2\pi R_0 H} \tag{13}$$

The elasticity of the graft was not dependent only on the elasticity moduli of the material but on its thickness as well as its radius (or length). For these reasons, we chose the elastic moduli coupled with the wall thickness HE and it was denoted D_θ:

$$D_\theta = HE_\theta = \frac{PR_\theta^2}{dr_0} \tag{14}$$

by similarity, the longitudinal elongation was represented by D_z as follows:

$$D_z = HE_z = \frac{dF}{dl} \cdot \frac{L}{2\pi R_0} \tag{15}$$

2. MATERIALS AND METHODS

2.1 Graft selection

Mechanical testing was performed on 36 models of polyester textile prostheses representing all classes of textile polyester grafts. Most of these grafts are widely used in clinical practice. Some are discontinued but were analysed in this study since they played a historic role in vascular surgery.

a) Woven grafts

Ten models were selected. Their characteristics and properties are summarized in Table 1. A woven fabric is formed when two sets of yarns are interlaced at right angles to each other. The longitudinal yarns are known as the warp and the transversal yarns are known as the weft. The warp yarns lie alternately over and under the weft yarns as shown in figures 1, 2 and 3.

b) Weft knitted grafts

Seven models were selected (Table 2). Weft knitted grafts are fabricated from one set of yarns that are interlooped as opposed to interlaced. Knitting is done through the raising and lowering of knitting needles. The simplest structure is that of the single jersey (Figure 4). As shown in figure 5, the two surfaces of the fabric are different.

c) Warp knitted grafts

Eight grafts were selected (Table 3). In this kind of fabric, one or more sets of yarn are knitted. The yarns are fed to a series of

T A B L E 1.- Characteristics and physical properties of woven prostheses.

Type	Manufacturer	Fiber content	Internal diameter (mm)	Stitch density weft/cm	Stitch density warp/cm	Bursting strength kPa	Water permeability ml/cm^{-2}/mm^{-1}	Crimps per cm
AVT high porosity	Advanced Vascular Technology, USA	Texturized	10	34	28	1907	1090	12.7
Circumferentially compliant velour	Vascular Products* Ireland, Eire	Polyester + polyurethane	8	-	-	-	1480	-
Cooley Verisoft	Meadox Medicals, USA	Flat + texturized	10	58	35	2630	199	8.7
Cooley low porosity	Meadox Medicals, USA	Flat + texturized	10	53	40	3240	72	13.0
DeBakey woven	Bard Cardiosurgery Division, USA	Texturized	10	56	34	4900	232	7.7
DeBakey soft woven	Bard Cardiosurgery Division, USA	Texturized	10	50	42	5688	217	10.6
Meadox woven double velour	Meadox Medicals, USA	Flat + texturized	10	40	34	4287	327	11.1
Indian	Sree Chita Tirunal Institute, India	Texturized	12	44	42	3173	225	7.9
Vascutek woven	Vascutek, UK	Texturized	8	54	44	3247	97	8.3
Prototype	VI Tech.	Texturized	8	36	30	2909	905	11.4

* This company is no longer in business.

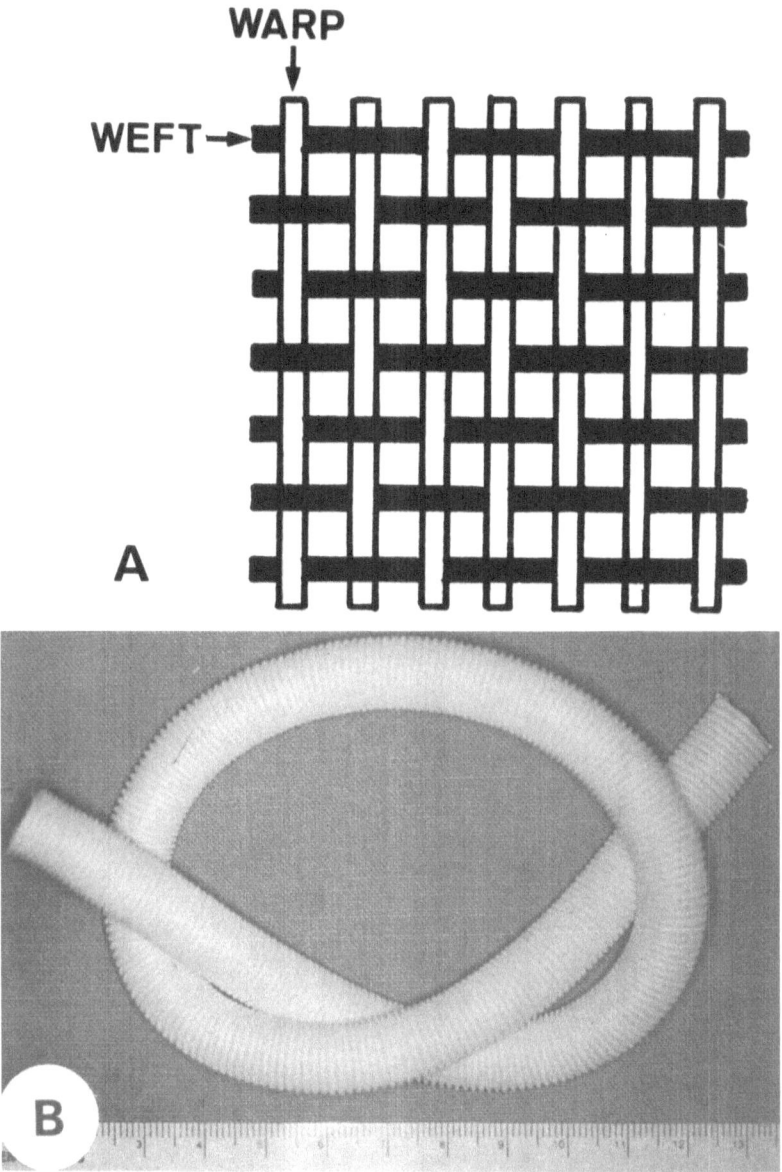

Fig. 1.- Plain woven structure. A: it consists of two sets
of yarn, called warp and weft, which are interlaced at right
angles to each other. The warp yarns lie alternatively over
and under each of the weft yarns. B: photomacrograph of the
Woven DeBakey graft.

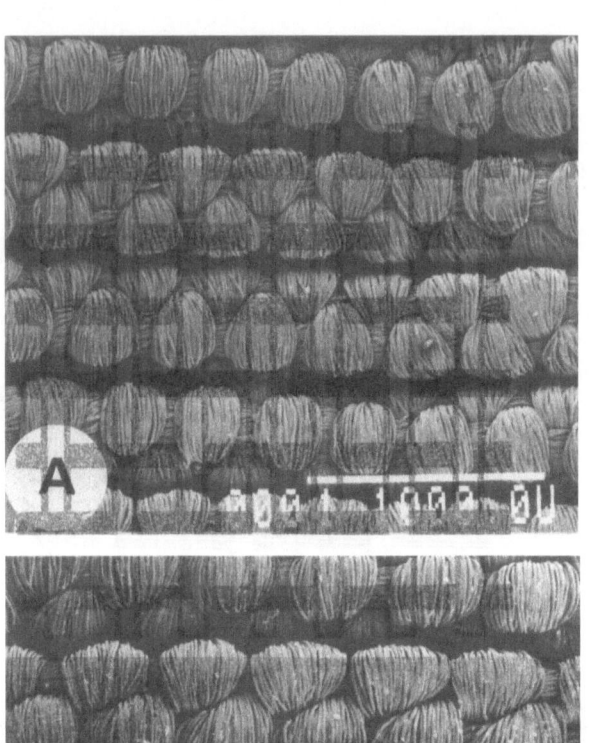

Fig. 1.— Plain woven structure. A: it consists of two sets of yarn, called warp and weft, which are interlaced at right angles to each other. The warp yarns lie alternatively over and under each of the weft yarns. B: photomacrograph of the woven Debakey graft.

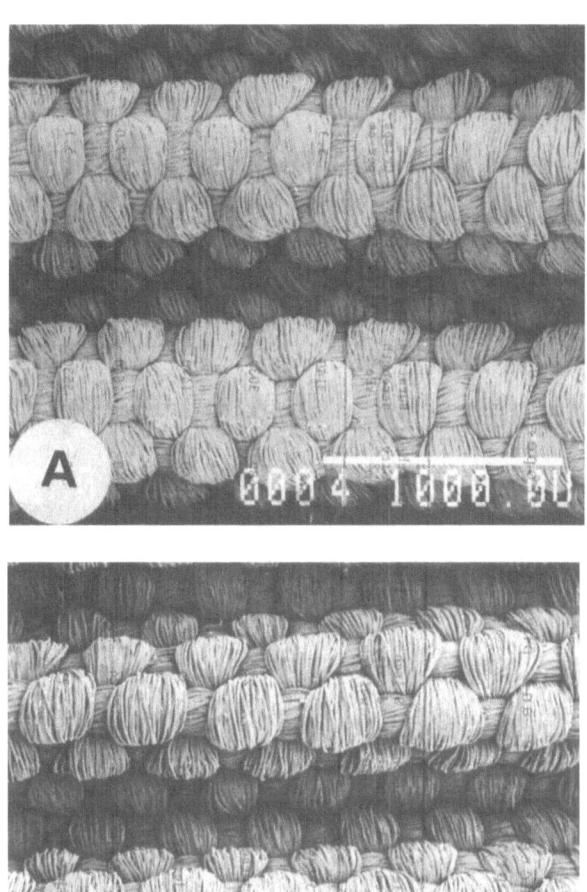

Fig. 3.- Photomicrographs of the Woven DeBakey soft graft. A: internal surface; B: external surface.

TABLE 2. – Characteristics and physical properties of weft knitted prostheses.

Type	Manufacturer	Fiber content	Fabric structure	Internal diameter (mm)	Stitch density weft/cm	Stitch density warp/cm	Stitch density /cm²	Bursting strength kPa	Water permeability ml/cm⁻²/mm⁻¹	Crimps per cm
Barone	Industrias HA Barone, Argentina	Texturized	Single jersey velour	8	16	21	324	3140	3660	5.3
DeBakey standard	Bard Cardio-surgery Division, USA	Flat	Single jersey	10	22	31	680	3030	2530	7.5
DeBakey ultra light weight	Bard Cardio-surgery Division, USA	Flat	Single jersey	12	32	42	1340	1440	4080	10.5
Lopor	Golaski Laboratories, USA	Texturized	Reverse single jersey	16	24	27	650	1640	2960	19.0
Mikrofroté	Vyzkumny ustav pletarsky, Czechoslovakia	Texturized	Single jersey	9	21	18	370	3530	2960	5.1
Milliknit	Golaski Laboratories, USA	Flat	Reverse single jersey	10	26	35	910	1890	5300	24.8
Vasculour D	Bard Cardio-surgery Division, USA	Texturized	Single jersey	10	20	26	520	1730	2250	10.1

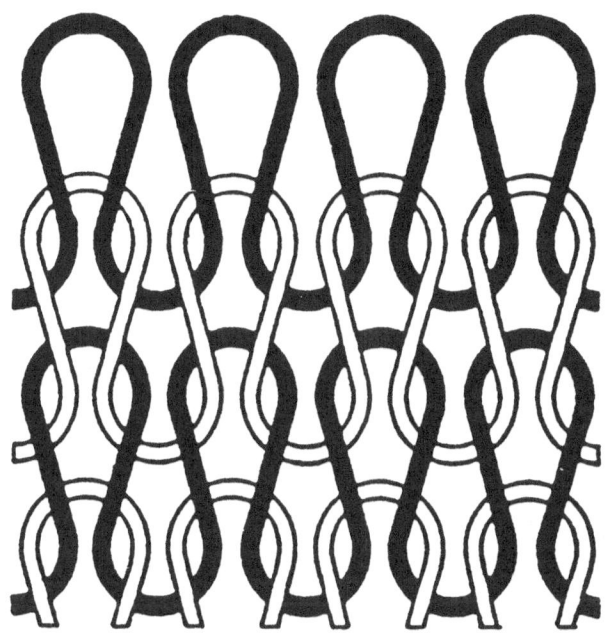

Fig. 4.- Loop diagram of a single jersey structure.
This knitted construction contains of sets of
yarns which are interlooped around each other
instead of being interlaced. When the yarns lie
predominantly in the transverse direction, it is a
weft knit structure.

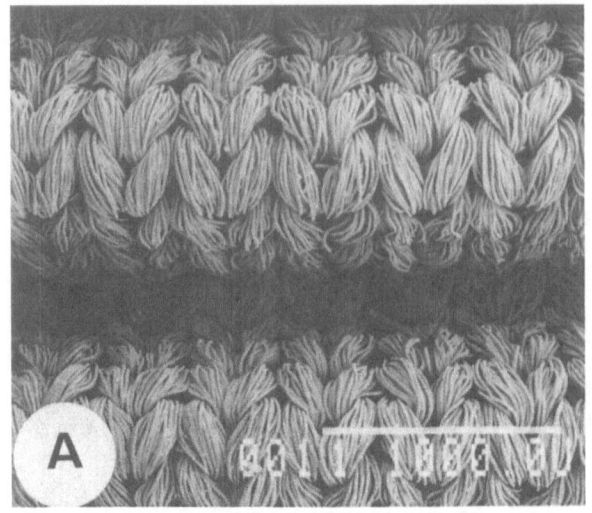

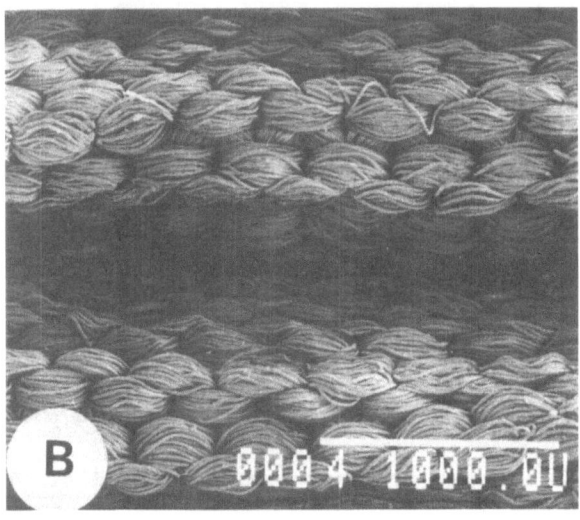

Fig. 5.- Photomicrographs of the knitted DeBakey
graft. A: internal surface. B: external surface.

TABLE 3. - Characteristics and physical properties of warp knitted prostheses.

Type	Manufacturer	Fiber content	Fabric structure	Diameter (mm)	Stitch density weft/cm	Stitch density warp/cm	Stitch density /cm^2	Bursting strength kPa	Water permeability ml/cm^{-2}/mm^{-1}	Crimps per cm
Cooley standard knit	Meadox Medicals, USA	Flat	2-bar reverse tricot	10	18	32	580	2770	1590	12
Cooley II	Meadox Medicals, USA	Texturized	2-bar reverse tricot	10	16	24	380	2110	2650	6.5
Sauvage Bionit	Bard Cardio-surgery Division, USA	Texturized	2-bar reverse locknit with open cord laps	10	14	24	336	2370	1620	8.4
Sauvage Bionit II	Bard Cardio-surgery Division, USA	Texturized	2-bar reverse locknit with open cord laps	10	14	28	392	1586	1330	6.3
Triaxial	Vascutek, UK	Texturized	2-bar reverse locknit with open cord laps	10	12	16	187	3190	1140	7.2
VP1200K	Vascutek, UK	Texturized	2-bar reverse locknit with open cord laps	10	14	15	210	2120	2000	8.7
Vasculour II	Bard Cardio-surgery Division, USA	Texturized	Inverse 2-bar reverse locknit with open cord laps	10	15	24	360	2420	2150	9.5
Weavenit	Meadox Medicals, USA	Flat	Inverse 2-bar tricot	10	26	33	860	1500	1670	14.7

needles where each yarn forms a loop. The loops are intermeshed and each yarn zigzags its way throughout the fabric (Figures 6, 7 and 8).

d)   Velour grafts

Two models were selected (Table 4). The velour grafts differ from the others in terms of surface characteristics in which the velour effect is achieved through the use of extra threads that are less frequently interlaced than the ground yarns. The Microvel[R] prosthesis developed by Meadox uses a half tricot structure where the texturized yarns are inlaid (Figures 9 and 10).

e)   Coated grafts

Four prostheses were selected (Table 5). The wall of the grafts were impregnated with bioerodible substances, i.e. albumin, collagen or gelatin. It was introduced to avoid preclotting prior to implantation. The luminal surface was thus smoother and less thrombogenic at implantation (Figures 11 and 12).

f)   Externally supported grafts

Five prostheses were tested (Table 6). In this case the grafts were either woven or knitted. The grafts were supported externally by a helicoidal monofilament which in some cases was welded to the outer surface of the graft (Figures 13 and 14).

2.2 Mechanical investigations

2.2.1 Dilation

The circumferential dilatation of arterial prostheses was performed using the OPTIDDIAC (24), a machine specially designed for this purpose (Figure 15). Although several experimental devices were also developed for the same purpose (11,21,28), the OPTIDDIAC had the advantage of being very simple, and measurements were recorded in the static state. The OPTIDDIAC system allowed a tubular prosthesis to be held between two clamps mounted on two walls and specially adapted to keep the graft in shape. One of the clamps was fixed and supplied the air pressure, while the other was mobile and applied a constant longitudinal tension on the specimen with a load supported by a sheave. This wall moved freely on a low friction track as it was mounted on 4 small ball-bearings. As the textile grafts were porous, a very thin tubular latex, a silicone rubber membrane, was inserted inside the prosthesis to allow a constant inner pressure. The diameter of the latex tube was wider than the graft in order to totally transfer the effect of the pressure on the prosthesis. In order to simulate in vivo conditions where there was no end plate on the vessel, the extremity of the latex membrane was closed and did not come into contact with the

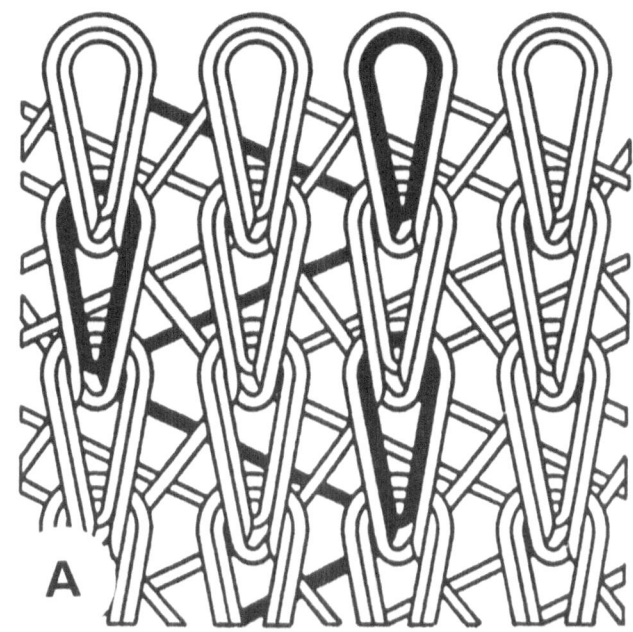

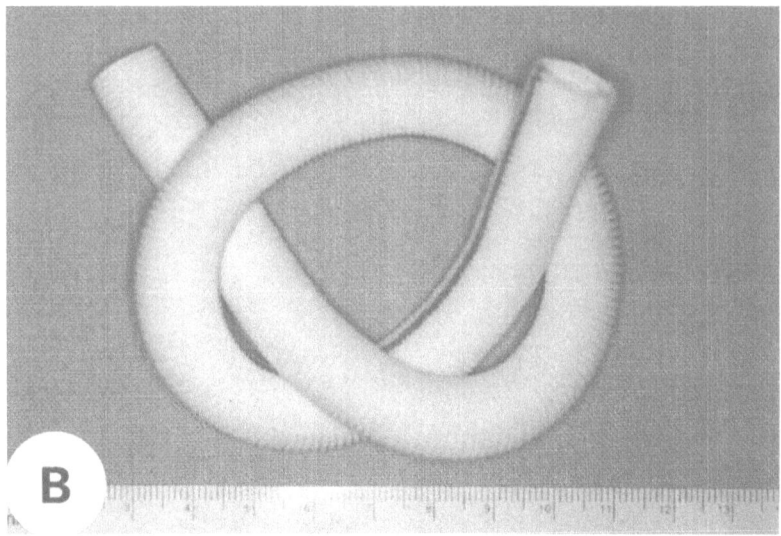

Fig. 6.- Knitted construction. A: loop diagram of a warp knit locknit structure. The yarns lie predominantly in the longitudinal direction. B: photomacrograph of the warp knitted Vasculour II.

154

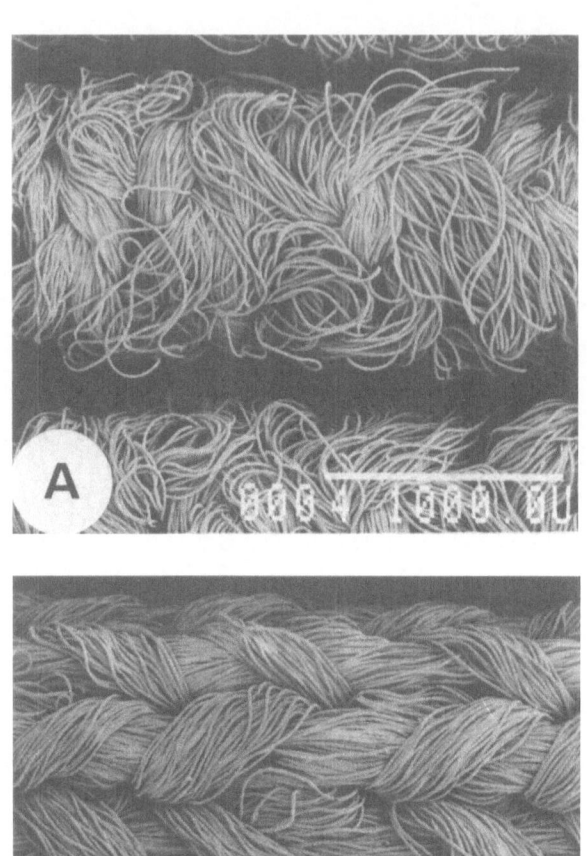

Fig. 7.- Photomicrographs of the wrap knitted
vasculour II graft. A: inside. B: outside.

Fig. 8.- Photomicrographs of the Weavenit graft. A:
inside. B: outside.

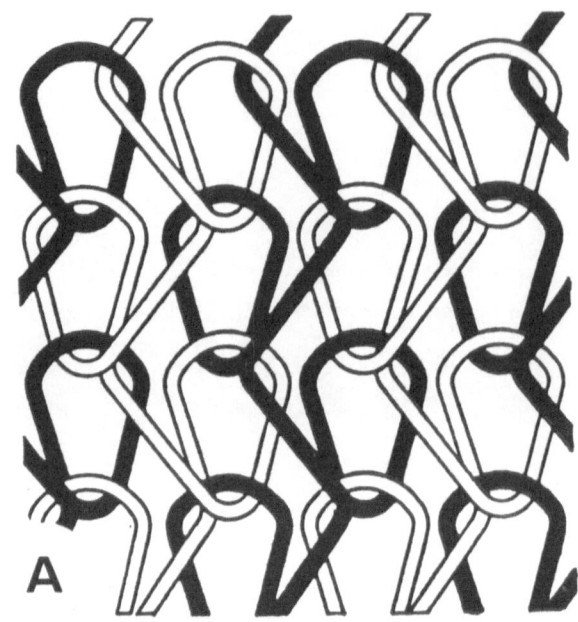

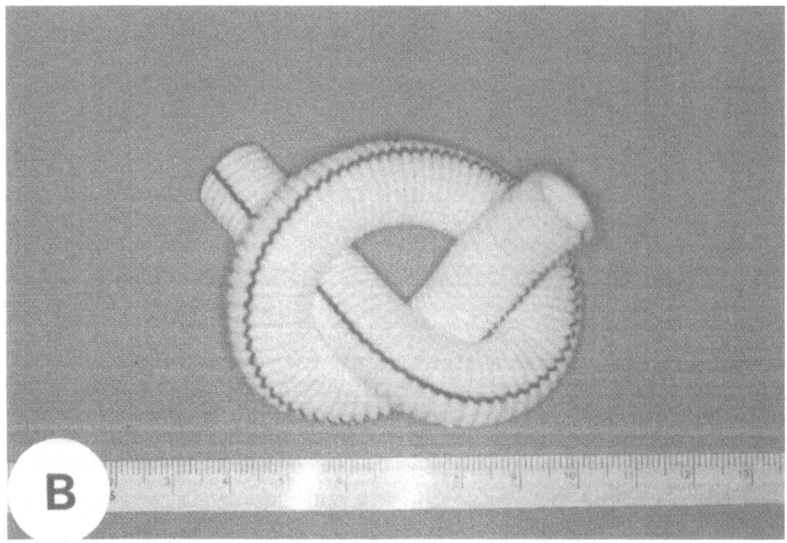

Fig. 9.- Half tricot structure. A: loop diagram. B: photo-macrograph of half tricot Microvel graft.

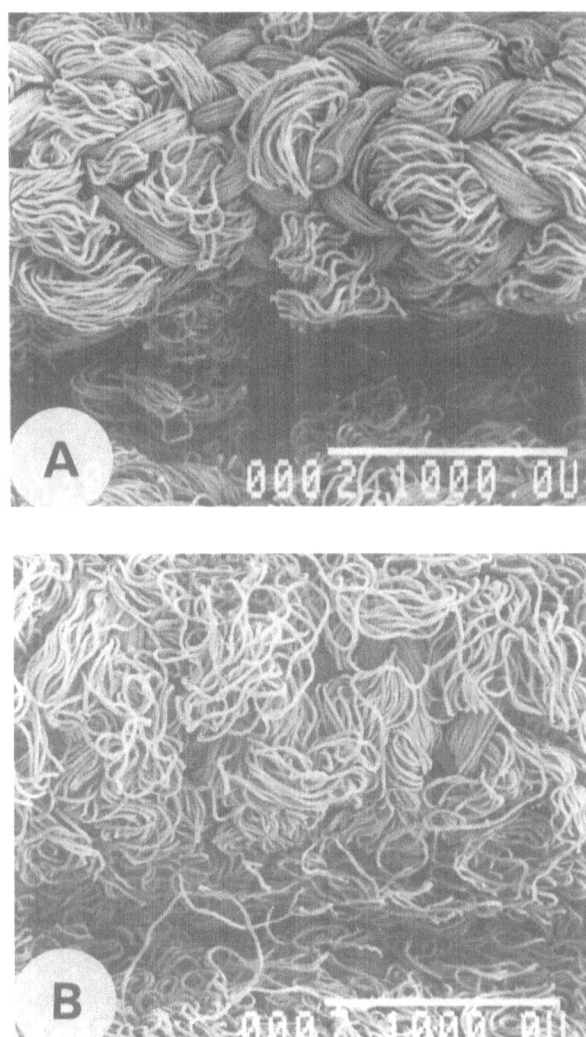

Fig. 10.- Photomicrographs of the half tricot
Microvel graft. A: internal surface. B: external
surface.

TABLE 4. - Characteristics and physical properties of velour prostheses.

Type	Manufacturer	Fiber content	Fabric structure	Diameter (mm)	Stitch density weft/cm	Stitch density warp/cm	Stitch density /cm²	Bursting strength kPa	Water permeability ml/cm²/mm⁻¹	Crimps per cm
Cooley double velour	Meadox Medicals, USA	Flat and texturized	2-bar reverse tricot	10	20	28	560	1160	1770	4.8
Microvel	Meadox Medicals, USA	Flat and texturized	Single tricot base + inlaid pile	10	22	32	700	1100	2400	6.5

TABLE 5. - Characteristics and physical properties of coated prostheses.

Type	Manufacturer	Coating product	Non coated support prosthesis	Diameter (mm)	Stitch density weft/cm	Stitch density warp/cm	Stitch density /cm²	Bursting strength kPa	Water permeability ml/cm²/mm⁻¹	Crimps per cm
Bard	Bard Cardio-surgery Division, USA	Albumin	Vasculour II	8	15	24	360	2377	10	4.5
Gelseal	Vascutek, UK	Gelatin	Triaxial	10	12	16	125	3700	1	3.7
Hemashield	Meadox Medicals, USA	Collagen	Microvel	8	22	32	700	1152	8.7	4.5
Tascon	Tascon Medical Technology*	Collagen	Sauvage Bionit	8	14	28	390	3067	5.9	3.4

* This company is now subsidary of Medtronic Inc., Minneapolis, MN, USA.

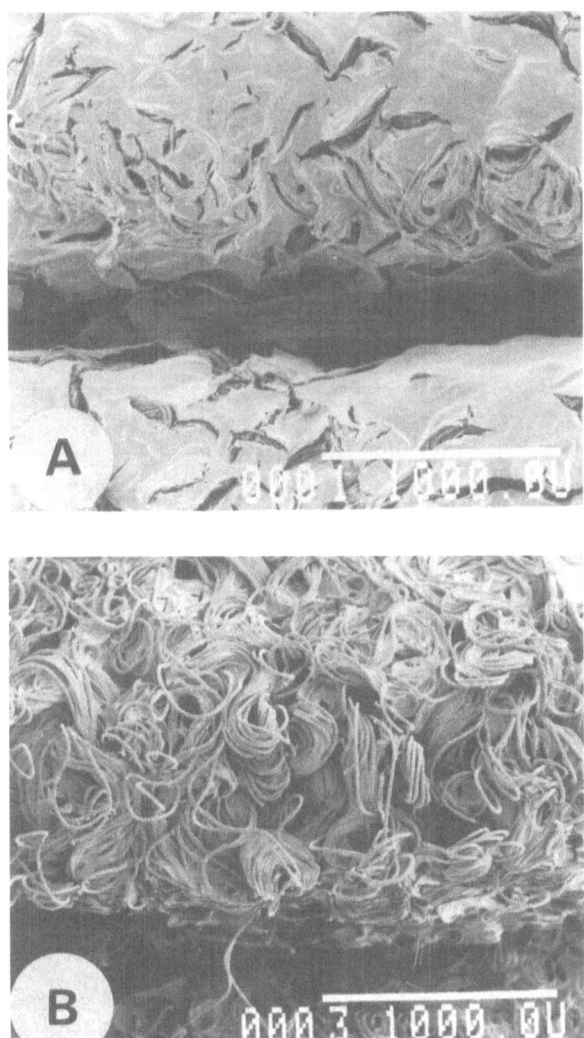

Fig. 11.- Photomicrographs of the Hemashield coated graft. A: internal surface. B: external surface.

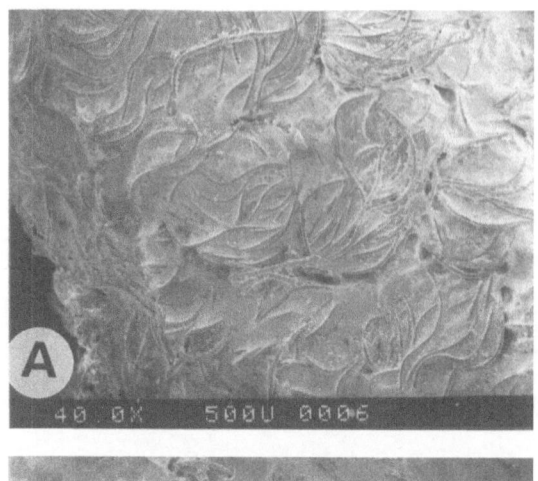

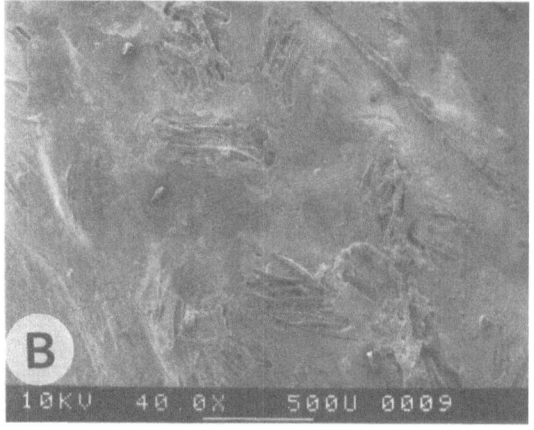

Fig. 12.- Photomicrographs of the Bard Albumin graft. A: internal surface. B: external surface.

TABLE 6. - Characteristics and physical properties of externally supported prostheses.

Type	Manufacturer	Support type	Non supported equivalent prosthesis	Diameter (mm)	Stitch density weft/cm	Stitch density warp/cm	Stitch density /cm²	Bursting strength kPa	Water permeability ml/cm⁻²/mm⁻¹
Atrium	Atrium Medical, USA	Helicoidal welded monofilament	Woven	8	46	43	-	3418	214
DK2-S	Vyzkunny ustav pletarsky, Czechoslovakia	Helicoidal unwelded monofilament	Weft knit	8	22	22	484	3250	1500
Sauvage EXS	Bard Cardio-surgery Division, USA	Helicoidal welded monofilament	Weft knit external velour	8	21	27	567	1750	3100
Sauvage EXS Bionit	Bard Cardio-surgery Division, USA	Helicoidal welded monofilament	Warp knit external velour	8	14	18	252	1349	2818
Vascutek ERT	Vascutek, UK	Helicoidal welded monofilament	Warp knit external velour	8	14	16	224	2506	1563

162

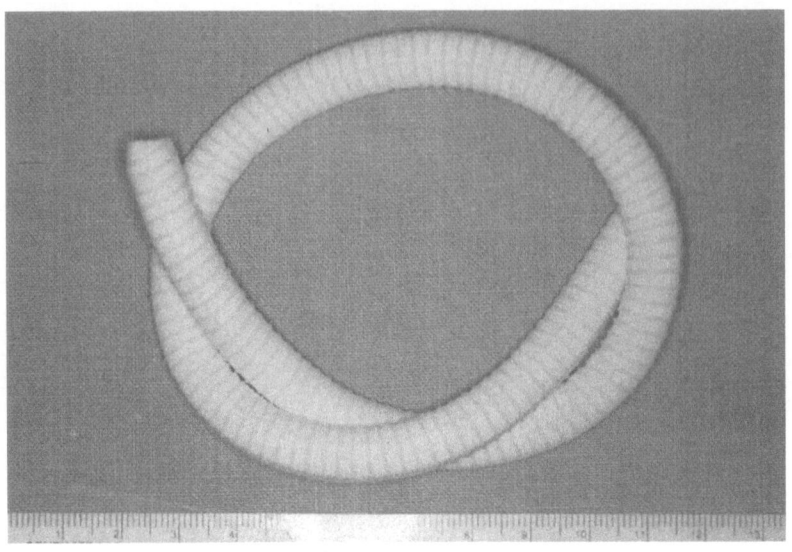

Fig. 13.- Photomacrograph of the Sauvage EXS Bionit
graft.

(24)

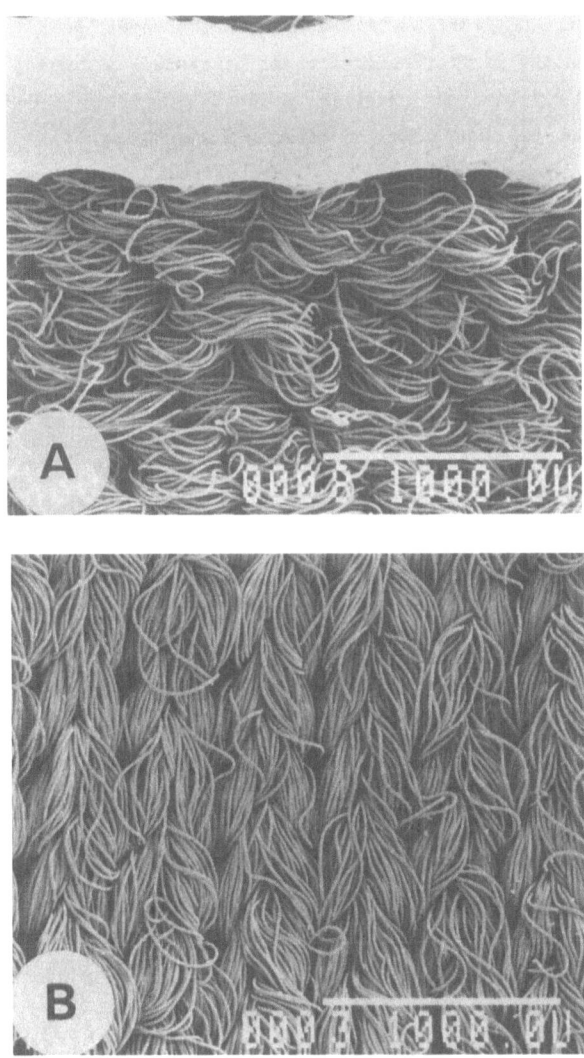

Fig. 14.- Photomicrographs of the Sauvage EXS
Bionit graft. A: internal surface. B: external
surface.

end-plate, so that only the radial component of the pressure force acted on the graft. The static air pressure was controlled by a mercury manometer. The inner pressure was increased from 0 to 340 mmHg (0 to 45 kPa) and the corresponding diameter was measured using a microscope which moved transversally or longitudinally to the nearest 0.05 mm by means of a cathetometer. For each graft, ten measurements were taken at ten different pressure values: 20, 40, 80, 120, 160, 200, 240, 280, 320 and 340 mmHg; and for three different longitudinal tensions. The approximate tension loads recommended by the ANSI/AAHI standard methods were (31): 114, 228 and 456 gf. These values correspond to 0.25, 0.50 and 1 lb.

2.2.2 Stretching

The stretching or longitudinal elongation of the grafts were measured using an Instron machine (Instron Corporation, 100 Royal Street, Canton, Mass. 02021), model 1130 Universal Testing Instrument. The intact graft specimen was held between two adapted clamps. One of them was fixed and the other was attached to the crosshead moving vertically at a speed of 10 cm/min (Figure 16). The geometry of the graft was preserved in order to reproduce the axial deformation in which simultaneous changes in length, thickness and circumference occured anisotropically. Also, the effect of the stretching tension was uniformly distributed at both ends. The tension load represented the elongational force which was recorded versus the crosshead displacement (representing the graft elongation) on the Instron chart recorder and on a HP 9000 Series 200 computer. However, prior to testing, a preconditioning technique was followed to obtain a reproducible stress-strain curve. Therefore, the graft was submitted to 5 loading-unloading cycles in order to reach the same stable hysteresis curve.

2.3 Data processing

The circumferential deformation, ϵ_θ, was expressed as the fractional change of the external radius compared to the reference radius, R_0, which was the radius when the prosthesis was not submitted to inner pressure. Since grafts collapsed when no internal pressure was applied, the reference diameter was measured at a pressure of 20 mmHg, which was sufficient to make the graft cylindrical. Also, the longitudinal deformation, ϵ_z, was determined as the fractional change in length compared to the reference length which was the length, L, of the graft in the absence of stress (30). Thus:

$$\epsilon_\theta = \frac{r_0 - R_0}{R_0} = \frac{dr_0}{R_0} \tag{16}$$

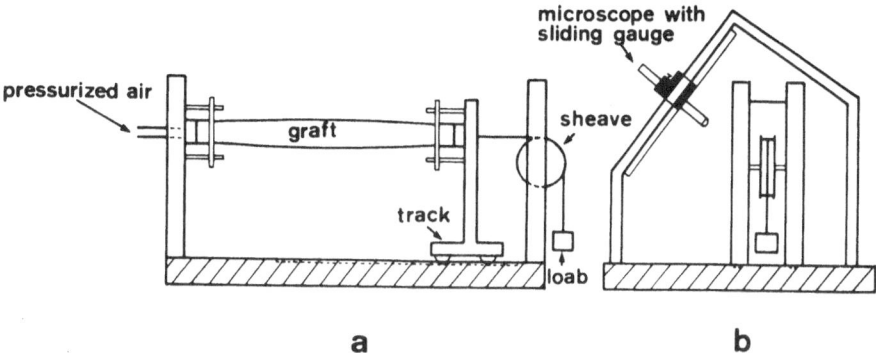

Fig. 15.- Schematic illustration of the dilation test apparatus (OPTIDDIAC). A: front view (without microscope). B: side view.

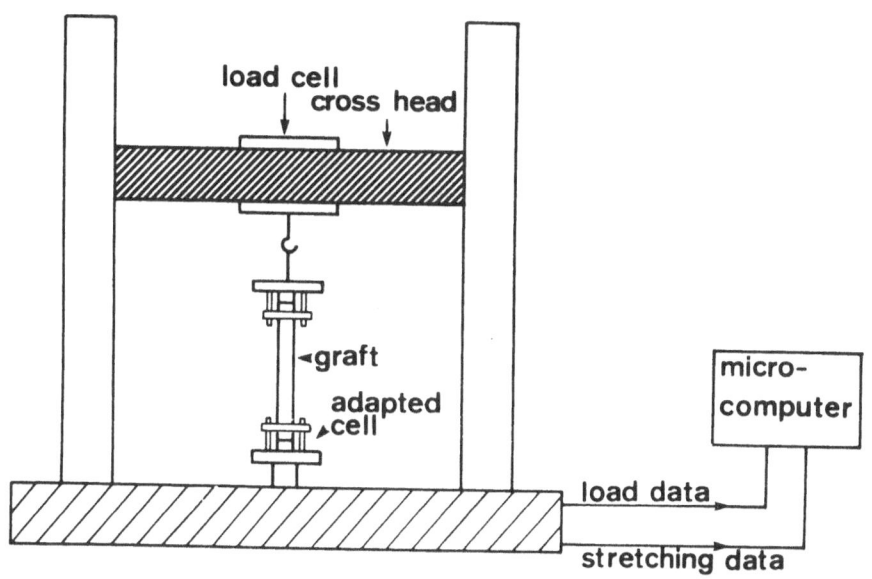

Fig. 16.- Schematic illustration of the longitudinal stretching test apparatus which uses a tensile tester coupled with a Hewlett Packard HP 9000 microcomputer.

$$\epsilon_z = \frac{dl}{L} \tag{17}$$

As the internal strain from the circumferential pressure was nearly linear for synthetic grafts, the circumferential deformation parameter was considered as the slope of the PR_0-ϵ_θ curve. D_θ was calculated by linear regression analysis of 10 points and was expressed in dynes/cm.

Since the term $L/2\pi R_0$ was known in the elongation test, the longitudinal deformation parameter, D_z, was determined by calculating the term dF/dl. As the computer recorded the elongation force versus the graft elongation, dF/dl was the slope of the curve which was computed at each point. D_z was expressed in dynes/cm.

By similarity to the circumferential diagram, D_z was calculated for the three values of the longitudinal tension: 114, 228 and 456 gf. The tangent dF/dl at 456 gf, in the force-longitudinal elongation curve, was almost perpendicular to the rising tension. Such tangent gave a very high value to the derivative dF/dl as well as the parameter D_z. For this reason, we prefered to calculate the third value of D_z at a tension of 342 gf (0.75 lb).

The results were presented in terms of D_θ or D_z versus the longitudinal tension. Data were then adapted using a polynomial spline function.

2.4 Statistical data analysis

Statistical analysis was performed on the circumferential deformation measurements in order to evaluate the accuracy of the dilation test results as well as the validity of hypotheses and assumptions considered in the present study, especially the use of the linear pressure-circumferential deformation relationship. As the material was complex, the form of relation $\epsilon_\theta = f(T,P)$ was unknown. In this case, the polynomial function which is frequently used is as follows:

$$\epsilon_\theta = f(P,T,PT,P^2,T^2,TP^2,T^2P,P^3,...) \tag{18}$$

Knowing the importance of each parameter in this expression will justify the use of the linear stress-strain relationship.

RESULTS

When the diameter variations were less than 0.05 mm, the elastic parameter could not be determined. In this case, this value was greater than the minimum value which could be detected by our measurement (it was estimated to be about 250×10^5 dynes/cm). The values approaching this limit were submitted to error since the variations on the graft diameter were small and were of the same order of magnitude as the reading error. Therefore, the results of some vascular pros-

theses were not presented accuratly, but as exceeded a maximum level of stiffness.

Dilatation

Figure 17 illustrates the elastic parameter versus the longitudinal tension for the woven grafts. At a constant tension of T = 456 gf, these grafts presented a large range of dilatability with an elastic parameter D_θ ranging from 10 x 10^5 to more than 250 x 10^5 dynes/cm. The four grafts, Cooley[R] low porosity, DeBakey[R] soft, Woven Vascutek[R], Cooley Verisoft[R], had very high rigidity which couldnot be measured with our experimental set-up. On the other hand, the four vascular grafts, Advanced Vascular Technology[R], Indian[R], CCV-VPI[R] and the woven prototype, were very elastic and showed a high dilatability when submitted to inner pressure.

The weft knitted grafts showed a high elasticity behaviour which seemed to be very sensitive to the longitudinal tension except for the Lopor[R] graft (Figure 18). However, the weft knitted grafts grew much stiffer with an increase in tension. At a longitudinal tension of 114 gf, the three grafts, Lopor[R], Mikrofroté[R] and Barone[R], presented the same elastic behaviour. Also, this behaviour was identical in the cases of the Barone[R] and the two DeBakey[R] grafts at 456 gf.

Warp knitted grafts were separated into two distinct groups (Figure 19). The first group included: Triaxial[R], Vascutek 1200K[R], Weavenit[R] and Cooley 2[R]. The other group comprised the Vasculour II[R], Knitted Cooley[R] and the two Sauvage Bionit[R]. The last two grafts were very close to each other in terms of rigidity at a longitudinal tension of 114 gf and were stiffer than all other models with a mean D_θ value of 42.2 x 10^5 dynes/cm. However, the grafts belonging to the first group were as elastic as the weft knitted grafts at a tension of 114 gf with a mean value of 16.6 x 10^5 dynes/cm.

The two velour grafts showed a high circumferential deformation, especially the Microvel[R] graft (Figure 20). These deformations were of the same order as that of the weft knitted grafts. However, the velour grafts did not seem to be sensitive to the longitudinal force. Indeed, D_θ changed between 14 x 10^5 and 17 x 10^5 dynes/cm for the Microvel[R], and between 27 x 10^5 and 30 x 10^5 dynes/cm for the Cooley Double Velour[R].

In the case of coated grafts (Figure 21), the Bard albumin[R], Hemashield[R] and Gelseal[R] presented a similar circumferential elastic parameter at 456 gf with a mean value of 48.4 x 10^5 $\pm$ 11.9 x 10^5 dynes/cm. They were not sensitive to tension, especially the Hemashield[R] graft. Nevertheless, the Tascon[R] graft became much stiffer when tension was risen; the elastic parameter at 456 gf was 2.5 times greater than that at 114 gf. In Table 7, we compared the circumferential elastic parameter values of the

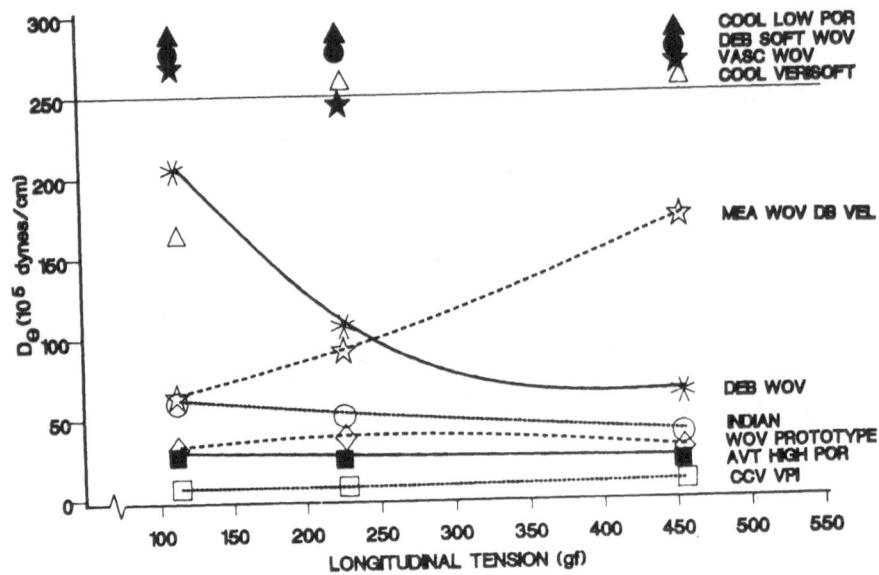

Fig. 17.- Woven grafts: circumferential elastic parameter versus longitudinal tension.

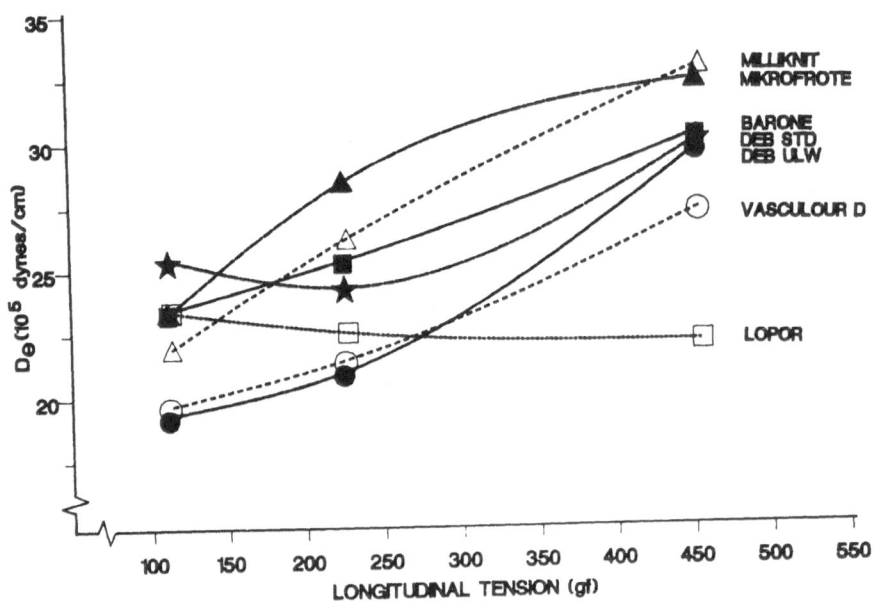

Fig. 18.- Weft knitted grafts: circumferential elastic parameter versus longitudinal tension.

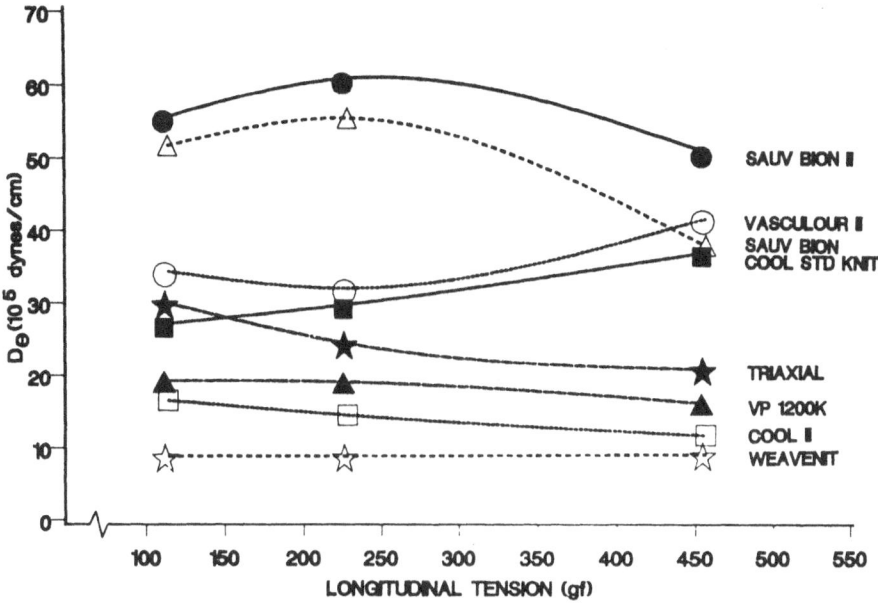

Fig. 19.- Warp knitted grafts: circumferential elastic parameter versus longitudinal tension.

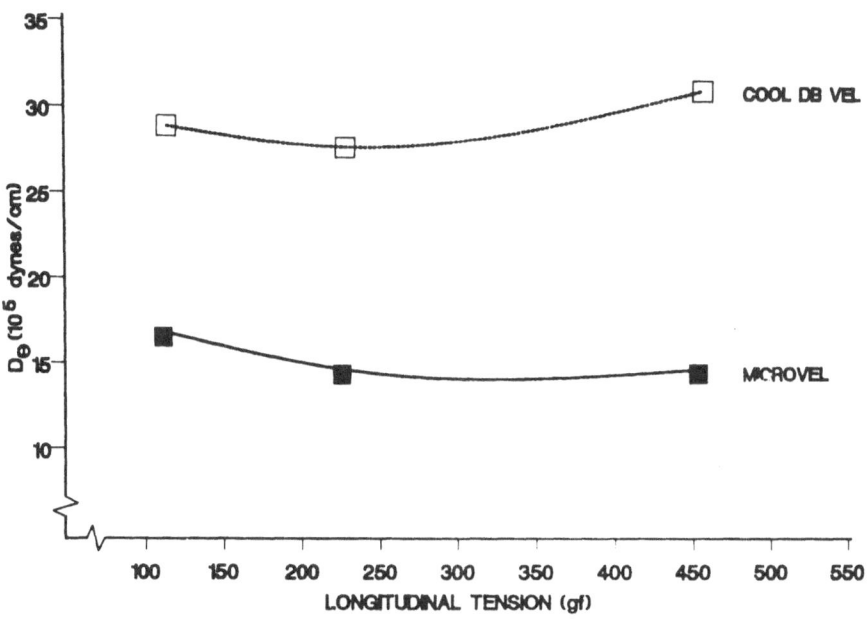

Fig. 20.- Velour grafts: circumferential elastic parameter versus longitudinal tension.

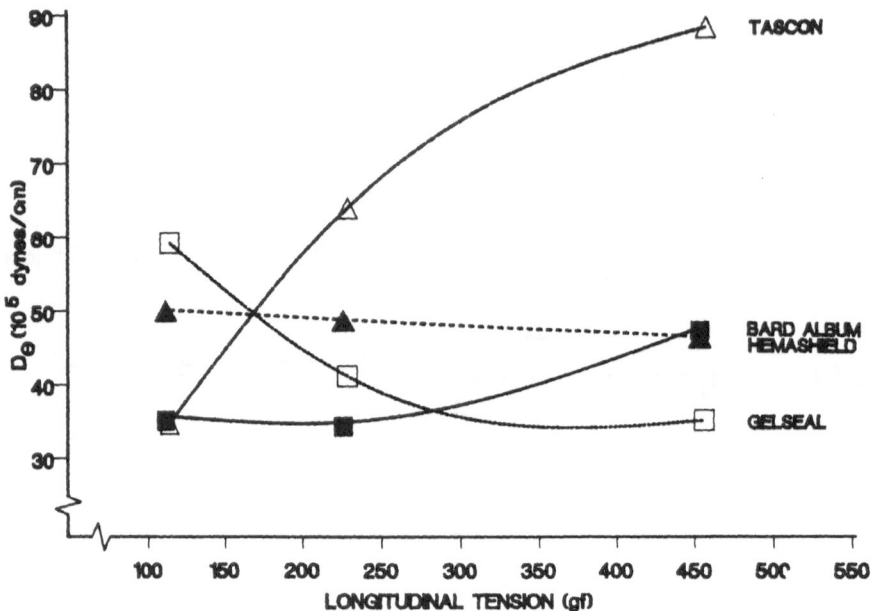

Fig. 21.- Coated grafts: circumferential elastic
parameter versus longitudinal tension.

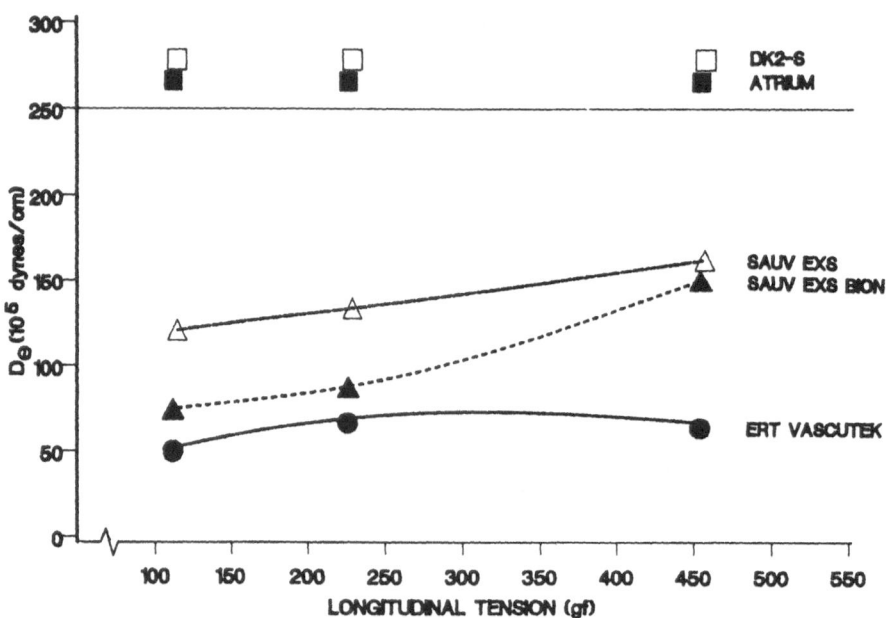

Fig. 22.- Externally supported grafts: circumferential elastic parameter versus longitudinal
tension.

TABLE 7.- Comparison of the circumferential elastic behaviour of the coated grafts and the textile structure.

| Prostheses | D_θ (10^5 dynes/cm) | | |
| | Longitudinal tension (gf) | | |
	114	228	456
Vasculour II	34.3	32.1	41.6
Bard Albumin*	37.7	34.9	48.0
Triaxial	30.0	24.5	20.9
Gelseal*	59.3	41.2	35.2
Microvel	16.7	14.6	14.6
Hemashield*	50.2	48.8	46.6
Sauvage Bionit	51.7	55.5	37.9
Tascon*	34.7	64.0	88.5

* Coated grafts

TABLE 8.- Comparison of the longitudinal elastic behaviour of the coated grafts and the textile structure.

| Prostheses | D_z (10^4 dynes/cm) | | |
| | Longitudinal tension (gf) | | |
	114	228	342
Vasculour II	11.1	28.0	39.0
Bard Albumine*	55.4	103.8	155.8
Triaxial	35.0	68.3	93.6
Gelseal*	22.9	50.9	85.6
Microvel	30.2	67.9	117.4
Hemashield*	24.12	64.08	112.0
Sauvage Bionit	13.2	28.9	46.8
Tascon*	43.3	89.8	133.5

* Coated grafts

coated grafts and their respective non coated prosthesis. The three grafts Gelseal[R], Hemashield[R] and Tascon[R], which were either a gelatin or collagen coated product, were less elastic than the non coated grafts. On the contrary, there was no significant difference between the Bard albumin[R] graft and the Vasculour II[R].

The two externally supported grafts DK2[R] and Atrium[R] had an excellent dimensional stability since our apparatus was not sensitive enough to detect any circumferential deformation (Figure 22). However, the DK2[R] graft showed an elasticity parameter of 250 x 10^5 dynes/cm at 114 gf, while the equivalent unsupported prosthesis (Mikrofroté[R] categorized in the weft knitted graft group) had an elasticity parameter of 23.4 x 10^5 dynes/cm. The other externally supported grafts exhibited high rigidity except for the Vascutek EXS[R] which dilates despite the presence of an external support.

Longitudinal stretching

The elastic parameter in the longitudinal direction was presented versus the longitudinal tension for each class of textile grafts: woven (Figure 23), weft knitted (Figure 24), warp knitted (Figure 25), velour (Figure 26), coated (Figure 27) and externally supported (Figure 28).

All groups of grafts showed the same longitudinal elongation behaviour within the same group. An increase in longitudinal tension resulted in a decrease in longitudinal distensibility. However, although the woven grafts showed a large range of dilatability, they behaved similarly in the longitudinal direction. Woven and knitted (weft knitted and warp knitted) had similar elasticity at the three longitudinal tension values. Only the Vascutek Triaxial[R] and the Mikrofroté[R] grafts were stiffer than the others.

Velour grafts were also less elastic than the previous mentioned grafts.

Table 8 illustrates the comparison between the coated grafts and non coated ones with regards to their longitudinal elastic behaviour. The coated grafts Bard Albumin[R] and Tascon[R] were less distensible longitudinally than their uncoated grafts Vasculour II[R] and Sauvage Bionit[R] respectively, while there was no difference between the Gelseal[R] and its skeleton Triaxial[R] as well as between the Hemashield[R] and the Microvel[R].

However, the most dimensionally stable grafts were the externally supported ones. Although DK2[R] and Atrium[R] prostheses were the least dilatable, they turned out to be the most stretchable. The Sauvage weft knit EXS[R] prosthesis was distinguished by its weak longitudinal elasticity.

Finally, Figure 29 allowed an easy comparison of graft dilatability and their elongability according to their textile construction and summarized

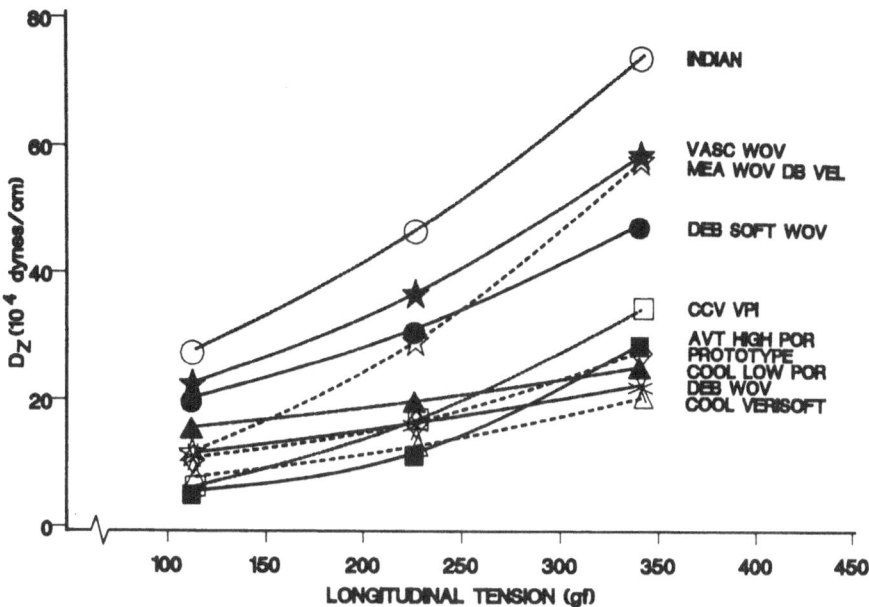

Fig. 23.- Woven grafts: longitudinal elastic
parameter versus longitudinal tension.

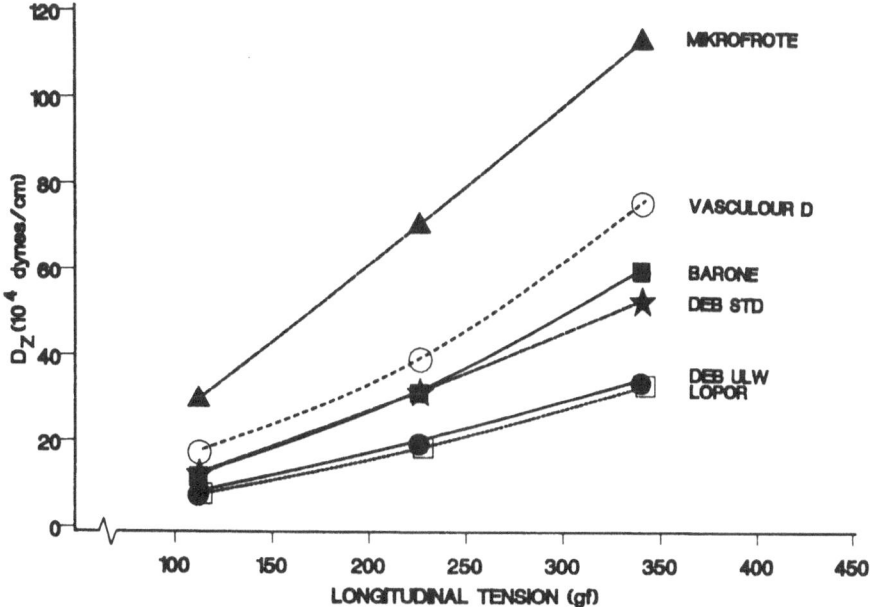

Fig. 24.- Weft knitted grafts: longitudinal elastic
parameter versus longitudinal tension.

174

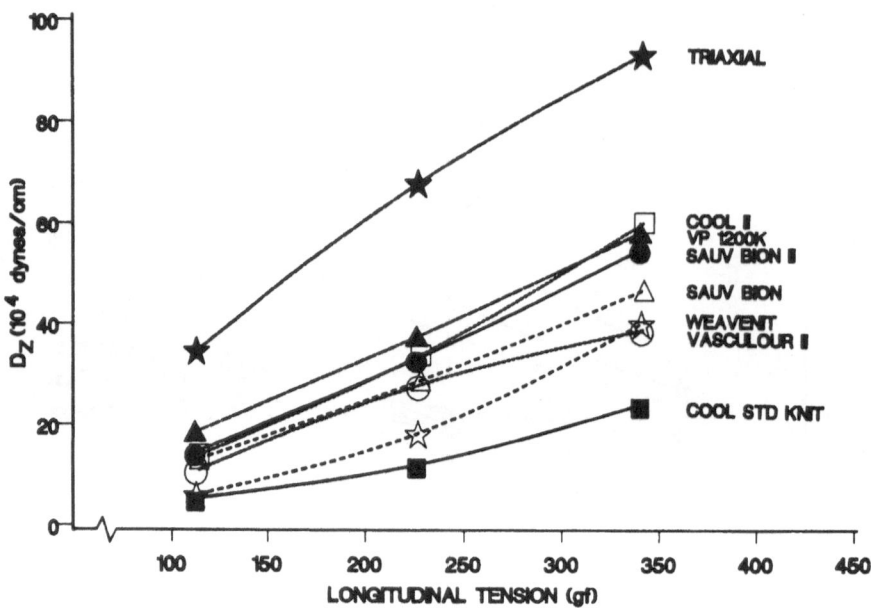

Fig. 25.- Warp knitted grafts: longitudinal elastic parameter versus longitudinal tension.

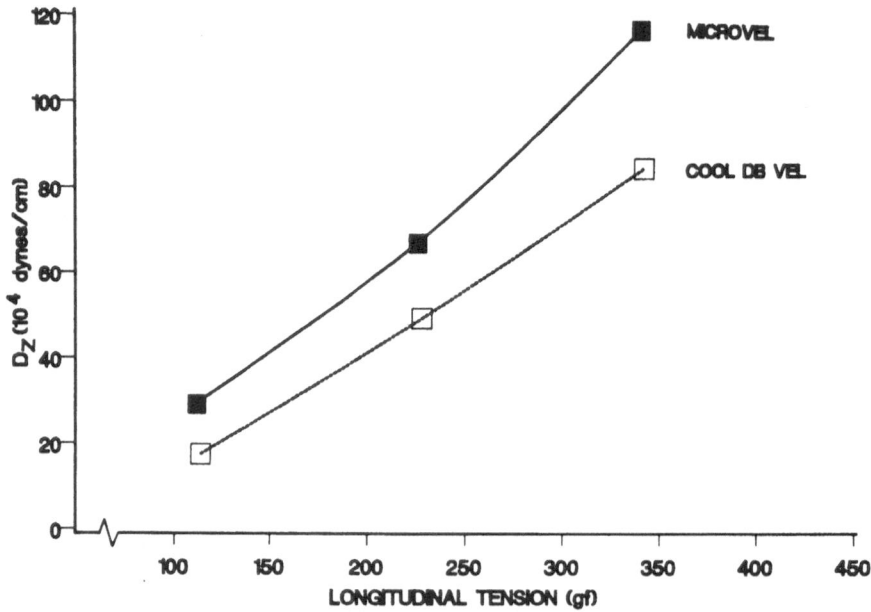

Fig. 26.- Velour grafts: longitudinal elastic parameter versus longitudinal tension.

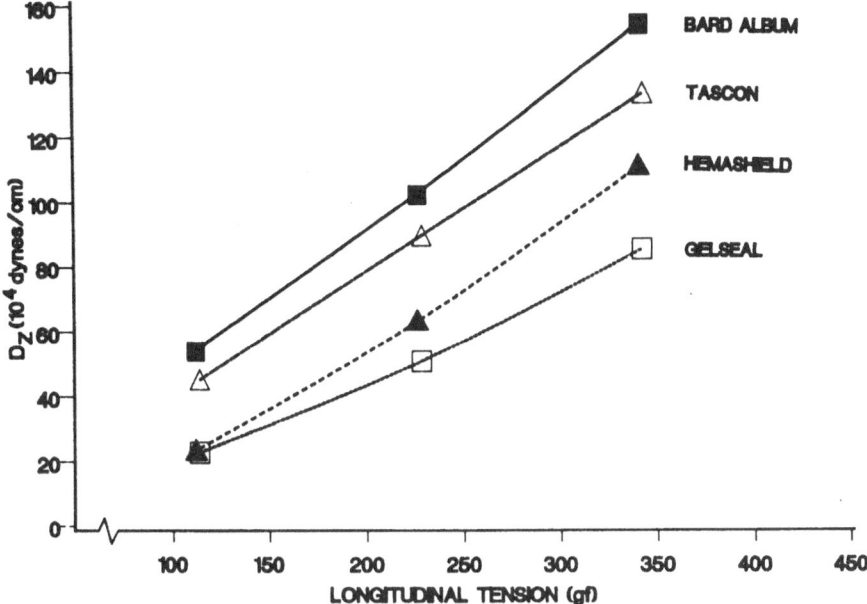

Fig. 27.- Coated grafts: longitudinal elastic
parameter versus longitudinal tension.

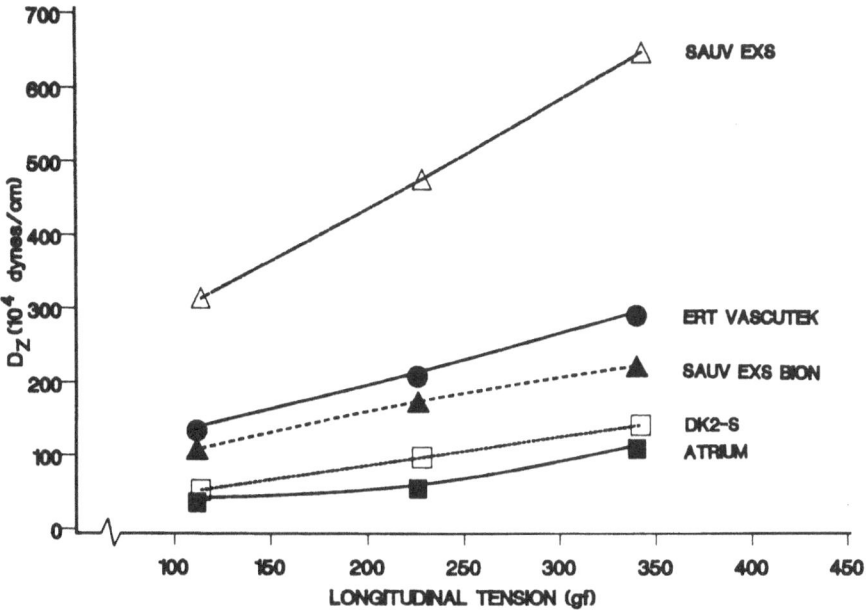

Fig. 28.- Externally supported grafts: longitudinal
elastic parameter versus longitudinal tension.

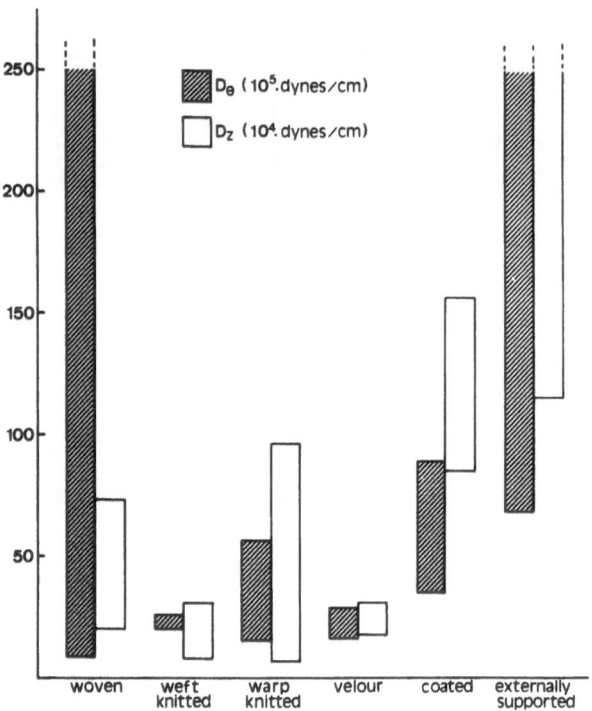

Fig. 29.- Comparison between circumferential and longitudinal deformability with regard to their fabric structure. The longitudinal tension corresponds to that recommended by the ANSI/AAMI standard methods.

all our observations. The weft and warp knitted prostheses exhibited high deformability in both longitudinal and radial directions. Velour grafts were as elastic as the weft knitted prostheses. Although the elongability of woven grafts was limited, they showed a large range of dilatability. In addition, coated grafts as well as externally supported grafts presented a high dimensional stability.

Statistical analysis

The use of elasticity coefficients was based on the assumption that the relationship between circumferential deformation and internal pressure or tension was linear. This assumption was checked by adapting equation (18) to the observed data for each graft and verifying that linear terms in P and T adequately described changes in circumferential deformation. The amount of variability in the observed means, which was accounted for by the various terms of a quadratic response surface, was expressed as a percentage of the variability among observed means at each combination of internal pressure and longitudinal tension (Table 9). It showed that the linear component of the pressure effect on circumferential deformation accounted for most of the variability among observed means at various values of P; this amount varied from 49 to 98% with only a few grafts under 80%. However, longitudinal tension (T) also affected circumferential deformation significantly for several prostheses (DeBakey[R], Woven[R], Lopor[R], Tascon[R], Meadox Woven Double Velour[R] and others). Although the amount of variation accounted for by the quadratic term in P, P^2, was relatively small for most prostheses (it ranges from 0.0 to 4.8%) this component had a significant effect on average circumferential deformation in 75% of the prostheses. The terms T^2 and TP were rarely significant.

DISCUSSION

Elastic behaviour of arteries and small diameter grafts was recently investigated by several authors who stated that the weak dilatability of synthetic grafts led to mechanical and hemodynamic trauma sufficient to stimulate intimal hyperplasia and thrombosis which then resulted in graft failure (1,19,20,30). However, this has to still be proved since the influence of compliance mismatch may depend intimately on the thrombogenic characteristics of the flow surface of the graft which could have a greater total effect than any compliance mismatch (22).

In larger diameter grafts, compliance mismatches between host arteries and grafts did not lead to such severe complications since the blood velocity is higher and the thrombotic threshold velocity was more difficult to reach; but dilation caused more damage to the graft encapsulation which was necessary for a

TABLE 9.- Percentage of variability accounted for by polynomial term.

Textile class	Model	P (%)	P² (%)	T (%)	T² (%)	TP (%)
Woven	AVT High Porosity	91.73	2.66	4.18	0.00	0.51
	Circumferentially Compliant Velour	94.57	2.27	1.64	0.02	0.90
	DeBakey Woven	49.02	0.88	39.08	0.06	8.89
	Meadox Woven Double Velour	59.15	2.14	28.01	0.40	7.50
	Indian	72.50	4.77	17.62	0.98	2.30
	Prototype	94.80	0.17	1.19	0.03	0.91
Weft knit	Barone	94.77	1.35	2.34	0.07	1.04
	DeBakey Standard	96.38	0.32	1.06	0.30	0.86
	DeBakey Ultralight Weight	63.78	1.10	31.88	0.00	2.35
	Lopor	95.98	2.41	0.23	0.93	0.00
	Mikrofroté	87.90	1.92	7.39	0.59	1.34
	Milliknit	85.08	0.99	9.52	0.05	2.52
	Vasculour D	92.18	0.02	5.64	0.05	1.74
Warp knit	Cooley Standard Knit	94.24	0.27	0.26	1.69	0.10
	Cooley II	94.86	2.60	0.22	0.15	0.98
	Sauvage Bionit	81.10	1.62	0.32	1.62	1.52
	Sauvage Bionit II	83.76	1.18	8.34	3.14	0.10
	Triaxial	-	-	-	-	-
	VP120OK	97.39	1.61	0.27	0.00	0.22
	Vasculour II	88.53	0.90	7.52	0.98	0.89
	Weavenit	98.43	0.01	1.00	0.00	0.06
Velour	Cooley Double Velour	96.38	0.84	1.45	0.82	0.20
	Microvel	98.23	0.82	0.19	0.33	0.05
Coated	Bard	-	-	-	-	-
	Gelseal	78.91	2.76	12.11	1.95	2.44
	Hemashield	98.77	0.00	0.24	0.08	0.03
	Tascon	54.77	0.04	29.72	4.94	0.05
Externally supported	Sauvage EXS	68.74	1.40	12.48	9.68	1.02
	Sauvage EXS Bionit	70.25	0.15	22.20	0.04	4.94
	Vascutek ERT	74.28	0.49	3.85	5.96	0.52

good healing process. The effective diameter of the prostheses should be close to the nominal diameter to prevent any break in, damage or rupture of both internal and external capsules in order to prevent any accumulation of thrombotic material (32) likely to facilitate occlusion and bacteremic colonization (33).

In the arterial system where the graft is subjected to a static pressure as well as to the repeated stress of pulsation, the elongation is unavoidable (34). Therefore, it is necessary for the surgeon to be aware of the difference between the nominal and the effective diameter of the graft which is subjected to a longitudinal tension and an inner pressure after implantation (32). This study was done in order to discuss the mechanical behaviour of Dacron arterial grafts and to evaluate and predict in situ deformations, especially dilation and elongation.

However, the AAMI standard graft dilation test consists of measurements of graft's diameter, pressurizing at the physiological pressure of 120 mmHg (31). This information is of interest but insufficient to characterize the graft elastic behavior.

Compliance has generally been used to assess circumferential deformation (1,13,22) and it was a significant parameter when reported together with the corresponding pressure range. However, the compliance measurements were performed at a constant graft length which did not correspond to the in vivo conditions. As a matter of fact, the graft in physiologic conditions is subjected to inner pressure, and complete longitudinal tethering was not be assumed prior to substantial tissue ingrowth and encapsulation. Therefore, the parameter D_θ was more significant in terms of in vivo deformation. However, the use of linear relationships between the pressure and circumferential pressure was justified by the polynomial regression since more than 81% of this deformation was due to pressure in most cases. Significance of the quadratic term P^2 in several graft models explained the light flattening of the pressure-circumferential deformation relationship at high pressure.

In general it was observed that woven grafts had a greater dimensional stability than knitted ones, however there was a large diversity. Some dilated more than knitted grafts and others were more rigid than some of the externally supported grafts.

The structure of woven grafts were simple, but led to different mechanical behaviours depending on the fiber used, the fiber density and other characteristics. The four very stiff grafts, Cooley low porosity[R], DeBakey soft[R], Woven Vascutek[R] and Cooley Verisoft[R] were introduced because of their low water permeability, for heparinized patients under extracorporeal circulation. To obtain this low water permeability, the texture was reduced to 40 stitches/cm leading to an excellent dimensional stability. On the other hand, the two grafts,

Advanced Vascular Technology[R] and CCV-VPI[R], which had very open structure, were not strong enough to resist the pulsatile blood pressure and flow. The other woven materials, even if difficult to handle and frayed easily at the edges were very popular because of their good mechanical stability and their bursting strength. However, the circumferential deformation of woven grafts was very much influenced by the longitudinal tension at which it was measured.

Weft knitted graft structure was more open as proved by the water permeability values. They were easier to handle than woven ones and did not fray at their edges but they lacked sufficient dimensional stability. The DeBakey ULW[R] and Milliknit[R] had greater water permeability than other weft knit grafts. They exhibited high dilatatility and low bursting strength, revealing a highly deformable graft. Such light-weight or thin-walled grafts were knitted with more needles than usual and with thinner fibers in order to reduce the wall thickness and to improve healing properties (35,36), which explains their low D_θ values and high dilatability.

In warp knitted grafts, the Weavenit[R] and Cooley 2[R] were also light weight products. However, clinical experiences reported that the healing process was not improved by this graft and some post-operative dilation, false aneurysms, suture line disruptions and interstitial bleeding occurred shortly after their implantation (37-44). The use of light-weight grafts is now discontinued in the USA. Light-weight warp knitted grafts were stiffer than weft knitted ones and the possible presence of a lost stitch in the structure when exposed to continuous cyclic stress did not lead to a run as in the case of a weft knitted structure (45). Compared to woven grafts they were easier to handle and had good surface quality and high water permeability. Warp knitted grafts are interesting materials and are replacing the earlier weft knitted models (46) but still require preclotting, and their implantation must be avoided when the graft is submitted to high mechanical stresses.

As far as the two velour grafts were concerned, the attempt at introducing trilobal fibers, in order to soften the construction, resulted in a decrease of the bursting strength and is now abandoned (47). The Microvel[R] model was of warp knitted structure, but it was knitted with half tricot stitch with only one set of yarn. The second yarn, which was used exclusively to form the loose-looped pile, was unable to contribute to strength and dimensional stability (48). The Cooley double velour[R] was stiffer than the above graft and thus more acceptable. Velour grafts have proved to have better healability than smooth walled grafts (49), but such material have a tendancy toward hyperplasia and stenosis (48), and concern for their high dilatability.

Although the coating of grafts, by either collagen or gelatin, maintained fibers together and improved the dimensional stability, the albumin did not improve the properties of the graft compared to the non coated support prosthesis, Vasculour II[R]. As demonstrated before, coatings improved the healing process of the grafts (50,51). However, the Tascon[R] graft induced both inflammatory reaction in the surrounding tissues and rapid pseudointimal accumulation, and the graft is no longer in use (50).

Externally supported grafts had good dimensional stability. The Vascutek[R] graft was more elastic than others due to the larger distance between rings of the external support.

Our study emphasized the role of longitudinal stress on circumferential elastic properties and especially on woven, coated and externally supported grafts. The longitudinal load recommended by the AAMI standard was 500 g for woven and non woven grafts, and 125 g for knitted grafts. This load was chosen to remove most of the graft crimps without damaging the structure of the specimen, but grafts behave so differently that this choice was arbitrary. Moreover, it is impossible for the surgeon to control the tension of the implanted graft. Therefore, it is preferable to specify the elastic behaviour of the graft at different possible longitudinal tensions. Although it is not satisfactorily understood, the elasticity in the longitudinal direction should also be considered because arteries are never in the relaxed state in vivo (52). At implantation, the longitudinal strain applied to the graft should be sufficiently high to avoid compression stress at elevated internal pressures but it should not exceed the arterial tethering force in order to reduce excessive loads on the host vessel at anastomosis (27). Furthermore, the graft should not stretch too much in order to keep its shape and avoid deformation by surrounding tissue.

The high longitudinal distensibility of woven, knitted and velour grafts was principally due to the presence of fine circular or helicoidal crimps (11,53). Longitudinal yarns were not be stressed until the crimp was straightened (53). This predominant role of crimping also explains why the elongabilities of woven grafts were very similar while their dilatabilities, more related to the manufacture weaving, were different.

CONCLUSION

This paper qualifies and quantifies the elastic behaviour of 36 grafts with regards to their type of textile structure as well as their individual characteristics. Since clinical reports revealed that circumferential deformation of weft knitted grafts led to serious complications, such as permanent dilatation, false aneurysms and interstitial bleeding, an inferior limit of dilation is required when manufacturing the graft. The results reported in this study should be an important criterion for the surgeon when selecting commercial Dacron graft. Furthermore, the success of graft implantation depends more on the adaptation of the prosthesis to the patient's characteristics than on the graft itself, and a good knowledge of the different grafts available is necessary to select the right one.

Finally, a compromise between good mechanical stability and bursting strength, on one hand, and good handling, healing and biocompatibility characteristics, on the other, has yet to be established. The harder the mechanical conditions which the graft are subjected to the stiffer and more dimensionally stable the graft will be, especially since mechanical stresses accelerate chemical degradation of the biomaterial.

ACKNOWLEDGEMENTS

This work was supported in part by the Medical Research Council of Canada and St-François d'Assise Hospital. The technical assistance of Yves and Jean-Pierre Gourdeau, Karen Horth, Suzanne Bourassa and Jacques Rodrigue was greatly appreciated. The grafts were kindly provided by Bard Cardiosurgery Division, Golaski Inc., Advanced Vascular Technologies, Vascutek Ltd., Sree Chita Tirunal Institute, Shanghai Chest Institute, Meadox Medical Inc., Vascular Products Ireland, Barone, Atrium and Vyzkunny Ustav Pletarsky. We are indebted to Drs. C. Poirier, C. Gosselin, G. Roy, T.J. Rao and A. Downs for their help and guidance.

REFERENCES

1. Turner RJ. Vascular graft development: an industrial perspective. In: "Modern Vascular Grafts", PN Sawyer edr, McGraw-Hill Book Company, 1987, pp 75-103.
2. Wesolow SA, Snyder RM. Textiles. In: "Handbook of Biomaterials Evaluation, Scientific, Technical and Clinical Testing of Implant Materials", AF Von Recum edr, Macmillan Publishing Company, 1986, pp 79-85.
3. Sauvage LR, Smith JC, Davis CC, Rittenhouse EA, Hall DG, Mansfield PB. Dacron[R] arterial grafts: comparative structures and basis for successful use of current prostheses. In: "Vascular Graft Uptade: Safety and Performance", HE Kambic, A Kantrowitz, P Sung edrs, Am. Soc. Test. Mat., 1986, pp 16-24.
4. King MW, Guidoin R, Gunasekera K, Martin L, Marois M, Blais P, Marrek JM, Gosselin C. An evaluation of Czechoslovakian polyester arterial prostheses. ASAIO J. 7: 114-133, 1984.
5. Guidoin RG, King M, Marois M, Martin L, Marceau D, Hood R, Maini R. New polyester arterial prosthesis from Great Britain: an in vitro and in vivo evaluation. Ann. Biomed. Eng. 14: 351-367, 1986.
6. Torché D, Lacombe J, King M, Guidoin R, Boyer D, Marceau D, Marois Y. An arterial prosthesis from Argentina: the Barone Microvelour[R] arterial graft. J. Biomat. Appl.
7. Guidoin R, Gosselin C, Roy J, Gagnon D, Marois M, Noël HP, Roy P, Martin L, Awad J, Bourassa S, Rouleau C. Structural and mechanical properties of Dacron prostheses as arterial substitutes. In: "Mechanical Properties of Biomaterials", GW Hasting, DF Williams edrs, John Wiley and Sons Ltd., 1980, pp 547-556.
8. Guidoin R, Gosselin C, Martin L, Marois M, Laroche F, King M, Gunasekera K, Domurado D, Sigot-Luizard MF, Blais P. Polyester prostheses as substitutes in the thoracic aorta of dogs. I. Evaluation of commercial prostheses. J. Biomed. Mat. Res. 17: 1049-1077, 1983.
9. Guidoin R, King M, Marceau D, Cardou A, DelaFaye D, Legendre JM, Blais P. Textile arterial prostheses: Is water permeability equivalent to porosity? J. Biomed. Mat. Res. 21: 65-87, 1987.
10. Fung YC, Fronek K, Patitucci P. Pseudoelasticity of arteries and the choice of its mathematical expression. Am. J. Physiol. 237: H620-H631, 1979.
11. Nahon D, Lee JM. A two-dimensional incremental study of the static mechanical properties of vascular grafts. Clin. Mat. 1: 177-197, 1986.
12. Dobrin PB. Mechanical properties of arteries. Physiol. Rev. 58: 397-460, 1978.
13. Lee JM, Wilson GJ. Anisotropic tensile viscoelastic properties of vascular graft materials tested at low strain rates. Biomaterials 7: 423-430, 1986.
14. Fung YC. Mechanical properties of blood vessels. In: "Biomechanics: Mechanical properties of living tissues", Springer-Verlag, New York, 1981, pp 261-301.
15. Weizsacker HW, Lambert H, Pascale K. Analysis of the passive mechanical properties of rat carotid arteries. J. Biomech. 16: 703-715, 1983.
16. Patel DJ, Fry DL. Longitudinal tethering of arteries in dogs. Circ. Res. 19: 1011-1021, 1966.
17. Taylor LA, Gerrard JH. Pressure-radius relationships for elastic tubes and their application to arteries: Part 1 - Theoretical relationships. Med. Biol. Eng. Comput. 15: 11-17, 1977.
18. Kinley CE, Marble AE. Compliance: A continuing problem with vascular grafts. J. Cardiovasc. Surg. 21: 143-170, 1980.
19. Baird RN, Kidson IE, L'Italien GJ, Abbott WM. Dynamic compliance of arterial grafts. Am. J. Physiol. 233: H568-H572, 1977.
20. Kidson IG, Abbott WM. Low compliance and arterial graft occlusion. Circulation 38: 1-4, 1978.
21. Abbott WM, Bouchier-Hayes DJ. The role of mechanical properties in graft design. In: "Graft Materials in Vascular Surgery", ed. Dardik H, Symposia Specialist Inc., 59-78, 1978.

22. Abbott WM, Cambria RP. Control of physical characteristics (elasticity and compliance) of vascular grafts. In: "Biologic and Synthetic Vascular Prostheses", eds. Stanley JC, Burkel WE, Lindenauer SM, Bartlett RH, Tucotte JG, Grune and Stratton, 189-221, 1982.

23. Steward SFC, Lyman DJ. Essential physical characteristics of vascular grafts. In: "Modern Vascular Grafts", ed. Sawyer PN, McGraw-Hill Book Company, 115-132, 1987.

24. Marceau D, Cardou A, Guidoin R, Gosselin C, King M. Etude de la déformation circonférentielle des prothèses artérielles en polytétrafluoroéthylène. Rev. Europ. Biotech. Med. 4: 114-117, 1982.

25. Megerman J, Abbott WM. Compliance in vascular grafts. In: "Vascular Grafting Clinical Applications and Techniques", CB Wright, RW Hobson II, LF Hiratzka, TG Lynch edrs, John Wright PSG Inc., 1983, pp 344-364.

26. Edwards WS, Tapp JS. Chemically treated nylon tubes as arterial grafts. Surgery 38: 61-70, 1955.

27. How TV, Annis D. Viscoelastic behavior of polyurethane vascular prostheses. J. Biomed. Mat. Res. 21: 1093-1108, 1987.

28. Papageorgiou GL, Jones NB. Physical modelling of the arterial wall. Part 1: Testing of tubes of various materials. J. Biomed. Eng. 9: 153-156, 1987.

29. Wesly RLR, Vaishnav RN, Fuch JCA, Patel DJ, Greenfield JC. Static linear and non linear elastic properties of normal and arterialized venous tissue in dog and man. Circ. Res. 37: 509-520, 1975.

30. Clark RE, Apostolou S, Kardos JL. Mismatch of mechanical properties as a cause of arterial prostheses thrombosis. Surg. Forum 27: 208, 1976.

31. National Standard for Vascular Grafts Prostheses, ANSI/AAMI VP20 1986, Association for Advancement of Medical Instrumentation, Arlington, USA.

32. Sanders RJ, Kempczinski RF, Hammond W, Diclementi D. The significance of graft diameter. Surgery 88: 856-866, 1980.

33. Goeau-Brissonnière JC, Pechère, R, Guidoin R, Noël HP. Colonisation expérimentale d'une prothèse artérielle en Dacron par un staphylocoque. J. Chir. 117: 397-401, 1980.

34. Pourdeyhimi B, Wagner D. On the correlation between the failure of vascular grafts and their structural and material properties: a critical analysis. J. Biomed. Mat. Res. 20: 375-409, 1986.

35. Harrison JH, Davalos PA. Influence of porosity on synthetic grafts. Ann. Surg. 82: 8-13, 1961.

36. Wesolowski SA, Fries CC, Karlson KE, DeBakey MC, Sawyer PN. Porosity: primary determination of ultimate fate of synthetic vascular grafts. Surgery 50: 91-96, 1961.

37. Deterling RA. Failure of Dacron arterial prostheses. Arch. Surg. 108: 13-14, 1974.

38. Cooke PA, Nobis PA, Stoney RJ. Dacron aortic graft failure. Arch. Surg. 108: 101-103, 1974.

39. Perry MO. Early failure of Dacron prosthetic grafts. J. Cardiovasc. Surg. 16: 318-321, 1975.

40. Lynn RB. Knitted Dacron ultralight weight grafts. A warning. Can. J. Surg. 22: 593, 1979.

41. Hayward RH, Korompai FL. Degeneration of knitted Dacron grafts. Surgery 79: 581, 1976.

42. Ottinger LW, Darling C, Wirthlin LS, Linton RR. Failure of ultralight weight knitted Dacron grafts on arterial reconstruction. Arch. Surg. III: 146-149, 1976.

43. Hussey HH. Arterial replacement. Failure of synthetic prostheses. J.A.M.A. 235: 848, 1976.

44. Blumemberg RM, Gelfand ML. Failure of knitted Dacron as an arterial prosthesis. Surgery 81: 493-496, 1977.

45. Guidoin R, Downs AR, Barral X, Marois M, Roy PE, King M, Gosselin C. Anastomotic false aneurysms with aortic Dacron grafts after 25 years. Ann. Vasc. Surg. 1: 369, 1986.

46. Couture J, Guidoin R, King M, Marois M. Textile Teflon arterial prostheses: how successful are they? Can. J. Surg. 27: 575-582, 1984.

47. Guidoin R, King M, Gosselin C, Blais P, Gunasekera K, Marois M, Cardou A. Les prothèses artérielles en polyester. Rev. Europ. Bioth. Med. 4: 13, 1982.

48. Guidoin R, King M, Blais P, Marois M, Gosselin C, Roy P, Courbier R, David M, Noël HP. A biological and structural evaluation of retrieved Dacron arterial prostheses. NBS Special Publication 601: 29-129, 1981.

49. Scott SM. A woven double velour Dacron vascular graft. Presented at the 8th Annual Meeting of the Society of Biomaterials, Anaheim, California, April 24-27, 1982.

50. Guidoin R, Marceau D, Couture J, Rao TJ, Merhi Y, Roy PE, De la Faye D. Collagen coatings as biological sealants for textile arterial prostheses. Biomaterials 8: 433-441, 1989.

51. Guidoin R, Marois Y, Rao TJ, Torché D, Marceau D, Tenney B, Duhamel RC, Lebowitz E, Walcott C. An albumin-coated polyester arterial prosthesis made ready to anastomose: in vivo evaluation in dogs. Clin. Mat. 3: 119-131, 1988.

52. Patel DJ, Fry DL. The elastic symmetry of arterial segments in dogs. Circ. Res. 24: 1-8, 1969.

53. Hasegawa H, Azuma T. Mechanical properties of synthetic arterial grafts. J. Biomech. 12: 509-517, 1979.

SELECTED DISCUSSION REMARKS:

von Recum: Originally the Gore Tex graft turned out to be very tissue compatible by a beautiful tissue ingrowth. But then by any reason the prostheses were coated with an outer wrapping. Now there is no tissue ingrowth and you have a seroma formation and you actually have a graft that totally relies on the sutures for continuation of the bloodflow.

Charara: In fact the external mesh of the Gore Tex model interfered with fibroblastic and tissue ingrowth into the graft wall. Then they added the wrapping up to increase the dimensional stability of the Gore Tex model. The external wrap is very stiff and doesn't have an open structure to allow this infiltration. This Gore Tex model is never healed even in the animal model.

Jerusalem: It is a very interesting phenomenon, there is a group in Groningen, Netherlands, they claim that the circumferential elasticity is essential for healing. It is generally the question of the requirement of the compliance of a vascular synthetic graft, because those vessels in which a graft is implanted are almost rigid without any compliance. There we may have a change from a rigid vessel to a graft with a good elasticity.

POLYURETHANES AND THEIR CYTOCOMPATIBILITY
FOR CELL SEEDING

J. Gerlach, H.H. Schauwecker, H. Planck*.

Chirurgische Universitätsklinik (Direktor: Prof. Dr. P. Neuhaus), UKRV/ Charlottenburg,
Freie Universität Berlin, Spandauer Damm 130, 1000 Berlin 19, FRG.
* Institut für Textil- und Verfahrenstechnik Denkendorf, FRG

Abstract: The present studies were done to investigate the cell behaviour on polymers as parameter for testing the cytocompatibility of biomaterials for cell coating. Endothelial cell kinetics and -morphology during initial adhesion, initial spreading and proliferation of the cells were examined as sensitive reaction of cells to the polymers. All investigated smooth polyurethanes had a similar cytokinetic and proliferation behaviour. Sprayed Tecoflex showed a faster initial adhesion and spreading of cells, but not a better cell proliferation than smooth Tecoflex. The results have shown that surface modifications have an influence on cell behaviour. Using bovine aortic EC, polyurethane as a basis for cell seeding can be used without any coating.

Introduction

Prostheses for the replacement of arteries give functionally good results in replacing large or middle diameter arterial vessels. However, a biological "restitutio ad integrum" of the blood contact surface was not described yet. An ingrowth of EC seems to be punctual without functional relevance[1,2,3].

To improve the patency rate of small diameter grafts, the coating of prostheses with EC layers was studied by several authors. The hope to reach a less thrombogenic inner surface is based on the use of active metabolic performances of the EC due to anti platelet active production of prostaglandins[4], EDRF[5,6,7] and other metabolites.

Most of these authors used Dacron[8,9,10,11,12,13,14,15] or Poly tetra fluorethylene (PTFE)[16,17,18].
Since polyurethanes (PU) showed a good blood compatibility and cell compatibility, they could be of interest as an alternative basis polymer for cell seeded vascular grafts.
To test the suitabilitiy of polyurethanes (PU) as a basis material for bioprostheses, six PU with smooth surfaces were tested on their abilities for endothelial cell seeding. In one PU, smooth surfaces were compared with porous surfaces. Cell kinetics and cell morphology during initial adhesion, initial spreading and proliferation of the cells were examined as a reaction of cells to their surrounding.

H. Planck M. Dauner M. Renardy (Eds.)
Medical Textiles for Implantation
© Springer-Verlag Berlin Heidelberg 1990

Materials

Polyurethanes were dissolved in tetramethylurea (TMU), dimethylformamid (DMF) or dimethylacetamid (DMAC) :

- *TECOFLEX EG-80A (Polyether urethane), TMU*
- *PELLETHANE 2363 80AE (Polyether urethane), DMAC*
- *PLATURAN 8300 (Polyester urethane), DMF*
- *ENKA PUR-1017 (Polyester urethane) ,TMU*
- *ENKA PUR-1059 (Polyester urethane) , TMU*
- *LYCRA 420 (Polyether urea urethane), DMAC*

To produce smooth surfaces, PU were cast into sheets with a thickness of about 1mm on cleaned Duran 50R glass (Schott, FRG). After curing for 6 h at 80 °C , samples of 5 x 20- mm were degassed for 48 hrs in a vacuum, then incubated in distilled water. Samples were not tacky, but smooth and without defects as determined by scanning electron microscopy (SEM).

The Tecoflex 80A with a porous surface with pores of ~10 μm was processed using a spraying method, described elsewere[19], by spraying the material onto a rotating mandril. From the resulting tubes, samples of 5 x 20- mm were cut.

Reference material was glass (Duran 50, Schott, FRG).

All samples were sterilised in ethylenoxyde and stored in vacuum for 20 days.

Method

Endothelial cells were isolated from bovine aorta using 0,05% collagenase 2 solution in PBS (Biochrom, FRG). Cell sheets of 10- 30 EC in suspensions were obtained after mechanical scratching with a rubber cell scraper, a single cell solution was prepared by pipetting the cell clumps several times with culture medium in the culture flasks. The method is described elsewere[20]. Cells were cultivated in Dulbeccos minimal essential medium (DME) with 5% FCS, streptomycin/ bacitracin and L-glutamine (all: Biochrom, FRG) in tissue culture polystyrol flasks (Falcon, USA).

To achieve an even distribution of cells on the samples, a rotation device was constructed, which allows to rotate glass tubes containing the samples in an incubator (5% CO_2 in air with humidity of 95% and temperature of 37°C). Samples of one measurement have been incubated together in a rotating tube with a diameter of 30 mm, made of glass (Duran 50, Schott, FRG), filled with 3 ml medium containing 4x105 endothelial cells/ml. All tubes of the investigation were simultaneous rotated 20 rph.

The experiments started at the same time with six of each smooth PU samples and six sprayed Tecoflex samples. Measurements of cytokinetics were performed by counting the initial adhesion at 15, 30, and 45 min (calculated as % of the maximal attached cells), measuring the spreading at 60, 90, and 180 min (calculated as % radius of the mean end radius of cells) and the proliferation time to confluenced growing on each cell layer (days). Measurements of cell morphology were done by using phase contrast microscopy (Zeiss IM 35, Germany) and SEM (Cambridge stereoscan 250 MK 2).

RESULTS

Table 1 shows the results in summary. All smooth polyurethanes had a similar cytokinetic and proliferation behaviour. Cells on all samples were grown to confluence at day 8 of incubation. In comparison to the glass tube surfaces, in which the samples were located, initial spreading on smooth Polyurethanes was faster. Corresponding to this results, there were no striking differences of cytomorphology on all smooth PU.
Sprayed Tecoflex showed a faster initial adhesion and spreading of cells, but not a better cell proliferation than smooth Tecoflex.

Table 1:

Cytokinetics on polyurethanes
(means, SD for the adhesion was between 1- 7 and SD for the spreading was between 5 -8).

Material	adhesion (%n)			spreading (%r)		
	15min	30min	45 min	60min	90min	180 min
Lycra 420 (DMAC)	16	37	80	57	88	92
Pellethane2363(DMAC)	14	38	81	55	89	93
Platuran 8300 (DMF)	14	38	80	56	88	92
Enka 1017 (TMU)	15	39	82	58	89	92
Enka 1059 (TMU)	15	38	82	58	88	91
Tecoflex 80A (TMU)	16	37	80	56	88	90
Tecoflex 80A sprayed	19	45	85	70	85	94

DISCUSSION

A cytocompatibility test was used to investigate the surface properties of polyurethanes for cell seeding. The cell type used for this test was adjusted to the requirements in which the materials will be used: PU for vascular prostheses were investigated with vascular endothelial cells.
To achieve an equal distribution of the cells over the whole surface of the samples, a special cell seeding device was used for application of EC on the samples with the same concentration of cells in suspension during the same period of time.
Under the investigated conditions, all polyurethanes showed similar suitabilities for seeding of endothelial cells. According to Fasol[21], solvents of polymers seemed to be without influence.

Endotheliasation of vascular prostheses prior to implantation could be a possibility to enhance the patency rate of small diameter grafts. In 1959, SZILAGYI[22] described a dependence between diameter of prosthesis and prognosis in the implantation of the graft: in aorto-iliacal position, he found an rate of 88% open prostheses after 24 months, but in femoro-popliteal position a rate of 56% open prostheses after 24 months. ECHAVE[23] reconfirmed these relations in 1979.

The critical diameter for the low complication use of arterial prosthesises was described by STANLEY[24] to be 4-5 mm. The critical length for the low complication use described HERRING[25] for Dacron-prostheses of 4 mm ID with 3,25 cm. SCHMIDT[26] reconfirmed these datas in the animal experimental use of PTFE.

1
DE BAKEY ME, ABBOT JP:
Endothelial lining of a human vascular prosthesis
Cardiovasc Res Cent Bull 3, 1964: 1-12

2
SAUVAGE LR, BERGER KE et al :
Interspecies healing of porous arterial prostheses
Arch Surg 109, 1974: 698-705

SAUVAGE LR, BERGER K:
Presence of endothelium in an axillary femoral graft of knitted
dacron with an external velour surface
Ann Surg 182, 1975: 749-753

3
WESOLOWSKI SA:
The healing of arterial prostheses - the state of the art
Thorac Cardiovasc Surg 30, 1982: 196-203

4
Moncada S, Vane JR:
Aracidonic acid metabolites and the interactions between
platelets and blood vessel wall
N Engl J of Med 300/ 1979: 507

5
Radomski MW, Palmer RM, Moncada S:
The anti aggregating properties of vasdcular endothelium:
interactions between prostacyclin and nitric oxide
Br J Pharmac 92/1987: 639-649

6
Radomski MW, Palmer RM, Moncada S:
Comparative pharmacology of endothelium- derived relaxing
factor, nitric oxide and prostacyclin in platelets
Br J Pharmac 92/1987: 181-187

7
ALHEID U, FRÖHLICH JC:
Endothelial - Derived Relaxing Factor from cultured human
endothelial cells inhibits aggregation of human platelets
Thromb. Res.47/5, 1987: 561-571

8
HERRING M, BAUGHMAN S:
Endothelial seeding of dacron and polytetrafluoroethylene
grafts: the cellular events of healing
Surg 96/4, 1984: 745-754

9
HERRING M, GARDNER A:

A single-staged technique for seeding vascular grafts with autogenous endothelium
Surg 84/4, 1978: 498-504

10
GRAHAM LM, BURKEL WE:
Immediate seeding of enzymatically derived endothelium in dacron vascular grafts
Arch Surg 115, 1980:1289-1294

11
GRAHAM LM, VINTER DW:
Endothelial cell seeding of prosthetic vascular grafts
Arch Surg 115, 1980: 929-933

12
BURKEL WE, VINTER DW:
Sequential studies of healing in endothelial seeded vascular prostheses: histologic and ultrastructure characteristics of graft incorporation
J Surg Res 30, 1981: 305-324

13
BELDEN TA, SCHMIDT SP:
Endothelial cell seeding of small-diameter vascular grafts
Trans Am Soc Artif Intern Org 28, 1982: 173-177

14
HUNTER TJ, SCHMIDT SP:
Controlled flow studies in 4 mm endothelialized Dacron grafts
Trans Am Soc Artif Int Org 29, 1983: 177-182

15
SHAREFKIN JB, LATKER CH et al:
Seeding of Dacron vascular prostheses with endothelium of aortic origin
J Surg Res 34, 1983: 33-43

16
HERRING M, BAUGHMAN S:
Endothelial seeding of dacron and polytetrafluoroethylene grafts: the cellular events of healing
Surg 96/4, 1984: 745-754

17
SEEGER JM, KLINGMAN W:
Improved endothelial cell seeding with cultured cells and fibronectin-coated grafts
J Surg Res 38, 1985: 641-647

18
RYAN US, WHITE LA:
Seeding dog endothelial cells on Gore-Tex grafts
Fed Proc 44, 1985:1661

RYAN US, OLAZABAL BM:
Endothelial seeding of filters, grafts, and tubes
J Tiss Cult Met 10/1, 1986: 61-65

19
Pat. Nr. DE 2806030

20
Gerlach J, Kreusel K.M., Schauwecker H.H., Bücherl E.S.
Endothelial cell seeding on PTFE vascular prostheses using a standardized nseeding technique
Int. J. Artif. Org. 12/4 1989:270-275

21
Fasol R., Zilla,P. et al.
Experimental in vitro cultivation of endothelial cells on artificial surfaces
Vol. 31 Trans Am Soc Artif Intern Organs 1985: 276-279

22
SZILAGYI DE : An elastic Dacron arterial substitute.
Surg Clin North Am 39, 1959: 1523-1538

23
Echave V, Koovnick AR: Intimal hyperplasia as a complication of the use of the polytetrefluoroethylene graft for femoro poplitela bypass. Surg 88, 1979: 903-910.

24
Stanley JC, Burkel WE et al: Biologic and synthetic vascular prostheses. NY, Grune& Stratton 1982: 495-509.

25
Herring M, Gardener A: Patency in canine inferior cava grafting: effects of graft material size and endothelial seeding. J Vasc Surg 1/6, 1984: 877-887.

26
Schmidt SP Hunter J et al: Small diameter vascular prostheses: two designs of PTFE end endothelial cell seeded and nonseeded Dacron. J Vasc Surg 2/2, 1985: 292-297

CELLULAR AND CYTOSKELETAL RESPONSE OF VASCULAR CELLS TO MECHANICAL STIMULATION

Peter C. Dartsch* and Eberhard Betz

Institute of Physiology I, University of Tübingen, Gmelinstrasse 5,
D-7400 Tübingen 1, Federal Republic of Germany

ABSTRACT

Smooth muscle cells from rabbit aortic media and endothelial cells from pig aorta were grown on hydrophilized and collagen coated silicone membranes which were subjected to cyclic and directional stretching and relaxing at a frequency of 60 per minute. The membranes were stretched with various amplitudes ranging from 2% to 20% (smooth muscle cells) and with an amplitude of 15% for endothelial cells. Cells on unstretched membranes in the same incubation chamber served as controls. In long-term experiments the stretching and relaxing of the membranes was continued for several days.

While the smooth muscle cells grown on unstretched membranes remained in random orientation in all experiments, the cells which underwent mechanical stimulation showed a high degree of orientation depending on the strength of the stimulus. The angle of cell orientation varied in direct relation to the stretching amplitude and became steeper with increasing intensity of the mechanical stimulus. For instance, by use of a stretching amplitude of 15%, smooth muscle cells oriented at angles of $\alpha = 76° \pm 8°$ ($\bar{x} \pm SD$) and $\alpha^* = 104° \pm 7°$ ($\bar{x} \pm SD$), respectively. In comparison, endothelial cells oriented at an angle of $\alpha = 89° \pm 12°$ ($\bar{x} \pm SD$) by use of a stretching amplitude of 15%, i.e. with their longer axis perpendicular to the stretch direction. Endothelial cells which were subjected to stretching elongated nearly four fold when compared with polygonally shaped cells grown on unstretched membranes. Short-term experiments demonstrated that a rearrangement of the intracellular actin filament system occurs prior to the orientation of the whole cell bodies. Rearrangements of other cytoskeleton components such as actin-binding protein caldesmon, microtubules and intermediate-sized filaments were also observed and are presented in detail.

The results indicate that periodic stretching and relaxing of the artery wall by blood pulsations seems to be an essential factor which accounts for the orientation of vascular cells within the vessel wall.

Key Words: *Cell Orientation - Cyclic Stretching - Endothelial Cell - Smooth Muscle Cell - Cytoskeleton*

---

*To whom correspondence should be addressed.

H. Planck M. Dauner M. Renardy (Eds.)
Medical Textiles for Implantation
© Springer-Verlag Berlin Heidelberg 1990

INTRODUCTION

The artery wall consists of three distinct layers: (1) the endothelial layer covering the luminal surface and serving as an effective mediator between blood and the underlying tissues; (2) the media with smooth muscle cells and matrix components such as collagen bundles, elastic lamellae and glycosaminoglycans. The media serves for adjusting artery diameter and wall tension to the demands; (3) the adventitia with fibroblasts/fibrocytes and connective tissue which give mechanical stability to the vessel wall.

The smooth muscle cells of the media are arranged in form of a helix which was termed «Schraubenzug» or «Faserschraube» [1, 2] to describe the helical orientation of the smooth muscle cells and their close association with parallely arranged bundles of collagen, elastic fibrils and elastic laminae. The helically arranged smooth muscle cells near the lumen are oriented in a steep angle, whereas the slope of the helix is flatter in the external regions [3]. Endothelial cells of the luminal surface of arteries are elongated and oriented in the direction of blood flow, but endothelial cells of large veins are polygonally shaped and not aligned [4,5,6,7].

In vivo, the pulsations of blood flow produce periodic oscillations in artery diameter resulting in a cyclic stretching and relaxing of the vessel wall. Moreover, the flow of blood causes an oscillating fluid shear stress. Medial smooth muscle cells are exposed only to cyclic stretching and relaxation of the artery wall, but endothelial cells are subjected to both cyclic stretches and relaxations and fluid shear stress. The orientation of endothelial cells in arteries with the axis of flow is thought to be mainly a result of hemodynamic forces, especially shear stress, acting on the vessel wall. However, in vivo research to monitor and evaluate the significance of mechanical forces is limited by the inability to measure and control all variables [8].

The aim of this in vitro study was to examine the response of cultured smooth muscle cells and endothelial to mechanical stimulation by the use of an apparatus which imitates the mechanical forces of periodic oscillations occuring during repeated stretching and relaxing of the artery wall by blood pulsation. In addition, we have examined the behaviour of intracellular cytoskeleton components such as actin filaments (stress fibers), caldesmon, microtubules and intermediate-sized filaments in response to mechanical stimulation.

MATERIALS AND METHODS

Cell Culture

Smooth muscle cells were obtained from rabbit thoracic and abdominal aorta either by the outgrowth of substrate-attached explants [9, 10, 11] or by enzymatic digestion as follows: Pieces of de-endothelialized media tissue were placed in a test-tube containing digestion medium. For each 100 mg of tissue (wet weight), the digestion medium consisted of 1,8 mg collagenase CLS III Worthington (Seromed, Berlin, FRG), 44 μl elastase (Boehringer Mannheim, Mannheim, FRG) and 1 mg trypsin inhibitor from soybean (Serva, Heidelberg, FRG) dissolved in 1 ml of serum-free culture medium buffered with 15 mM Hepes. This mixture was incubated for 160 min. at 37°C in a shaking water bath. Therafter, 20% fetal calf serum was added to the mixture which was then centrifuged for 15 min. at 170 g. The pelleted cells were washed twice with culture medium and finally resuspended and seeded into cell culture dishes.

Endothelial cells were isolated from the aorta of Goettinger minipigs as follows: The thoracic aorta was excised, cut longitudinally and fastened with the lumen upside onto a specially constructed metal frame to prevent mixture with other cells of the aortic wall. The endothelial layer was covered with a dispase grade II solution (Boehringer Mannheim, Mannheim, FRG) for 20 to 30 min. at 37°C under sterile conditions. Endothelial cells were then removed by carefully aspirating the enzyme solution into a pipette. After addition of 20% fetal calf serum cells were centrifuged for 15 min. at 170 g and finally transferred into cell culture dishes coated with lathyritic collagen type I from rat skin (Boehringer Mannheim, Mannheim, FRG).

At confluency, smooth muscle cells and endothelial cells were routinely subcultured by trypsin-EDTA treatment. Cells were grown in medium 199 (smooth muscle cells) or a mixture of DMEM/Ham F 12 (2:8, v/v; endothelial cells) supplemented with 10% to 15% fetal calf serum and standard amounts of penicillin and streptomycin. Cells were incubated in a humidified atmosphere of 7% CO_2 and 93% air at 37°C. All cell culture reagents were purchased from Gibco BRL, Eggenstein, FRG.

Stretching Apparatus and Elastic Membranes

The apparatus (Fig. 1) which produces a wide range of cyclic and directional stretches and relaxations comparable to those which may occur in the artery wall has been described in detail elsewhere [12, 13, 14]. Briefly, a synchronous motor drove a steering eccentric disk which was formed so that its rotations imitated the diameter changes of the abdominal aorta during its volume pulsation including its typical incisura. The rotations of the steering eccentric disk were transferred to a Teflon piston-rod which moved back and forth and so stretched and relaxed the two inner elastic membranes mounted in the incubation chamber. The outer membranes were held stationary and served as corresponding controls. Extensible and transparent silicone membranes, which served as elastic substrata, were produced by polymerization of Sylgard 184 silicone elastomer (Dow Corning, München, FRG) as described [12]. Before the cells were seeded, the surface of the silicone membranes was hydrophilized in oxygen plasm (Wöhlk-Contact-Linsen, Kiel, FRG) and coated with lathyritic collagen type I from rat skin. Hydrophilizing as well as collagen coating proved necessary in order to facilitate a firm attachment of the cells to the substratum. In a series of control experiments, it was checked that hydrophilizing as well as collagen coating did not exhibit surface structures which might induce the orientation of cells by contact guidance [15, 16, 17].

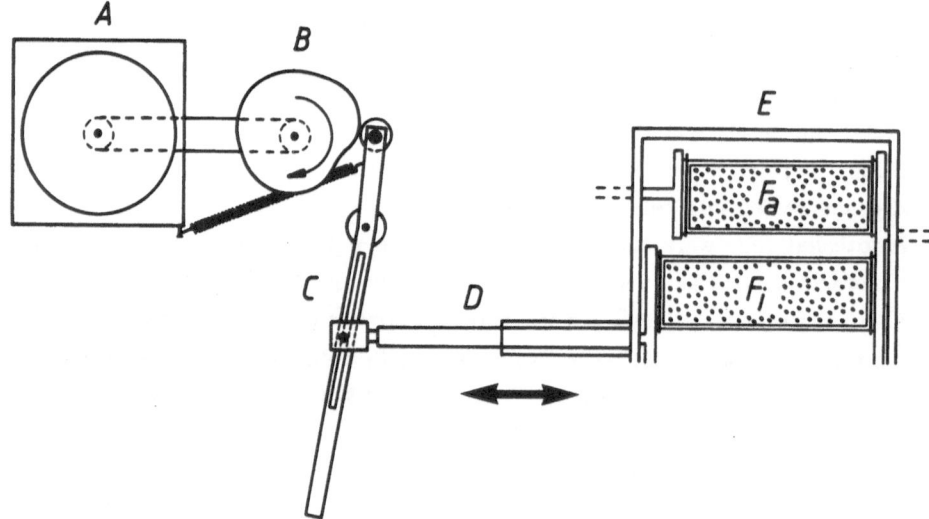

Fig. 1. Apparatus for cyclic and directional stretching and relaxing of silicone membranes. Only one pair of membranes is shown. A, synchronous motor with a constant speed of 50 rpm; B, steering eccentric disk to imitate the time course of the volume pulse of the abdominal aorta; C, rotary lever; D, Teflon piston-rod; E, incubation chamber; F_a, unstretched outer membrane; F_i, inner membrane which was stretched and relaxed at a frequency of 60 times per minute. The direction of stretching is marked by the arrows.

The silicone membranes were 50 µm thick and 20 x 60 mm in size = 12 cm^2 growth area. The lateral borders were thickened to 150 µm to avoid narrowing of the central part of the membrane during stretching. Both the shorter sides had fastenings made of a silicone tube with a V_2A-wire enclosed. To ensure a firm cell attachment, silicone membranes were hydrophilized in oxygen plasm and coated with lathyritic collagen type I prior to cell seeding.

Protocol of Stretching Experiments

Smooth muscle cells were seeded in primary cultures on pretreated silicone membranes at a density of 5×10^5 cells/cm^2. Subcultured smooth muscle muscle cells and endothelial cells of second through third passage (in vitro age of approximately 10 population doublings) were seeded at a density of 1×10^5 to 2×10^5 cells/cm^2. Cells were allowed to attach and spread for 48 hours. Cells were then subjected to continuously performed cyclic and directional stretching and relaxation at a frequency of 60 per minute. Dobrin [18] has suggested that most vessels appear to undergo about 8% to 10% oscillation in external diameter in the exposed state and about 15% oscillation in internal diameter in the unexposed state. Therefore, in a series of experiments, smooth muscle cells were stretched with amplitudes ranging from 2% to 20%. Endothelial cells were stretched 15% beyond their resting length.

In long-term experiments, the stretching and relaxing of the membranes was continued for several days (see results) until the degree of cell orientation was stable and no longer changed. In short-term experiments, smooth muscle cells were exposed to cyclic stretching for 3, 6 and 12 hours with an amplitude of 10%. Cells on unstretched membranes in the same incubation chamber served as corresponding controls.

The experiments were terminated by washing the cells in two changes of phosphate buffered saline (PBS), pH 7.4, and fixation. For phase contrast microscopy and staining of stress fibers and caldesmon, cells were fixed with 3.5% formaldehyde in PBS for 10 min. at room temperature. For staining of microtubules and intermediate-sized filaments cells were fixed with methanol for 6 minutes at -20°C. In all experiments, cells were fixed at a mean extension of the elastic membranes in order to avoid possible cell alterations and cytoskeletal changes by the extremes of either maximal elongation or complete relaxation of the membranes.

Evaluation of Cell Viability

Cell viability of cells on silicone membranes was checked with fluorescent dyes [19, 20] as follows: Cell cultures were washed twice with PBS and incubated for 2 minutes at room temperature with PBS containing 6 µg/ml fluorescein diacetate and 3 µg/ml ethidium bromide (both compounds from Sigma Chemie, Deisenhofen, FRG). After removal of dyes and another two washes with PBS cells were examined with an inverted microscope using epifluorescence illumination. Viable cells which can hydrolyse fluorescein diacetate exhibit a green fluorescent staining of cytoplasm when using blue excitation, whereas the nuclei of dead cells show a red fluorescent staining by use of green excitation.

Fluorescent Staining of Intracellular Cytoskeleton

The pattern of intracellular actin filaments (stress fibers) in smooth muscle cells and endothelial cells was visualized by fluorescent staining with TRITC-phalloidin (Sigma Chemie, Deisenhofen, FRG) as previously described [13, 21]. Microtubules and intermediate-sized filaments were stained by indirect immunofluorescence with monoclonal antibodies against α- and β-tubulin (Amersham Buchler, Braunschweig, FRG) and vimentin (Sigma Chemie, Deisenhofen, FRG) as described [21, 22, 23]. Polyclonal antibodies against caldesmon were a kind gift of Dr. V.P. Shirinsky, Moscow, USSR, and polyclonal antibodies against smooth muscle myosin and smooth muscle tropomyosin of Dr. U. Gröschel-Stewart, Darmstadt, FRG. Antibodies against smooth muscle α-actin were purchased from Progen Biotechnik, Heidelberg, FRG. FITC- and TRITC-labeled goat anti-mouse IgG and goat anti-rabbit IgG (secondary antibodies) were purchased from Miles Scientific, München, FRG and Dianova, Hamburg, FRG.

Measurement of Cell and Cytoskeleton Orientation and Cell Elongation

The angle of cell orientation (= α and α*, respectively) to the direction of stretching was determined by measuring the longer axis of the cells of at least five arbitrarily selected microscopic fields at phase contrast. The angle of actin filament orientation (= γ) was determined likewise by measuring the direction of the filaments in the cells using

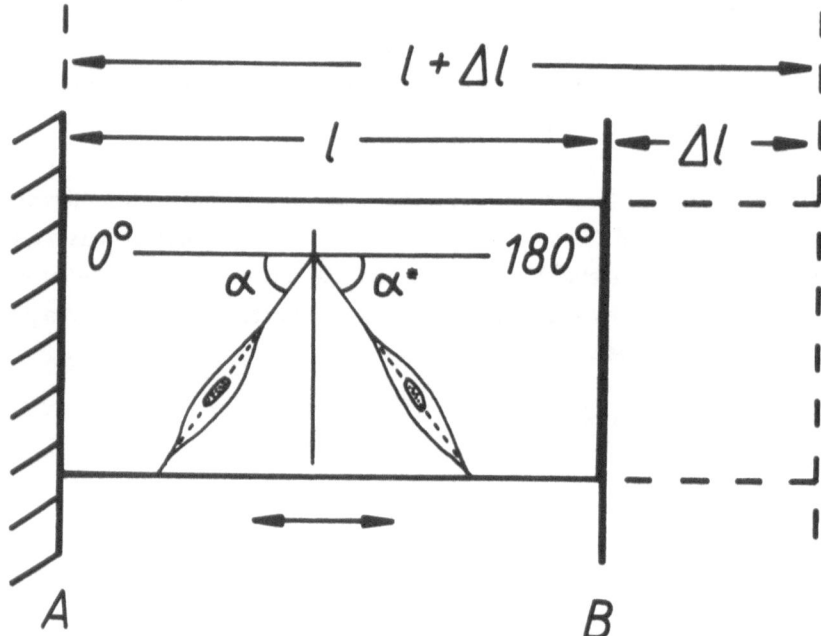

Fig. 2. Schematic drawing of cell and actin filament orientation on silicone membranes which were subjected to cyclic and directional stretching and relaxation. The direction of stretching is marked by the arrows. A, fixed end of the membrane; B, variable end of the membrane; l, length of the membrane relaxed to initial tension; l + Δl, total length of the membrane stretched to maximum; Δl/l, relative change of length of the relaxed membrane (= stretching amplitude); α, angle of cell and actin filament orientation; α*, complementary angle of cell and actin filament orientation.

epifluorescence illumination. Fig. 2 gives a schematic presentation of cell and actin filament orientation. The factor of cell elongation (= f) of endothelial cells was determined by measuring the width of the cells in the stretch direction in relation to cell length perpendicular to the stretch direction.

RESULTS

Cell Identification

The smooth muscle cells were identified by morphological and immunological criteria as follows (Figs. 3,4): (1) spindle shape; (2) «hill and valley» growth pattern after reaching confluency; (3) positive reaction with antibodies against smooth muscle α-actin, smooth muscle myosin, smooth muscle tropomyosin and caldesmon. Endothelial cells were identified by the following criteria (Fig. 5): (1) polygonal cell shape; (2) «cobblestone» growth pattern after reaching confluency; (3) positive staining of cell margins with silver nitrate of confluent cultures; (4) marginal actin filament network. Cultures containing smooth muscle cells and endothelial cells, were rare because of the precautions in the preparation of the vessel wall. Such mixed cultures were rejected from stretching experiments.

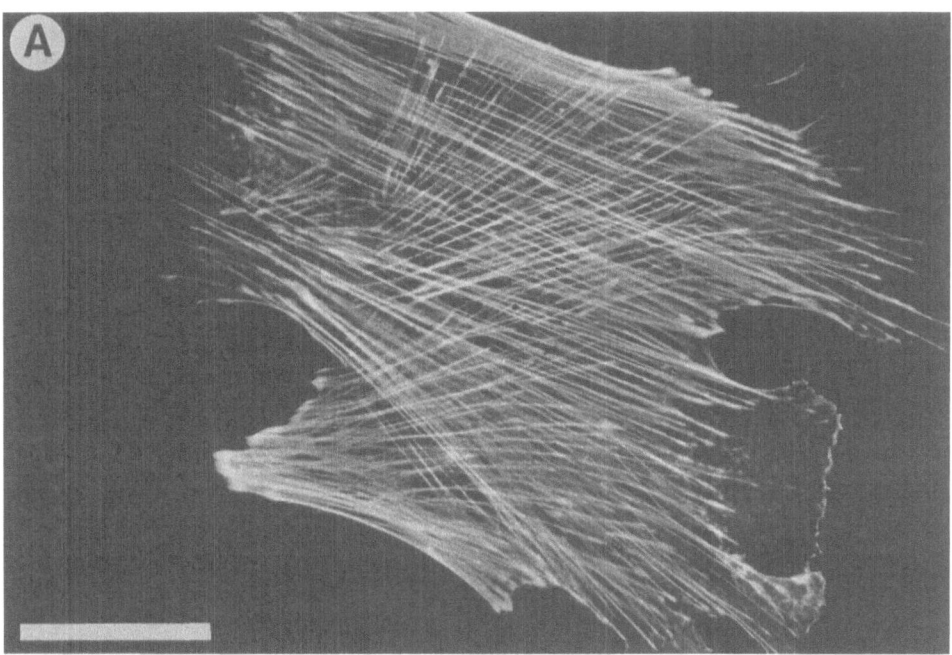

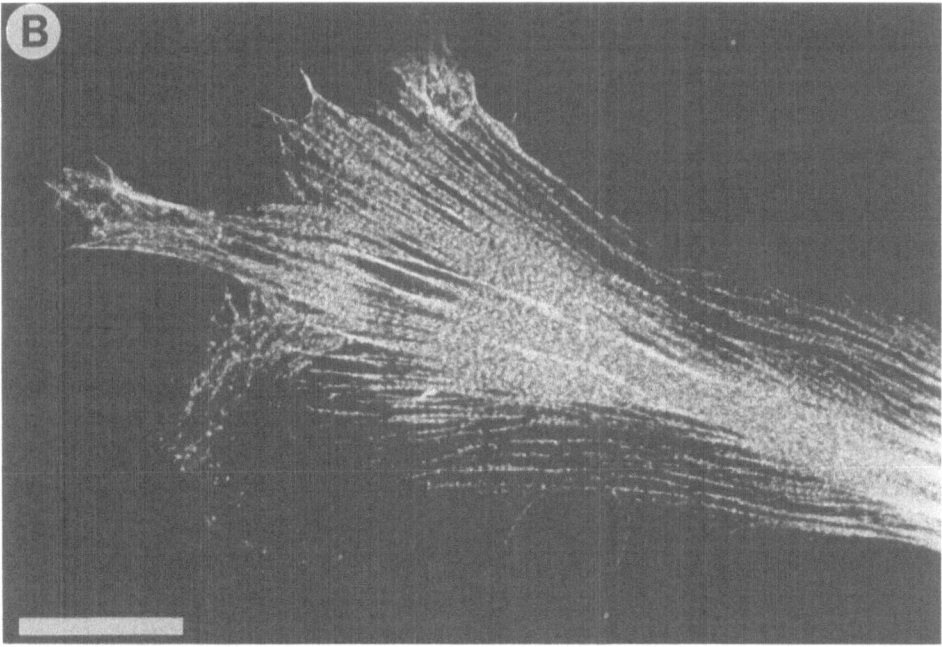

Fig. 3. Identification of smooth muscle cells from rabbit aortic media. (A) Positive reaction with antibodies against smooth muscle α-actin. Bar - 50 μm. (B) Positive reaction with antibodies against smooth muscle myosin showing an intracellular distribution in form of myosin aggregates. Epifluorescence microscopy. Bar - 30 μm.

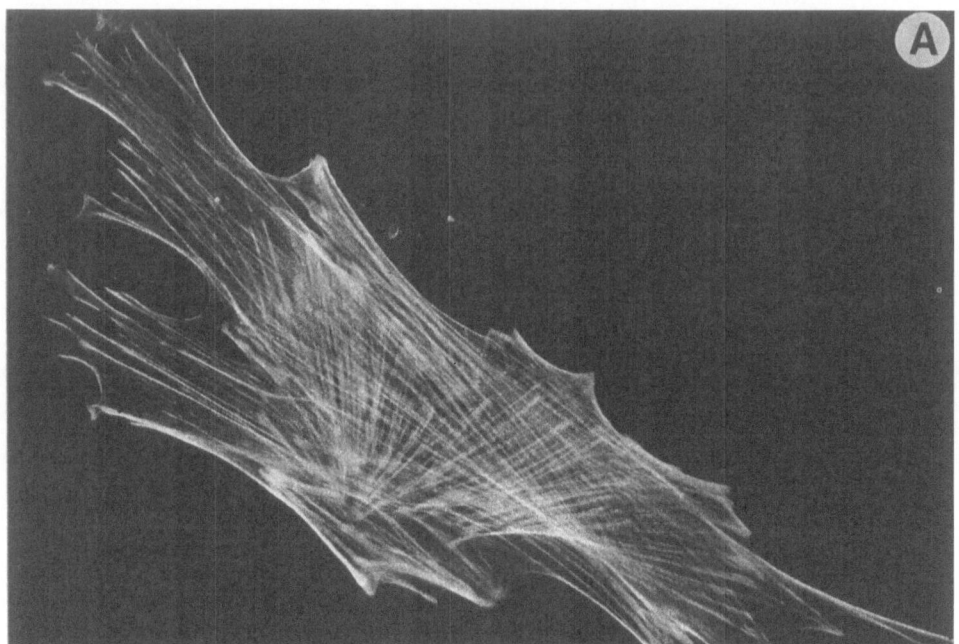

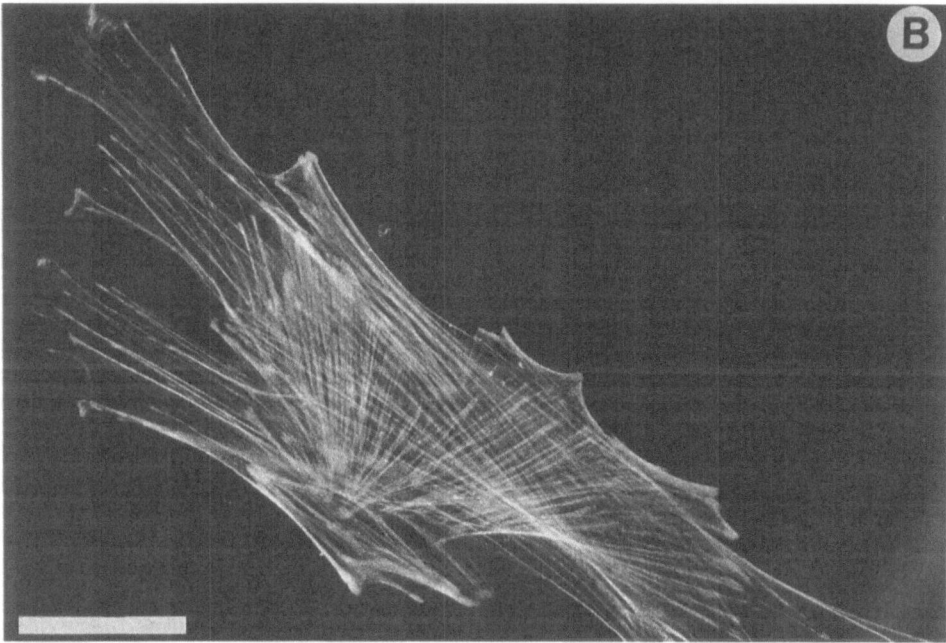

Fig. 4. Double fluorescence staining of a smooth muscle cell with TRITC-phalloidin (A) and antibodies against caldesmon (B). Note the co-localization of both actin-containing stress fibers and caldesmon and their distribution in form of a three-dimensional network. Bar - 50 μm.

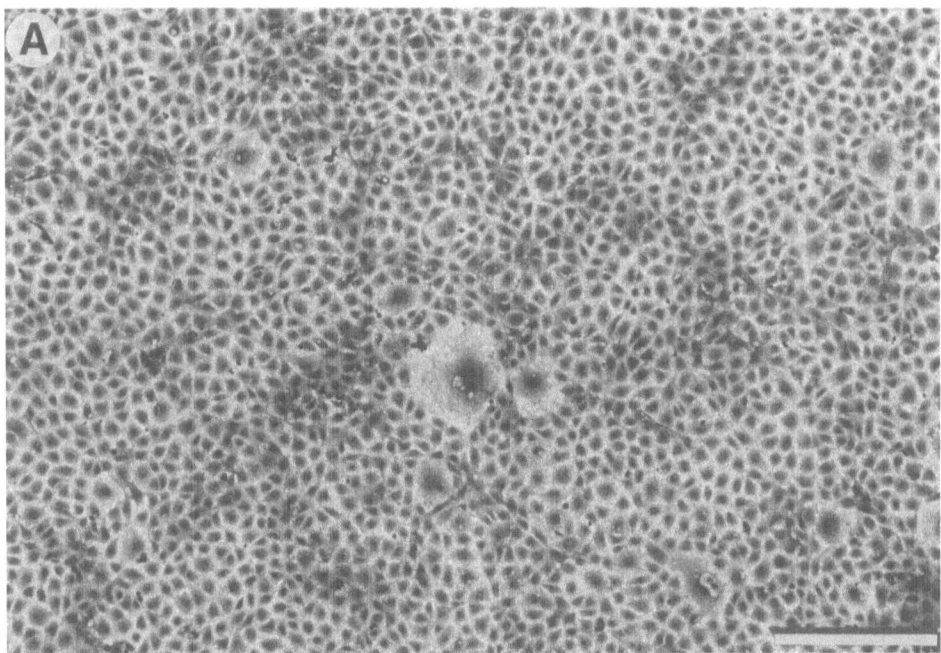

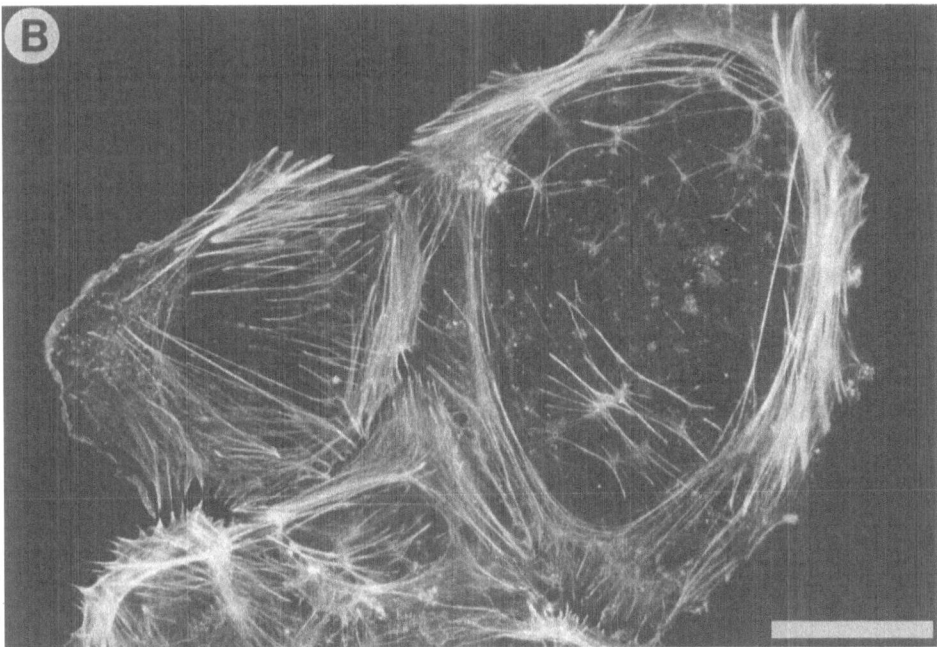

Fig. 5. Identification of endothelial cells from pig aorta. (A) Characteristic « cobblestone » growth pattern. Coomassie giemsa staining. Bar - 500 µm. (B) Marginal actin filament network. Epifluorescence microscopy. Bar - 50 µm.

Smooth Muscle Cell Orientation

At the beginning of each stretching experiment, the well-spread smooth muscle cells were randomly oriented. As shown in table 1, with the exception of the cells which were stretched with an amplitude of 2%, the angle of cell orientation after 5 to 14 days of continuously performed periodic stretching became steeper with increasing stretching amplitudes. In agreement with previous studies [12] we observed that a certain minimum of mechanical stimulus was required to induce cell orientation. When a stretching amplitude less than 3.5% was used, the smooth muscle cells remained randomly oriented. On the other hand, a maximum stretching amplitude of 20% resulted in a cell orientation perpendicular to the stretch direction. For example, figure 6 shows the orientation of smooth muscle cells after 6 days of cyclic stretching and relaxations with an amplitude of 10%. The polar representation of this cell alignment (Fig. 7) clearly demonstrates the distribution at two complementary angles α and α*. In all experiments presented here, no detachment of cells was observed.

With the cell densities used for the experiments, the stretched cells maintained their orientation even after cessation of further stretching. The cells which underwent cell division aligned themselves parallel to the cell arrays, probably by contact guidance of synthesized extracellular matrix material.

In all stretching experiments, cells on unstretched membranes remained in random orientation. On a very local basis it was a common finding that in unstretched controls the cells showed a visible «orientation», i.e. some rows of cells had the same angle of orientation. But a short distance away, another row of cells was aligned in an entirely different direction.

Rearrangement and Orientation of Smooth Muscle Cell Cytoskeleton Components

The well-spread smooth muscle cells growing on unstretched membranes revealed a relatively uniform pattern of actin distribution (Fig. 8B). The staining showed long,

Stretching Amplitude	Cells	Angle of Cell Orientation ($\bar{x}$ ± SD)	
2%	Cells of passage 2	Cells remain in random orientation	
3.5%	Cells of passage 3	α = 61° ± 15°	α* = 120° ± 12°
5%	Cells of passage 3	α = 60° ± 13°	α* = 119° ± 16°
10%	Cells of passage 1	α = 67° ± 16°	α* = 116° ± 15°
15%	Cells of passage 3	α = 76° ± 8°	α* = 104° ± 7°
20%	Cells of passage 1	α = 88° ± 9°	

Tab. 1. Orientation response of cyclically stretched rabbit aortic smooth muscle cells as a function of the stretching amplitude. The given data represent mean values of at least two independent experiments each. Note that α and α* are complementary angles to the stretch direction.

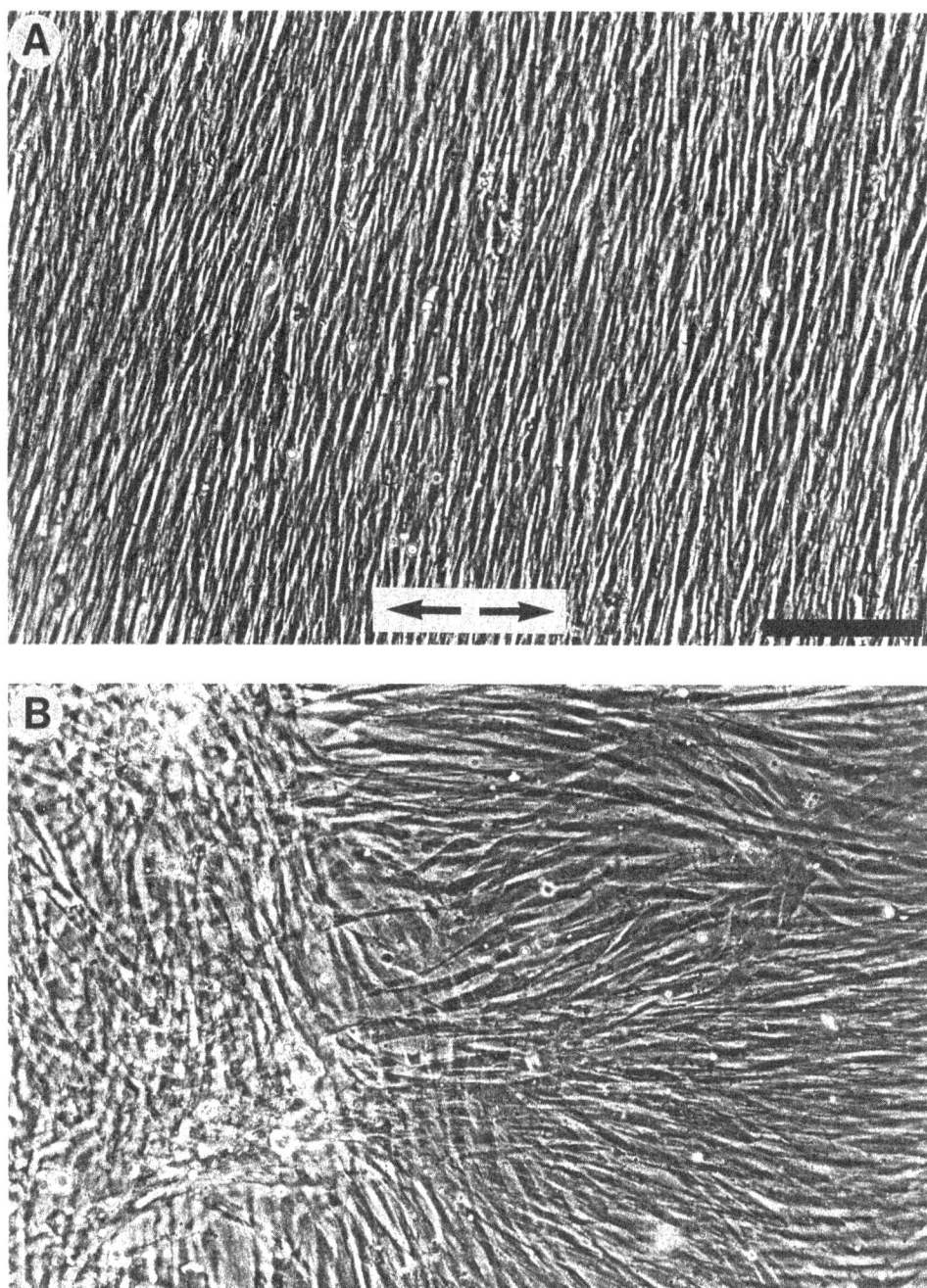

Fig. 6. (A) Orientation of rabbit aortic smooth muscle cells after 6 days of cyclic and directional stretching with an amplitude of 10%. Only the orientation at an angle α is depicted. (B) The cells on unstretched membranes remained in random orientation. The direction of stretching is marked by the arrows. Phase contrast microscopy. Bars - 200 μm.

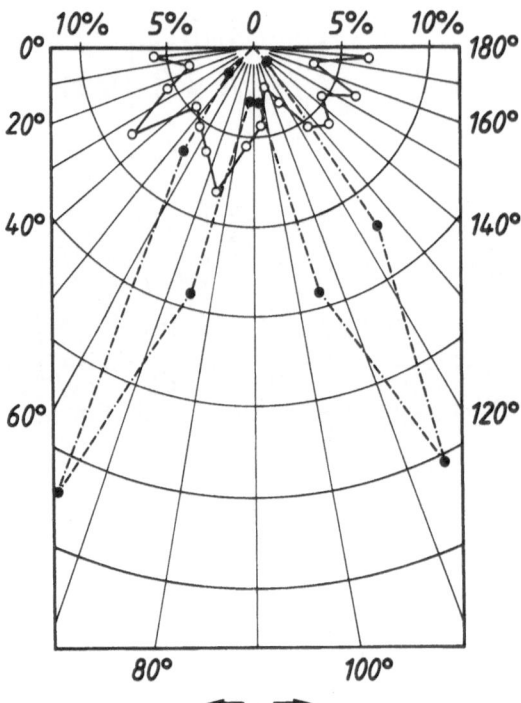

Fig. 7. Polar representation of the orientation response of smooth muscle cells to cyclic stretching and relaxation. The cells were stretched with an amplitude of 10% for 6 days (●—●). The cells on unstretched membranes remained randomly oriented (O—O). Note the nearly symmetrical distribution of stretched cells at the two complementary angles α and α*. The direction of stretching is marked by the arrows.

straight and non-interrupted filaments in parallel alignment along the longer axis of the cells. Frequently, a second layer of actin filaments transversed the other ones and giving the impression of a three-dimensional network. The cells which were exposed to cyclic stretching for only 3 to 12 hours exhibited several striking differences in actin filament architecture (Fig. 8A). The vast majority of the cells showed one single layer of straight and non-interrupted actin filaments in a clearly parallel alignment that was not found in corresponding controls. In contrast to unstretched smooth muscle cells, this parallel alignment of filaments showed a preferential orientation at the same angles as observed for the whole cell bodies after several days of continuous stretching.

Since the intracellular actin filament rearrangement and orientation took place within some hours, whereas the orientation of the cell bodies occurred after days, we consequently observed a number of cells with parallel arranged actin filaments which, however, ran transverse to the longer axis of the cells (Fig. 9A). Moreover, in early stages of stretching we observed smooth muscle cells with strongly fluorescent branchings of actin filaments (Fig. 9B), possibly indicating their depolymerisation and reorganization as a response to the mechanical stimulus.

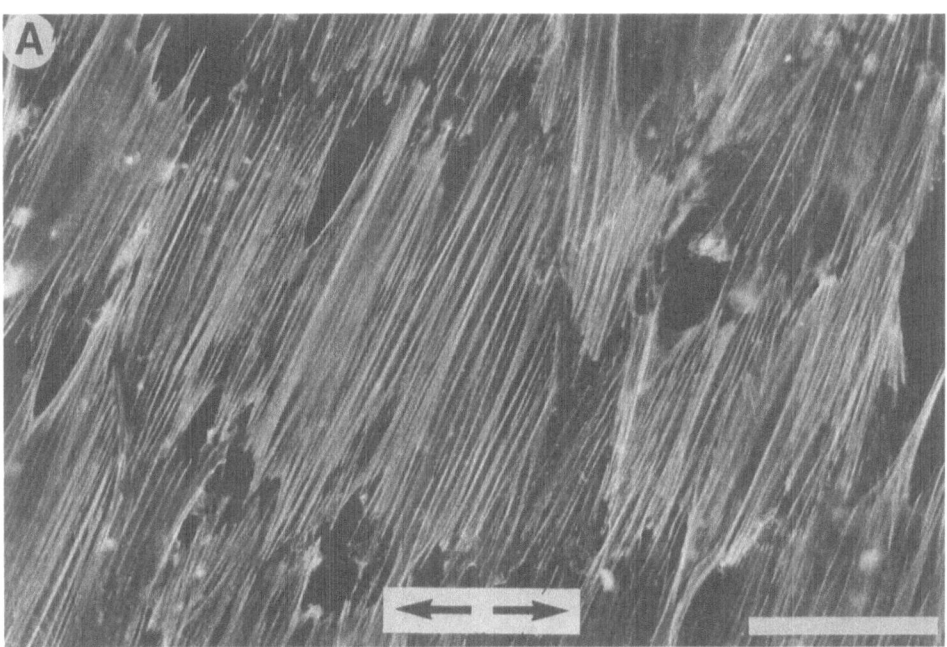

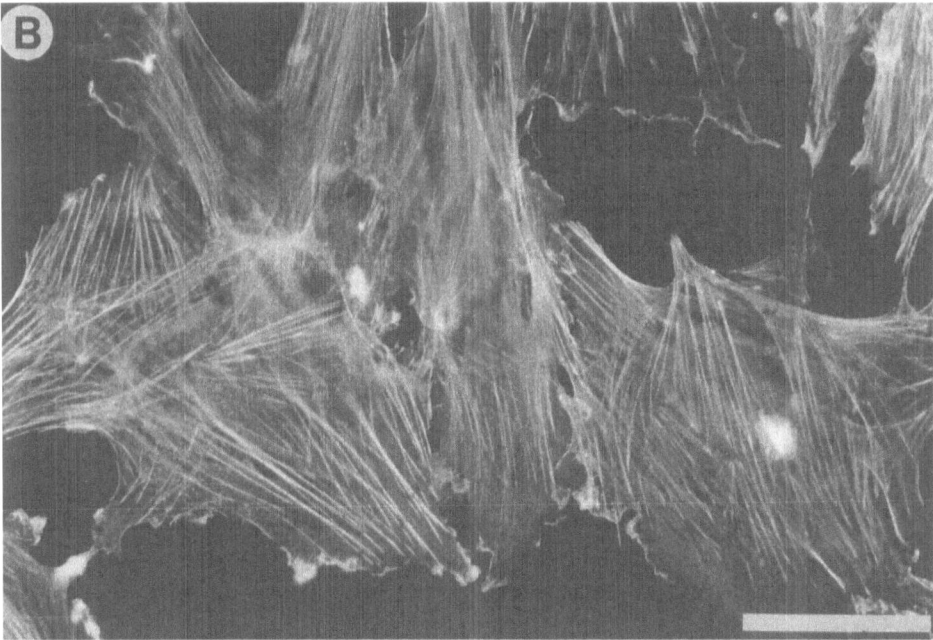

Fig. 8. (A) Orientation of actin filaments (stress fibers) in arterial smooth muscle cells after 12 hours of cyclic stretching with an amplitude of 10%. Note the parallel alignment of filaments. (B) Actin filaments in randomly oriented cells on unstretched membranes. Note the three-dimensional network of filaments. The direction of stretching is marked by the arrows. Epifluorescence microscopy. Bars - 50 μm.

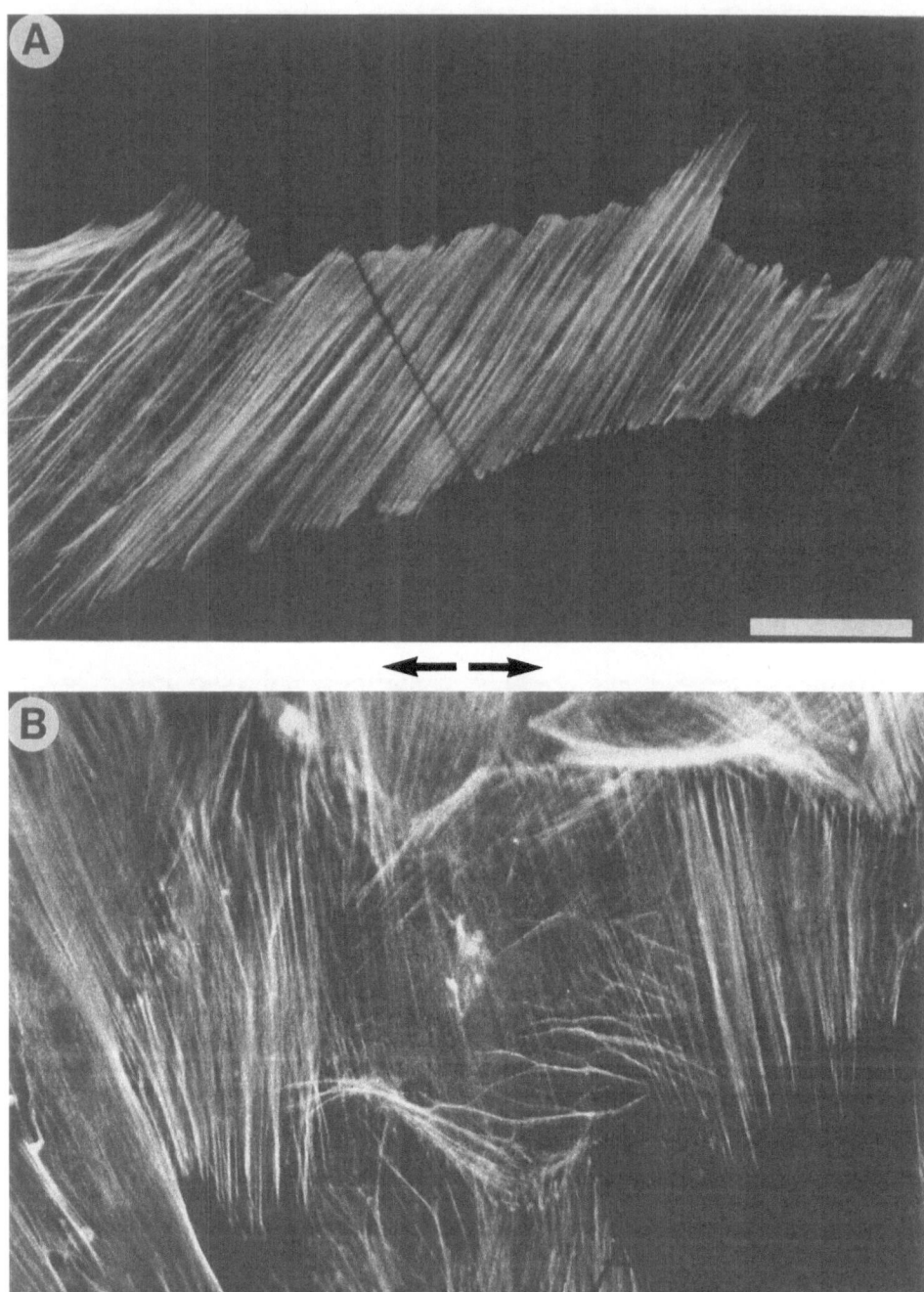

Fig. 9. (A) Smooth muscle cell with one layer of parallel arranged actin filaments running transverse to the longer cell axis. Epifluorescence microscopy. Bar - 25 μm. (B) Strongly fluorescent branchings of actin filaments in an early stage of stretching. Epifluorescence microscopy. Bar - 30 μm. The direction of stretching is marked by the arrows.

Caldesmon, an actin- and calmodulin-binding protein which has been described to play a role in the regulation of smooth muscle contraction [24, 25, 26, 27], is distributed along the stress fibers. In stretched smooth muscle cells, caldesmon remained bound to F-actin and, therefore, underwent the same rearrangements as described for actin filaments (Fig. 10). Vimentin is the only type of intermediate-sized filament found in rabbit aortic smooth muscle cells. As described by Franke et al. [28, 29], in unstretched cells, vimentin was observed to be organized as wavy fibers extending through the cytoplasm in a radial arrangement and a particularly abundant network around the nucleus. In contrast, the vimentin filaments in stretched cells showed the same parallel alignment and orientation as observed for actin filaments. The microtubule network in smooth muscle cells which were exposed to cyclic stretching did not exhibit a distinct orientation, although a remarkable number of microtubules were parallely aligned and oriented at the same angles as described above.

Endothelial Cell Elongation and Orientation

At the beginning of each stretching experiment the endothelial cells which were attached and completely spread out on the elastic membranes exhibited a polygonal shape and were randomly distributed. In the course of the continuous periodic stretching of the membranes with an amplitude of 15%, endothelial cells elongated and oriented perpendicular to the direction of stretching (Fig. 11 A). The factor of elongation of stretched cells was $f = 6.8 \pm 1.3$ ($\bar{x} \pm$ SD). Endothelial cells on unstretched membranes remained randomly distributed and polygonally shaped (Fig. 11 B) with an elongation factor of $f = 1.8 \pm 0.8$ ($\bar{x} \pm$ SD). This means that the endothelial cells responded to mechanical stimulation by elongating nearly four fold. Despite this fact, endothelial cells which were subjected to cyclic stretching and relaxing remained attached and viable. Besides this pronounced elongation, cells oriented perpendicular to the stretch direction (Fig. 12). The angle of cell orientation was $\alpha = 89° \pm 12°$ ($\bar{x} \pm$ SD). Endothelial cells on unstretched membranes remained randomly oriented (Fig. 12).

Rearrangement and Orientation of Endothelial Cell Cytoskeleton Components

The well-spread endothelial cells growing on unstretched membranes showed a relatively uniform pattern of actin distribution as already described (Figs. 5 B and 13 B). Endothelial cells which were subjected to cyclic and directional stretching and relaxations for 3 days exhibited several differences in actin filament orientation and architecture (Fig. 13 A). The marginal actin filaments became more pronounced and revealed a more intense fluorescence staining. In correlation to cell elongation and orientation, the main portion of marginal actin filaments were oriented perpendicular to the stretch direction. Often, parallely aligned actin filaments were observed running along the longitudinal cells axis. The angle of actin filament orientation was $\gamma = 90° \pm 9°$ ($\bar{x} \pm$ SD; Fig. 12).

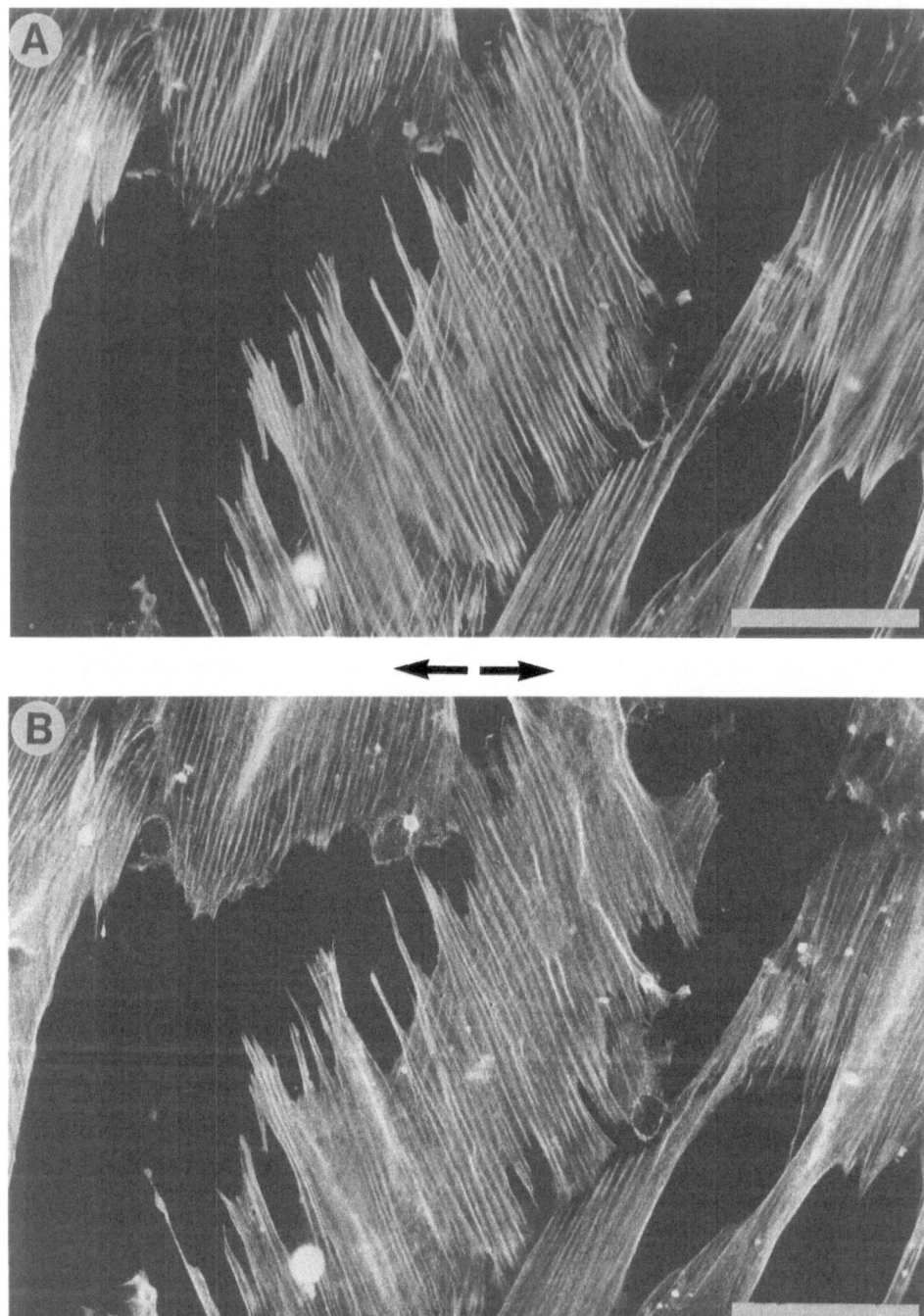

Fig. 10. Double fluorescence staining of stretched cells against actin (A) and caldesmon (B). Note that caldesmon is co-localized to F-actin and remains bound even after cyclic stretching. The direction of stretching is marked by the arrows. Epifluorescence microscopy. Bars – 50 µm.

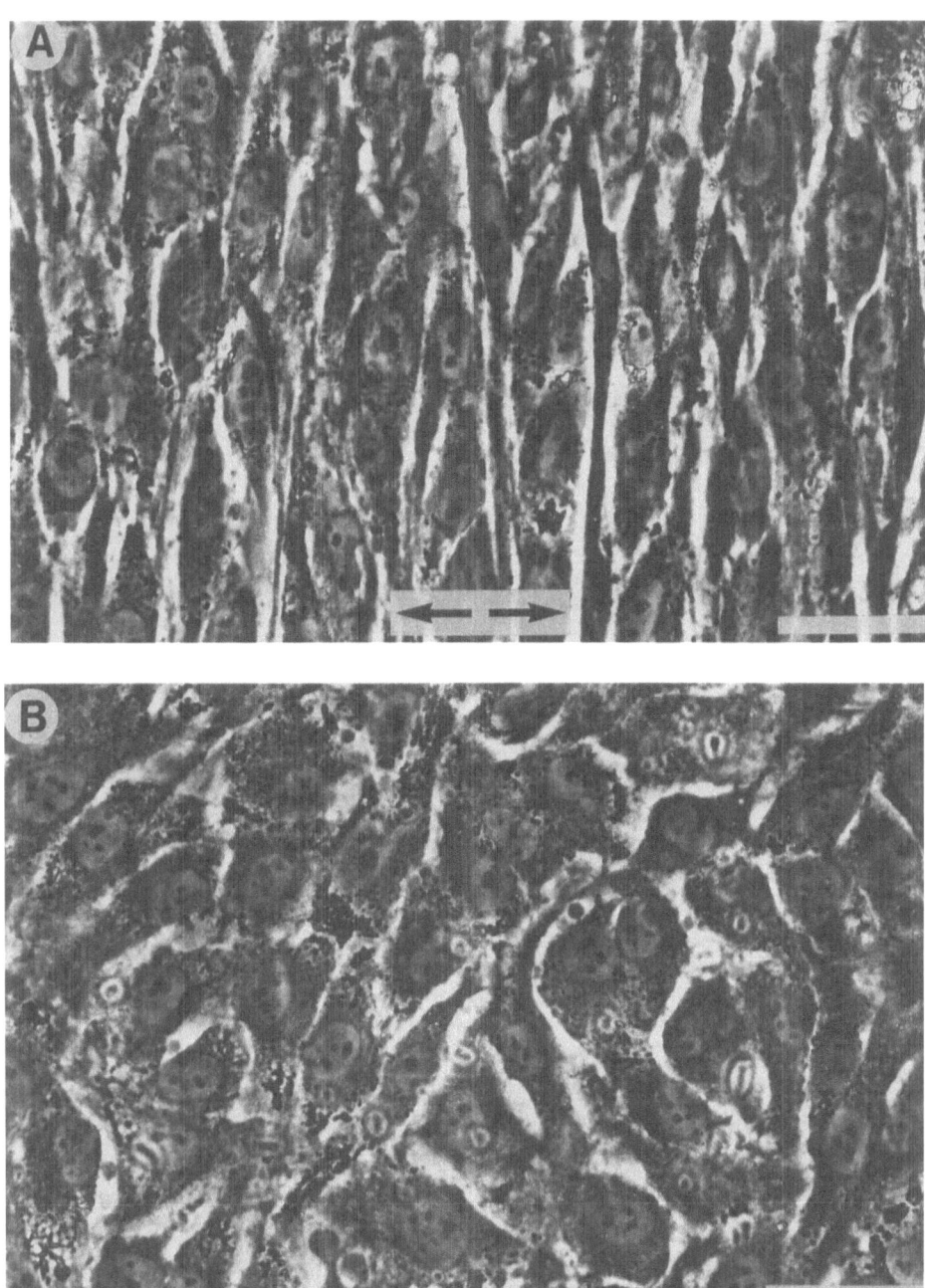

Fig. 11. (A) Elongation and orientation of endothelial cells perpendicular to the direction of stretching after 3 days of cyclic stretching and relaxations with an amplitude of 15%. (B) Cells on unstretched membranes remained randomly oriented and exhibited a polygonal shape. The direction of stretching is marked by the arrows. Phase contrast microscopy. Bars – 200 μm.

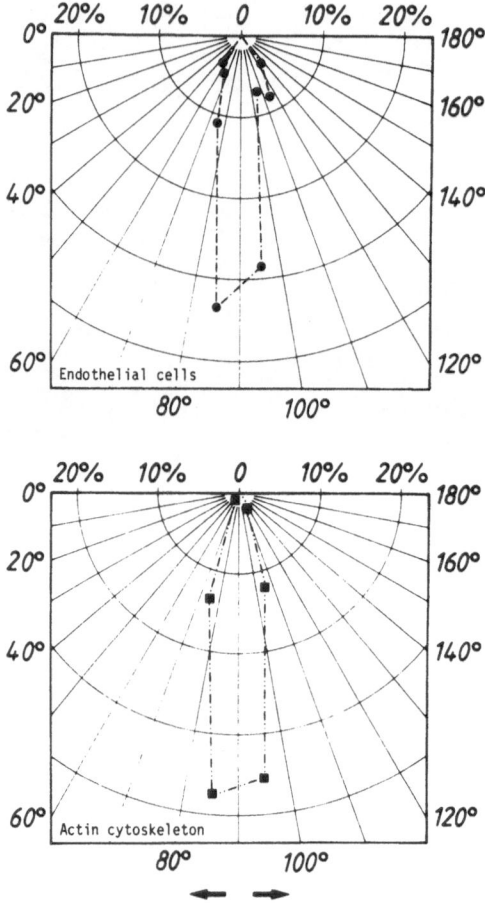

Fig. 12. Polar representation of the orientation response of endothelial cells and actin filaments to cyclic and directional stretching and relaxations. The cells were stretched with
an amplitude of 15% for 3 days. Note the distribution of both cells and actin filaments
perpendicular to the stretch direction which is marked by the arrows.

In contrast to actin filaments, the microtubule network in endothelial cells which
underwent mechanical stimulation did not exhibit a distinct orientation pattern, but
we observed a noticeable number of microtubules parallely aligned perpendicular to the
stretch direction (Fig. 14 A). In comparison, microtubules in unstretched cells (Fig. 14 B)
originated around the nuclei in areas termed «microtubule organizing centers» and ran
radially through the cytoplasm terminating near the cell periphery. The distribution of
microtubules in elongated and oriented cells was observed to be highly asymmetric
after 3 days of mechanical stimulation. Microtubules in stretched cells were no longer
uniformly distributed over the cytoplasm, but seemed to be accumulated or condensed
in small cytoplasmic areas. Moreover, the circular structures of «microtubule organizing centers» around the nuclei were dramatically reduced to fluorescent focal regions.

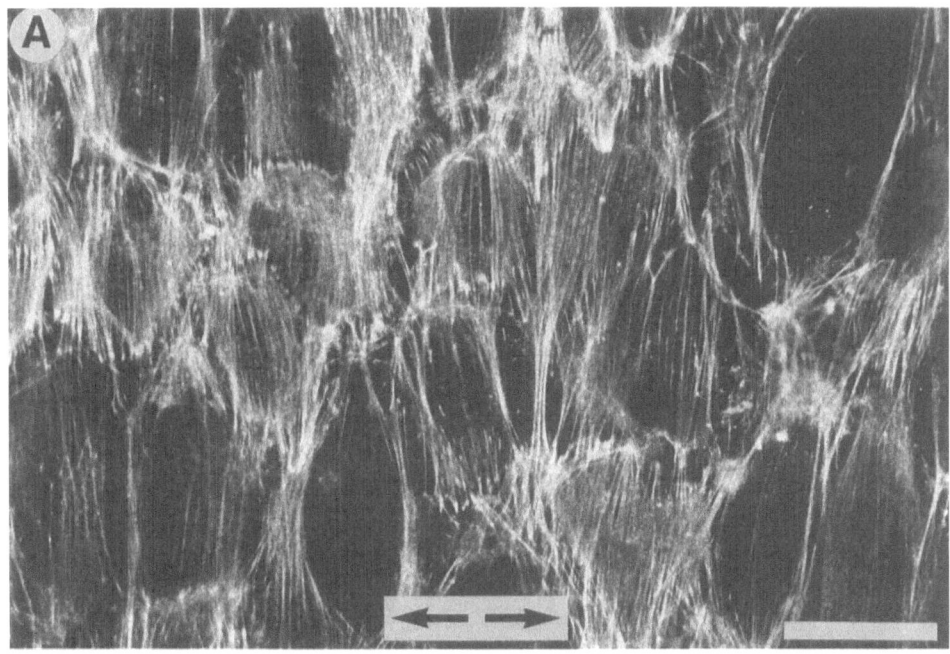

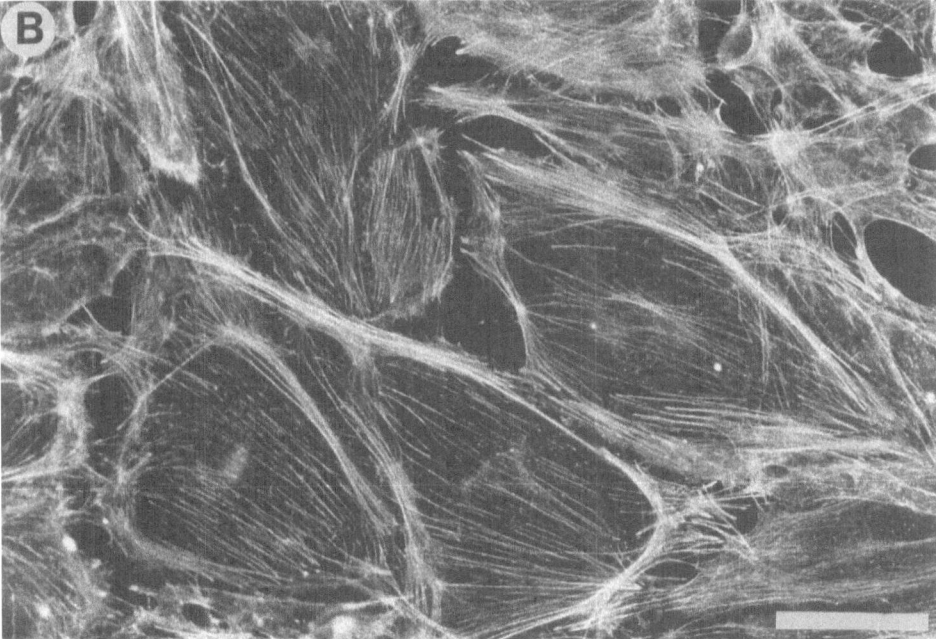

Fig. 13. (A) Strongly aligned actin filaments in endothelial cells after 3 days of cyclic and directional stretching with an amplitude of 15%. (B) Actin filaments in randomly oriented endothelial cells on unstretched membranes. The direction of stretching is marked by the arrows. Epifluorescence microscopy. Bars – 30 μm.

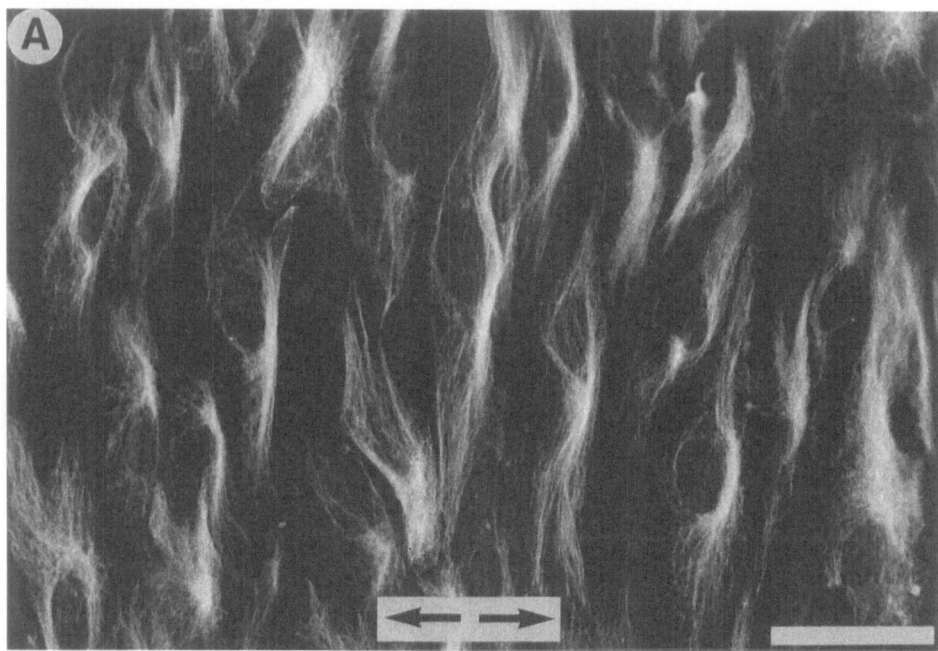

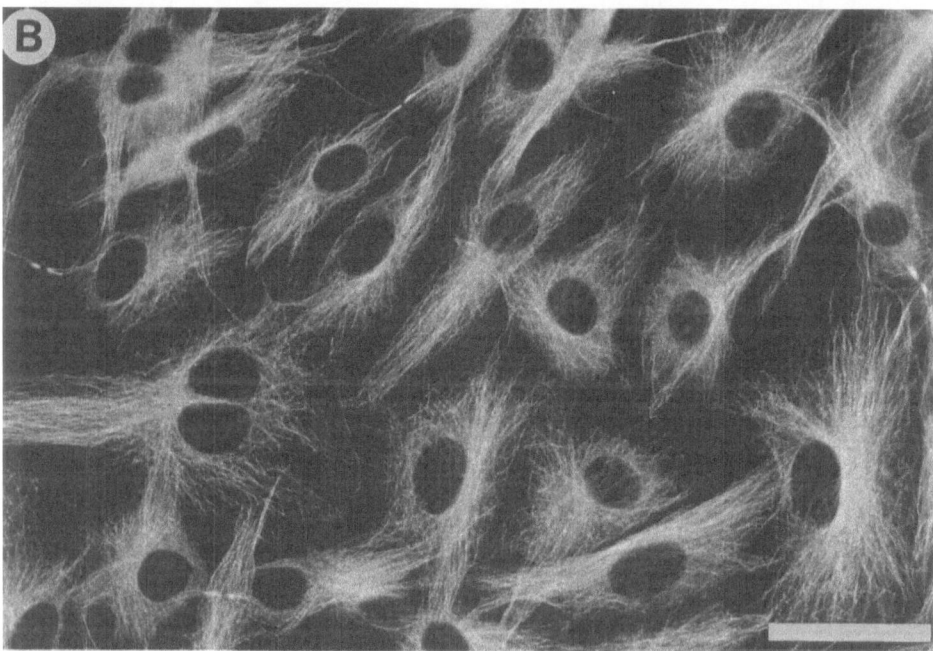

Fig. 14. (A) Distribution of microtubules in elongated and oriented endothelial cells after 3 days of cyclic stretching and relaxations. (B) Distribution of microtubules in randomly oriented endothelial cells on unstretched membranes. The direction of stretching is marked by the arrows. Epifluorescence microscopy. Bars – 75 μm.

DISCUSSION

The present study demonstrates that cultured smooth muscle cells and endothelial cells show a distinct orientation when subjected to cyclic and directional stretches and relaxations comparable to those which are induced in the artery wall by blood pulsation. The response to mechanical stimulation is reflected by cell orientation with preceding alterations and rearrangements of cytoskeleton components such as actin filaments, microtubules and intermediate-sized filaments.

In vivo, the majority of the smooth muscle cells within the media is helically oriented at an angle of 50° to 70° to the extension of the artery wall by blood pulsation. The helices may have a different pitch and vary in different segments of the same artery. The layers of the helix near the endothelium are usually steeper than the layers at the adventitial site. As the helices with a higher pitch cross those with a lower one, a twisted structure is formed [3, 30, 31] which gives not only mechanical stability to the artery wall but enables the levelling of the force of alternating blood pressure [1, 2, 32-36]. The angles of smooth muscle cell orientation obtained by our in vitro-system are comparable to the pitches of the smooth muscle cell layers in the media of the artery wall indicating that our apparatus represents a suitable model for the in vivo situation. Moreover, the induction of smooth muscle cell orientation by cyclic stretching provides evidence that the diameter oscillations by blood pulsation are a main factor which accounts for the arrangement of smooth muscle cells in vivo.

In contrast to medial smooth muscle cells, it appears at first glance that the endothelial lining of vessel walls in vivo is governed only by shear stress resulting from the flow of blood. However, study of aortic valves has shown that endothelial cells covering both sides of aortic valve leaflets are arranged rectangularly to the direction of flow [37], suggesting that shear stress is not the only factor influencing endothelial cell morphology and alignment. With these perspectives, a number of investigations have been undertaken to demonstrate the effect of hemodynamic forces on cultured endothelial cells in vitro. One of the first in vitro studies of flow response was that of Krueger et al. [38] who introduced a specially designed cell culture chamber in which the flow conditions over living cells could be precisely defined. The first major effort in this area was that of Dewey et al. [39, 40] who have developed a cone-plate apparatus for studying the dynamic response of vascular endothelial cells to controlled levels of fluid shear stress. They reported that the endothelial cells which were exposed to shear forces underwent a time-dependent change in cell shape from polygonal to ellipsoidal and became oriented with flow. Additional studies have demonstrated the response of endothelial cells to steady flow in vitro by changes in cell shape [41-43], cytoskeletal [44-48] and metabolic alterations [49-55].

Only little attention has been focussed on the effects of repeated stretching and relaxing of the vessel wall by blood pulsation on the endothelial lining in vivo. To our knowledge, there is only one other report which describes the response of endothelial cells to cyclic stretching [56]. In this study it was demonstrated that endothelial cells elongate and orientate perpendicular to the axis of deformation of the substrate. This is in agreement with the observations presented here. In our stretch-system, endothelial cells which were subjected to continuous cyclic and directional stretches and relaxations, showed a pronounced cell elongation within 3 days. Moreover, endothelial cells did not only elongate but oriented also perpendicular to the stretch direction. This agrees with in vivo observations, because the distending forces of blood pulsation, which cause a periodic stretching and relaxing of the artery wall, are perpendicular to the direction of blood flow and endothelial cell alignment. In addition, our findings also agree with flow experiments in which fluid shear stress as the only hemodynamic force caused an endothelial cell elongation and orientation [41-43, 49, 53, 57, 58].

Besides cell orientation, the other important result of this study are the alterations of cytoskeleton components. Especially in the short-term experiments with smooth muscle cells, we observed a rearrangement of the intracellular actin filament system as a response to cyclic stretching and relaxations which may be due to fast depolymerization - polymerization sequences of F-actin. It seems likely that such alterations in F-actin distribution induce cell orientation, because the response of actin cytoskeleton occurs prior to the orientation of the whole cell bodies.

Also in endothelial cells, a stretch-induced orientation of actin filaments was observed. For comparison, endothelial cells which were exposed to fluid shear stress, exhibited alterations in actin filament organization such as an increase in number and staining intensity of central straight and coarse fibers [44, 46-49]. The portion of marginal actin filaments was observed to be diminished. For example, Franke et al. [46] reported that the shear-induced stress fibers were more or less randomly distributed with respect to the direction of fluid flow. Under the experimental conditions of our study, the actin filament network of endothelial cells was strikingly altered. In contrast to shear stress experiments, we observed a pronounced rearrangement of actin filaments with a distinct orientation of central fibers perpendicular to the stretch direction and an overall increase of staining intensity. This pattern of actin filament distribution which apparently differs from the results of shear stress experiments, is in agreement with the alignment of stress fibers in arterial endothelial cells in situ which are characterized by orientation of actin filaments parallel to the longer cell axis and thus parallel to the direction of blood flow [44, 59, 60].

The distribution of microtubules and centriolar regions in endothelial cells of pig aortas in situ was described by Rogers and Kalnins [61]. They showed that individual microtubules ran parallel to the longer axis of each cells, appeared to be helical and formed a criss-cross pattern throughout the cytoplasm. The centriolar region was located at one end of the nucleus, in 60-80% of the cells in the region located towards the heart [62]. Under our experimental conditions of cyclic stretching, the microtubule network of endothelial cells was altered after exposure to mechanical stimulation exhibiting an highly asymmetric distribution of microtubules which also seemed to be accumulated or condensed in small cytoplasmic areas. A significant parallel alignment of microtubules along the ells axis was not observed, neither in endothelial cells nor in smooth muscle cells. The existence of strongly fluorescent focal regions in stretched cells at one side of the nucleus instead of hollow rings in unstretched cells indicates that «microtubule organizing centers» which are associated with the centrioles change their position in the perinuclear region during cyclic stretching. A preferential location of «microtubule organizing centers» at one end of the nucleus was not observed, probably because the dynamic process was not finished after this relatively short experimental period of continuous stretching and relaxations. On the other hand, the location of «microtubule organizing centers» in stretched endothelial cells might reflect the state of actively migrating cells in respect of their alignment perpendicular to the stretch direction as it was shown that «microtubules organizing centers» of migrating endothelial cells are positioned in front of the nucleus towards the direction of movement [63, 64].

The present study provides evidence that stretching the artery wall by blood pulsations results in an orientation response of both smooth muscle cells and endothelial cells of the artery wall. Moreover, the two different phenomena, fluid shear stress of blood flow and periodic stretching of the artery wall by blood pulsations, appear to be the main factors which account for the arrangement of the two different cell types within the vessel wall.

Acknowledgements

This work was supported by the Bundesministerium für Forschung und Technologie (grants 03-8566-3 and 0318 846A) and by the Ministerium für Wissenschaft und Kunst des Landes Baden-Württemberg, Forschungsprojekt No. 26.

The authors are indebted to Dr. V.P. Shirinsky, Moscow, USSR, for providing caldesmon antibodies and to Dr. U. Gröschel-Stewart, Darmstadt, FRG, for smooth muscle myosin and smooth muscle tropomyosin antibodies.

REFERENCES

1 *Schultze Jena BS* (1939) Über die schraubenförmige Struktur der Arterienwand. Gegenbaurs Morph Jb 83: 230-246
2 *Staubesand J (1959)* Anatomie der Blutgefäße. I. Funktionelle Morphologie der Arterien, Venen und arterio-venösen Anastomosen. In: Ratschow M (ed) Angiologie. Thieme, Stuttgart, pp 23-82
3 *Rhodin JG* (1980) Architecture of the vessel wall. In: Bohr DF, Somlyo AP, Sparks HV (eds) Handbook of physiology, section 2: the cardiovascular system, Vol. II: vascular smooth muscle. American Physiological Society, Bethesda, Maryland, pp 1-31
4 *Langille BL, Adamson SL* (1981) Relationship between blood flow direction and endothelial cell orientation at arterial branch sites in rabbits and mice. Circ Res 48: 481-488

stress on lipoprotein endocytosis. In: Biology of the arterial wall – Satellite Meeting Siena, CIC Edizioni Internazionali, Rome, pp. 183-188

56 *Ives CL, Eskin SG, McIntire LV* (1986) Mechanical effects on endothelial cell morphology: an in vitro assessment. In Vitro Cell Developm Biol 22: 500-507

57 *Dewey CF Jr, Bussolari SR, Gimbrone MA Jr, Davies PF* (1981) The dynamic response of vascular endothelial cells to fluid shear stress. J Biomech Engineering 103: 177-185

58 *Ives CL, Eskin SG, McIntire LV, DeBakey ME* (1983) The importance of cell origin and substrate in the kinetics of endothelial cell alignment in response to steady flow. Trans Am Soc Artif Intern Organs 29: 269-274

59 *Drenckhahn D* (1983) Cell motility and cytoplasmic filaments in vascular endothelium. Prog appl Microcirculation 1: 53-70

60 *Wong AJ, Pollard TD, Herman IM* (1983) Actin filament stress fibers in vascular endothelial cells in vivo. Science 219: 867-869

61 *Rogers KA, Kalnins VI* (1983) Comparison of the cytoskeleton in aortic endothelial cells in situ and in vitro. Lab Invest 49: 650-654

62 *Rogers KA, McKee NH, Kalnins VI* (1985) Preferential orientation of centrioles toward the heart in endothelial cells of major blood vessels is reestablished after reversal of a segment. Proc Natl Acad Sci USA 82: 3272-3276

63 *Gotlieb AI, McBurnie May L, Subrahmanyan L, Kalnins VI* (1981) Distribution of microtubule organizing centers in migrating sheets of endothelial cells. J Cell Biol 91: 589-594

64 *Gundersen GG, Bulinski JC* (1988) Selective stabilization of microtubules oriented toward the direction of cell migration. Proc Natl Acad Sci USA 85: 5946-5950

SELECTED DISCUSSION REMARKS:

Zekorn: The alteration of endothelial cells by shear forces was referred to induce class II antigens. So it might be interesting from an immunological standpoint to investigate the class II expression of the endothelial cells prior and after the alteration in your apparatus.

Charara: Have you any idea concerning the wall shear stresses and the orientation of the endothelial cells?

Dartsch: Under the light of the results presented here I think, that for the endothelial cells not only the fluid shear stress but also the stretching and relaxing of the arterial wall are two main factors which account for endothelial cell orientation.

PERIGRAFT REACTION OF VASCULAR
PROSTHETIC GRAFT AND THERAPEUTIC
MANAGEMENT

C.Pallua, H.J.Meinecke and W.Hepp

Chirurgische Klinik und Poliklinik,
Universitätsklinikum Rudolf Virchow,
Standort Charlottenburg, Spandauer
Damm 130, D-1000 Berlin 19

INTRODUCTION

Synthetic vascular prostheses which are biologically inert and constant in their
chemical and physical characteristics have been proven reliable for arterial vascular replacement. Incompatibilities have been observed extremely seldom during
their millonfeld application (1,3,4,5). The perigraft reaction, the socalled
biological incompatibility of synthetic vascular prostheses was first described
by KAUPP et al. (3) in 1979, until now over 300 times by different authors (1,2,
3,4,5,6,7,8,9). It occurs most often after extraanatomical positioning (6,9).
We will discuss our three own cases.

PATIENTS, TREATMENT, RESULTS

A 66-years old female patient was presented because of a swelling in the medial
position in the right distal thigh and proximal lower leg. Ten weeks ago she
received a femoropopliteal bypass in an other hospital. Here a brief description
of the vascular history of the otherwise healthy patient:
In 1979 implantation of a supragenual femoropopliteal venous bypass in reversed
technique because of an occlusion of the femoral superficial artery. The postoperative course was without complications and the following seven years the
patient was free of symptoms.

In 1986 an occlusion of the femoropopliteal bypass linked to a stage II b. Further
a long distant occlusion of the femoral superficial artery and a long distant
stenosis of the supragenual popliteal artery was observed. Now, also in an other
hospital, an infragenual femoropopliteal PTFE-bypass (Vitagraft) was placed in
orthotopic position. Once again the postoperative course was inconspicuous.

In April 1987 the patient suffered repeated painful occlusion of the bypass.

H. Planck M. Dauner M. Renardy (Eds.)
Medical Textiles for Implantation
© Springer-Verlag Berlin Heidelberg 1990

This time an occlusion of the total popliteal artery was found, hence insertion of a new distal femoropopliteal synthetic prosthesis (Impraflex) with subcutaneous transplant positioning. The postoperative course was inconspicuous until the patient observed a swelling 8 weeks after operation and then she came for consultation.

Macroscopically we saw a fluctuating painless tumor of about 4x10 cm without local or systemic signs of infection. The subcutaneously placed bypass could be palpated. It was dislocated and "swimming graft". Angiographically an inconspicuous aortoiliacal segment was seen with free bypass perfusion and a discrete distal anastomotic stenosis. The lower leg arteries were, with the exception of the anterior tibial artery long distant perfused. In the computer tomography of the upper and lower leg an extensive paravasate was seen by means of intravenous radiopaque liquid in the aerea of the middle and distal bypass segment with regular formation of cysts in the popliteal region till to the distal anastomosis. The proximal anastomosis showed a good incorporation. No pathogenic agents were found in the amber coloured liquid of the cysts, but plenty of eosinophil granulocytes. There was suspicion of hyperergic-allergic genesis in the opinion of the pathologist.

In the corrective operation of 2.11.1987 you could see the old graft in the distal and middle segment long distantly not incorporated in the healing process and swimming in a filled to bursting utricular cyst of amber coloured liquid, which reached the distal anastomosis. The PTFE-prosthesis was subtotally exchanged against an EXS-Sauvage-prosthesis. The central suture was done handwidth beneath the old proximal anastomosis, end-to-end technique, in the well incorporated segment of the old prosthesis. The new graft was placed subfascial in the thigh and then pulled through the fossa poplitea. The distal anastomosis was placed just above the truncus tibiofibularis. The surrounding of the old prosthesis was cut away as much as possible. The histological examination of the cyst wall showed a sclerotic swollen connective tissue, which was interspersed with vascular non specific granulation tissue poor on cells. Especially perivascularly there were some mixed cells of irritated infiltrats with a high pool of eosinophil granulocyts. One can say that an allergic genesis is possible. The postoperative course was inconspicious, the unlimited distance of walking was also after 18 months of control unchanged.

In the next case an 80-year-old patient is presented. She is diabetic and takes insulin. She was here for the first time in June 1988 with arterial occlusive disease of the right leg stage IV a. By angiographic means there was seen a generalized arteriosclerosis, an occlusion of the right femoral superficial artery in Hunter's canal, a stenosis of the beginning of the superficial femoral artery, an occlusion of 2 cm length of the popliteal artery and filiform stenoses of the arteries of the upper leg and multiple stenoses in the pelvic region. Macroscopically a necrosis was seen in the region of the big right toe.

After a dilitation of the right external iliac artery in August 1988 the patient was operated on and received a lateral tibial anterior bypass on the right side with a PTFE-prosthesis. First the postoperative course was inconspicious, the wound of the toe healed. In October a painless swelling appeared in the region of the distal medial thigh. In the computer tomography one could see a liquidlike space in the surroundings of the bypass. Angiographically the bypass was normal.

On strong suspicion of allergic course in October she was operated on and a subtotal bypass exchange with autologous vein was performed. Intraoperatively a cyst with swimming graft was found, the anastomoses were completely intact. Therefore they were not touched. The postoperative course was inconspicuous and the patient is up to now without complaints.

DISCUSSION

The so-called "perigraft reaction" occurs in literature in about 2.3 o/oo to 9.5 o/oo (3,7). In VOLLMAR's opinion (9) the incidence is up to 10 o/oo, since diagnosis has been improved. Both materials, Dacron double velour and PTFE were affected to the almost same percentage (9). In 75 % it occurs after extraanatomical positioning (6,9).

Various factors for this reason are under discussion: physicochemical irritation of the surrounding tissue by the fabrics itself, mechanical trauma by continuous movement of the graft in the tissue bed, i.e. insufficient tissue fixation resulting in gaps and exsudation of fluid round the prosthesis, and also a latent or manifest renal insufficiency (2,3,7,9).

Often the diagnosis of admission is lymphcyst, but this is more unlikely. Most of the perigraft reactions have a relatively long latency period, about 28 months in the literature (9), therefore a lymphcyst must be excluded.

This rare complication can be very dangerous. Repeated graft occlusions of unknown morphologic genesis can be caused by a perigraft reaction (9). But also lethal complications can not be excluded. Thus it was reported about a patient, who suffered from a perigraft reaction concerning the total aortobifemoral prosthesis (46 months after the operation) and who refused the exchange operation. This patient died because of a hemorrhagic shock. This was due to a central anastomotic bleeding 5 1/2 years after the implantation of the prosthesis resp. 20 months after the diagnosis "perigraft reaction" (6).

Concerning differential diagnosis the deep wound infect is possible. The latter is much more dangerous and its therapy differs completely. Therapy of the "peri-graft reaction" includes total or partial replacement of the affected portion of the graft and cyst wall with substitution by a prosthesis of a different synthetic material. The anastomosis must be done in a region where the old graft is absolutely well incorporated. In case of doubts a complete bypass exchange must be performed (2,6,7,9).

The incidence is rare, but the vascular surgeon must realize this complication. He must recognize it in time and must know how to handle it.

References:

1 Bolton W, Cannon JA (1981) Seroma formation associated with PTFE vascular grafts used as arteriovenous fistulae. Dialysis and Transplant 10:60
2 Dale WA, Blumenberg RM, Gelafnd ML (1983) Peri-graft seromas. J Cardiovasc Surg 24:372
3 Kaupp HA, Matulewitcz TJ, Lattimer GL, Kremen JE, Celani VJ (1979) Graft infection or graft reaction? Arch Surg 114:1419
4 Martinez RR, Vincente LC, Ferrer FD, Grau LJ, Mulet MJ (1982) Periprosthetic cyst formation: An unusual complication of polytetrafluoroethylene prosthesis implantation. Tex Heart Inst J 9:221
5 May J, Harris J, Patrick W (1979) Polytetrafluoroethylene (PTFE) grafts for haemodialysis: Patency and complications compared with those of saphenous vein grafts. Aust NZJ Surg 49:639
6 Paes E, Vollmar JF, Mohr W, Hamann H, Brecht-Krauss D (1988) Perigraft reaction: Incompatibility of synthetic vascular grafts? New aspects on clinical manifestation, pathogenesis, and therapy. World J Surg 12:750
7 Szilagyi DE, Smith RF, Elliot JP, Yrandecic MP (1972) Infection in arterial reconstruction with synthetics grafts. Ann Surg 176:321
8 Vollmar JF, Hesse G, Mohr W (1982) Infektion oder Unverträglichkeit von Kunst-stoffprothesen? Akt Chir 17:19
9 Vollmar JF, Gildner NW, Mohr W, Paes E (1987) Perigraft-reaction after implantation of vascular prostheses. Pathogenesis, clinical picture and treatment. Inter Angio 6:287

The Historical Development of Sutures Comparing the Manufacturing Process, Handling Characteristics and Biocompatibility

Kniepkamp, H.

B. Braun Melsungen AG, D - 3508 Melsungen

Suture materials are well known over many centuries. From the ancient egyptians it is known, that they used ants for skin-closure. After the bite of the ants, the head was broken up and this is the first example of a skin-closure with clips, which is nowadays very common in use.

After that period not many things happened. Hippokrates and Galen have given a lot of instructions how to close a skin wound, which were still in use up to the 19th century. One thing, that couldn't be solved until the beginning of the 19th century, was the suture of the small intestine. Due to the high infection of this injury most of the patients died. Nister from the UK tried to solve this problem. He developed a certain kind of disinfectant: He used the carbolic acid by treating not only the suture material but also the tissue and the surrounding in the OP-room. Up to that time the mortality due to infection rate in the hospital was more or less 80 % of all interventions in the small intestine. After this treatment with carbolic acid the infection rate in the hospital was going down to 20 %.
Unfortunately Nister treated for example the catgut material, a resorbable suture material from beef or sheep intestine, only on the outer surface with carbolic acid. During the resorption of this material the bacteria, which were - deriving from the beef or sheep intestine - in the catgut itself, were dissolved into the body and infection started again.
1906 a surgeon from Kassel, who had worked a lot about gelatine derived from bones, found, that you must sterilize or disinfect the material just before processing. Therefore he started to process the raw material of catgut with untwisted lamellaes with jodine, so that the whole material was purified. This was the first step for a sterile suture material.

In the thirties the surgeons used catgut as a resorbable material, silk and linen as nonresorbable materials. After the development of nylon by Carothers in the USA, also synthetic nonabsorbable materials were introduced into the surgery. This was started with the Nylon and followed by a patent from the BASF who had developed a so called "pseudo-monofilament", a structure in between a monofilament and a multifilament, which consists of a core of twisted nylon-filaments sheathened with the same material.

Another major step for the development of suture materials was the development of the atraumatic needle. This means, the steel, from which the needle was produced, received a drilled hole at the end, which allows to fix with mainly the same diameter the needle and the thread. By this way you can avoid a lot of traumata, which are caused by a so called surgical needle by passing the needle with the thread through the tissue.
After developing the polyesters, which mainly consist of polyethyleneterephthalate, this material was introduced as suture material.

H. Planck M. Dauner M. Renardy (Eds.)
Medical Textiles for Implantation
© Springer-Verlag Berlin Heidelberg 1990

Beginning at the late sixties there was a new development: The synthetic absorbable suture materials from polyglycolic- and polylactic acid. This development led to much better results in surgery, especially concerning resorbable suture materials. At the moment the use of Dexon and Vicryl - two synthetic resorbable materials in a braided configuration - supersedes more or less the use of catgut.

There are different possibilities to classify suture materials. First of all they can be divided in synthetic and natural suture materials.
Synthetic suture materials, which are commonly used in surgery, are polyester, polypropylene, polyamide, polyethylene, polyglycolic acid, polyglycolic acid-polylactic acid-copolymer, polydioxanone and polyglycolic acid-polycarbonate-copolymer.
Natural suture materials are catgut, silk and linen, very traditional suture materials.

Another possibility is to distinguish between absorbable sutures (fig. 1) and nonabsorbable sutures (fig. 2).

Fig. 1: Absorbable suture material   Fig. 2: Nonabsorbable suture material

Absorbable sutures are catgut, polyglycolic acid, polyglycolic acid-polylactic acid-copolymer, polydioxanone and polyglycolic acid-polycarbonate-copolymer.
Nonabsorbable suture materials, which are unresorbed or thought to be unresorbed in the body for a long period of time, are silk, polyester, polypropylene, polyamide, polyethylene and steel.

Apart from the raw material different appearances of suture materials result from the manufacturing process: the monofilament, the braid, the pseudo-monofilament and the twist. All these four appearances have different advantages and disadvantages and the surgeon has to choose, which material should be used according to these advantages and disadvantages and the field of indication.

First of all a monofilament can be made only from synthetic material and due to the manufacturing process, where the polymer is extruded, the surface is very smooth. According to that there are some properties which are similar for all monofilaments. Resulting from the very smooth surface, there is a very minimized tissue drag with a low trauma. The ability to make or to place a knot in the depth of the body is quite perfect, but on the other side you have a low knot security. Due to the massive structure of a monofilament the capillarity of the suture material is minimized or zero, which is a very essential point, especially for suture materials used in the intestine or for skin-closure. Using a suture material with a high capillarity could cause bodyfluids coming out of the wound or infection penetrating the wound. The flexibility of a massive material is certainly less than with a braided material.

In surgery the materials used for monofilaments are:
Polyester, which has a high knotpull tensile strength, very good flexibility and the material will not be degraded in the body over a long period of time.
Polypropylene is also used as monofilament. The material has an excellent tissue drag, the knot security is a little bit less. But it is also very stable for implantation over a long period of time.
Polyamide, which has a similar chemical structure as the collagenous material, will be degraded within the body within a certain period of time. The degradation cannot be compared with a real resorption, where the loss of tensile strength is more or less independent from the size of the material and the site of implantation. With polyamide the degradation is also depending on the implanted mass and is depending on the site in the body.
Polydioxanone and polyglycolic acid-polycarbonate-copolymer are two synthetic absorbable monofilaments with a very long resorption time. So knotpull tensile strength with these materials is gone within 60 days, the material can be found in the body for 180 days.
The tissue reaction of a monofilament is very minimal, due to the smooth surface, too. Within these synthetic nonabsorbable materials, there are quite no differences in tissue compatibility.

The next appearance is the braid: Different yarns, consisting of multifilament fibres, are braided together. Due to the fact, that the different fibres are crossing each other, you cannot untwist the material. The number of yarns is depending on the size, which is desired for the suture, 8 to 16 yarns can be used within one braid.
Caused by this manufacturing process there are some properties, which are the same for all braids.
Considering the way of manufacturing a braid, it is obvious, that the surface of a braid will be certainly rougher than that of a monofilament. Due to the rougher surface the tissue drag will also be higher than that of a monofilament.

The knotability due to the rough surface is certainly not so good as with a monofilament, causing problems to place the knot in the depth of the body. But on the other side the knot security is very high. The capillarity is present to a certain extent, the flexibility is very good and the material is easy to handle.

Materials like polyester are mainly used for braids. Polyamide is also used for braids as well as the very famous silk, which makes - comparing the meters or the kilometers sold all over the world - the highest number even today. Unfortunately the silk is a very hydrophilic material in contrast to the synthetic absorbable materials. This can cause problems concerning the capillarity and therefore concerning wound infection.

The polyglycolic acid and the polyglycolic acid-polylactic acid-copolymer, the synthetic resorbable materials, are resorbed within 21 and 28 days respectively in the body, independent of the site of implantation.

Fig. 3 shows a histological picture of a braided polyester material. The tissue reaction is much higher than with a monofilament. Therefore the biocompatibility of different suture materials can only be compared if the materials have the same surface structures.

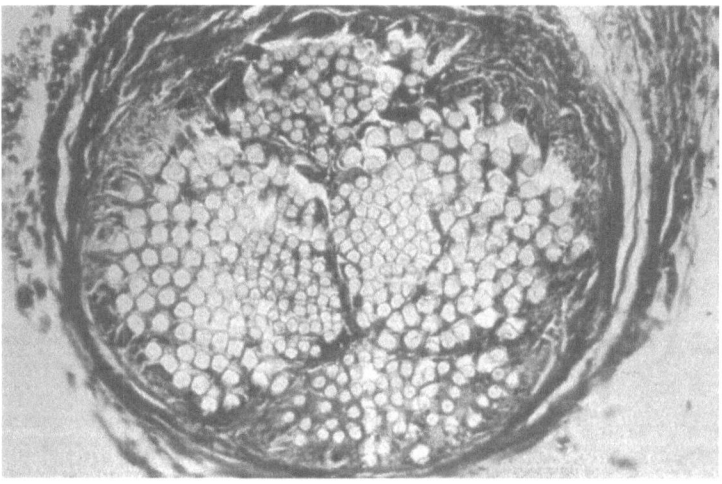

Fig. 3: Histology of an implanted braided polyester suture material

Fig. 4 - 6 show the histological evaluation of an implanted polyglycolic acid suture material. Fig. 4 is the situs after 10 days with a very low inflammatory reaction around the suture material. After 50 days (fig. 5) diminishing of the original diameter of the multifilament can be seen, after 90 days (fig. 6) quite all of the suture material is resorbed with a very low reaction around the suture material.

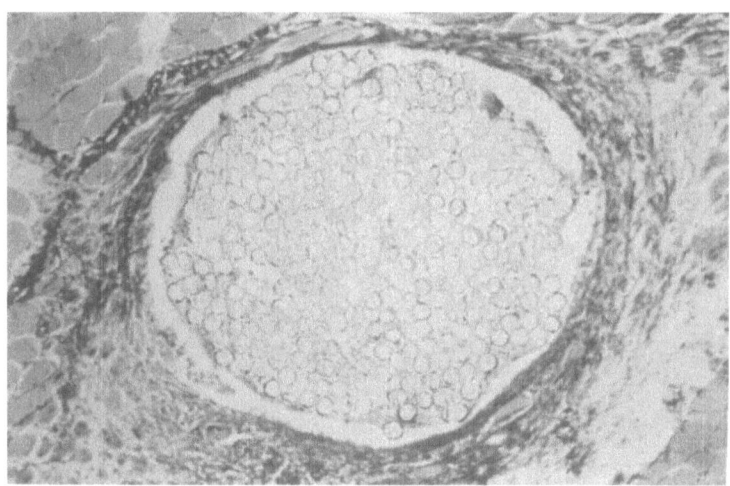

<u>Fig. 4:</u>   Histology of an implanted polyglycolic acid suture material after 10 days of
implantation

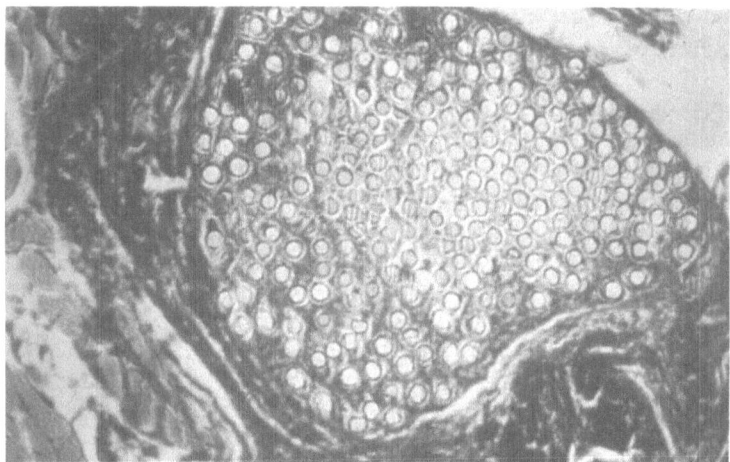

<u>Fig. 5:</u>   Histology of an implanted polyglycolic acid suture material after 50 days of
implantation

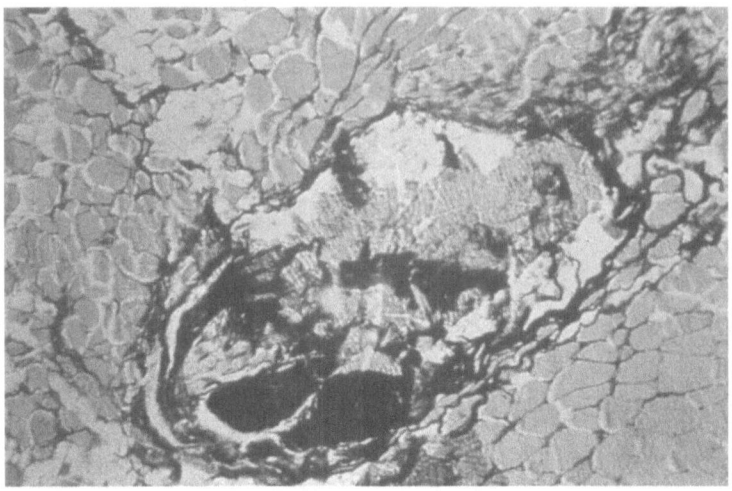

<u>Fig. 6:</u> Histology of an implanted polyglycolic acid suture material after 90 days of
implantation

The surface of a braid can be improved by coating the material with a lubricant, which reduces the friction between thread and thread during the knotting operation. The surface will be smoother, the tissue drag will be lower, the knotability will be better, but certainly the knot security will be lower.

The next appearance of suture materials is the pseudo-monofilament, a very old suture structure. Around a core of twisted material a sleeve of the same material is extruded (fig. 7). One property due to this manufacturing process is a relatively smooth surface. Due to that fact the tissue drag is relatively low, the knotability is good in contrast to the material which is made in nylon, the knot security is not as well as with the braided materials, the capillarity is quite zero if the coating of the material is not damaged, the flexibility lies in between braids and monofilaments.

The last appearance, a very traditional one, is the twist, where several yarns are twisted together to achieve the diameter and strength. Due to this manufacturing process twists have a relatively rough surface and therefore they have a pronounced tissue drag, the knotability is relatively good, the knot security very high. But due to the materials, which are used for this appearance, silk and linen, the capillarity is relatively high. The flexibility is quite the same as with braids. The silk is used as a twist only in a very thin size for ophthalmic suture, which is called the Virgin Silk. The linen is traditionally used, but today due to the high tissue reaction more or less abandoned. Steel is used also as a twist in a multifilament configuration but due to the handling properties steel is more or less abandoned in surgery.

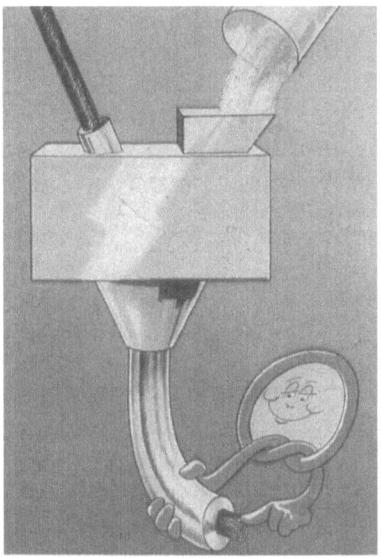

Fig. 7: Manufacturing of a pseudo-monofilament

Drawing the conclusions, there are a lot of well known suture materials and the surgeon has to choose the material for a certain indication. In soft tissues like mucosal structures a monofilament should be used, but the surgeon has to consider, that the handling properties are less than with braids. If a very smooth and very flexible material is needed, a braided material should be used. With this material the surgeon has to accept that the tissue drag is not as good as with a monofilament structure. At the moment in sutures the trend is more or less going to resorbable ones, whenever a resorbable suture material can be used. Due to the development of the monofilament resorbable suture materials there is a certain extension in indication for the resorbable suture materials.

SELECTED DISCUSSION REMARKS:

Gogolewski: Just a comment: In Groningen, we have been producing structurized monofilaments by using the so called "melt fracture effects". You can produce fibers with a very regular surface, which looks more or less like a threaded monofilament. You get a surface, which doesn't cause any problems as far as tissue drag is concerned, yet you improve the knot strength and of course the stability of the knots.

Aud.: What kind of lubricants are used for monofilaments?

Kniepkamp: At the moment the lubricants are used for braids only. A very old lubricant is the paraffin oil. But that will be applied in the hospital directly before use.
For a non-resorbable suture material you can use silicone for example or some polyesters or a combination of vinylacetate and polypropylene.
For resorbable sutures the coating must be resorbable, too. Otherwise you can have problems relating the resorption.

TEST METHODS FOR SURGICAL SUTURES

H. Planck, O. Weber, C. Elser, M. Renardy, K. Mayr, M. Milwich

Institut für Textil- und Verfahrenstechnik, D-7306 Denkendorf

Summary

Only few and unsufficient standards describing surgical suture properties exist in literature. Several test methods characterising suture, suture surface and handling behaviour were worked out, to help the surgeon choosing the fitting material for a needed purpose.

Introduction

Discussions with surgeons and producers of surgical sutures showed, that there are only few and unsufficient standards, which classify and describe the properties of surgical sutures. They are, as written down in USP or EP (Chapter 21):
1) Knot-pull tensile strength
2) Diameter
3) Sterility
4) Needle attachment
5) Extractable color
6) Analysis of soluble chromium compounds (absorbable sutures).
Only tensile strength and diameter characterise directly the handling of sutures, whereas various surgeons mentioned that a surgical suture should have high tensile strength, long degradation time (30 to 60 days), good knot run down, knot security and good tissue drag (influenced by suture surface characteristics).

To measure suture properties, exact standardized testing methods and test devices should be developed with following aims:

- reproducible test-results by automation
- classification and comparison of surgical sutures
- specifying needed properties to help the surgeon to choose the right suture
  for a specific purpose to minimize operation and healing risks.

Testing was made both under "dry" and "wet" conditions. Some test methods showed in this special regard no significant differences in the results, so that in these cases it will be sufficient to test under dry conditions.

Testing material was braided polyester and polyglycolicalcohol.

H. Planck M. Dauner M. Renardy (Eds.)
Medical Textiles for Implantation
© Springer-Verlag Berlin Heidelberg 1990

1. Knot-pull tensile strength

USP or EP standards classify a surgical suture by limits on average diameter and average knot-pull tensile strength which are laid down in USP chapter 21.

Knot-pull tensile test is sufficiently described in USP in terms of the surgeons-knot and the shape of the clamps which fix the suture (Fig. 1), but it lacks the data of the test speed.

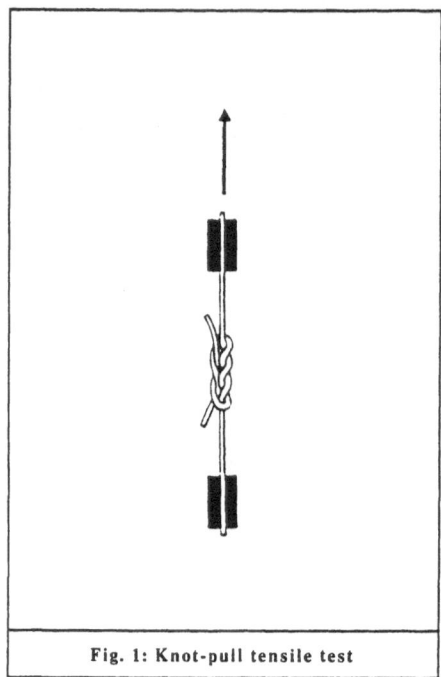

Fig. 1: Knot-pull tensile test

2. Diameter

Because of unexact methods measuring the diameter, often times a too strong surgical suture is used during an operation to be on the safe side in regard of yarn strength.

Thus too much thread is in the body, causing more foreign body reaction with delayed wound healing. Too thick threads also cause too big stitching holes.

Several diameter testing devices being on the market were tested and had various disadvantages.

Therefore a new mechanical diameter testing device was designed with the following abilities:

- test stamp with definite lowering speed
- microprocessor for evaluation and digital display - printing mean value, standard deviation and histogramm
- automatic scanning of a selectable number of tests along the thread
- after turning the thread for 90 degrees, the scanning is repeated automatically (on exact the same spots) to get exact results of roundness
- dead weights for exact load at measuring.

With the developed diameter tester (Fig. 2) we obtained reproducible test results of diameter and roundness (including standard deviation and variance).

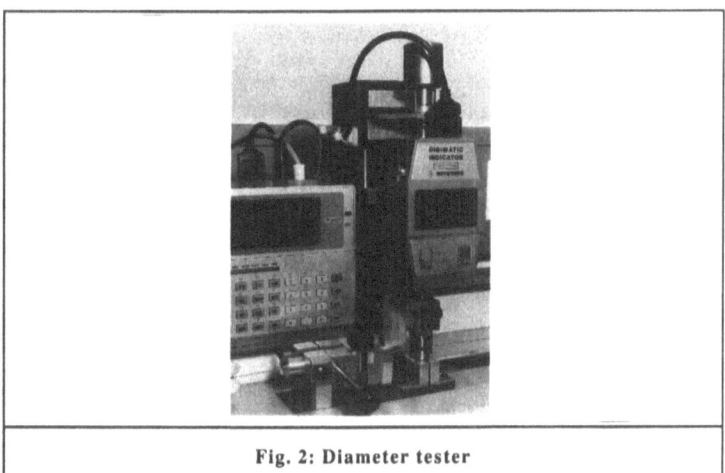

Fig. 2: Diameter tester

3. Bending stiffness

Bending stiffness determines in general, how well it is to work with a suture.
A slack suture is formed easier into a knot and gives a secure knot, which doesn't move or open under stress. A stiff suture offers more resistance in making a knot and opens easier under strain.
Therefore a suture should be slack!
Several stiffness testing devices being on the market came out to be unexact, not automated, therefore too slow and too dependent of human concentration abilities!
Thus a new stiffness tester (Fig. 3) was designed, where a yarn loop is pressed unto measuring bolts of a precision scale by the circling movement of the clamp which fixes the suture ends.

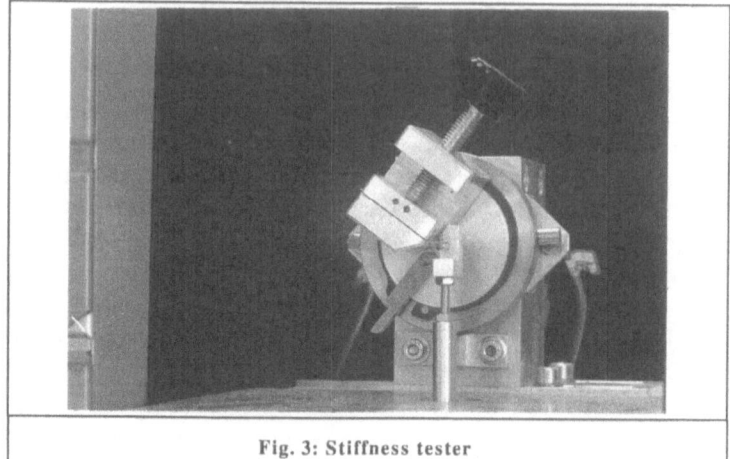

Fig. 3: Stiffness tester

4. Surface roughness

A rough surface of a surgical suture causes high friction resistance from the biological tissue when pulled through, thus injuring it. A rough surface can also damage a surgeons operation gloves. On the other hand offers a rough surface a better knot stability and knot security because of higher friction between corresponding surfaces.

A suture with a smooth surface (which depends upon the filament material, the braid procedure and the coating material) needs less force to form a knot and offers a better knot run-down.

Good test results gave the Hommel roughness tester where a leight-weighted, rounded measuring tip (Fig. 4) runs along the moving thread, the yarn causing an up and down motion of the tip which is recorded on-line with a plotter.

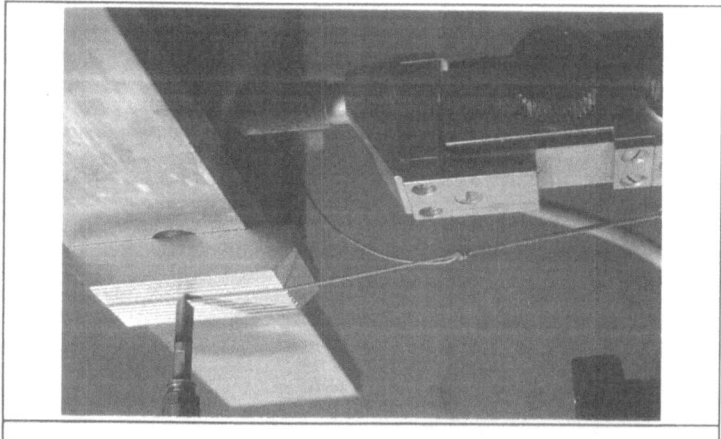

Fig. 4: Measuring tip of roughness tester

The analysis of the records obtained is performed in the Hommel evalution equipment and issues several different characteristic values of roughness.

5. <u>Knot run-down</u>

Knot run-down depends on
- suture stiffness
- surface roughness
- coating material

A slack and smooth suture needs less force to form a knot and place it, thus the tension on the biological tissue is small. A knot run-down measuring assembly was developed (Fig. 5), where both ends of the yarn are loaded with dead weights so that the loop can move within its limits.

A double loop was used for testing, a double loop being the first part of a surgeons knot.

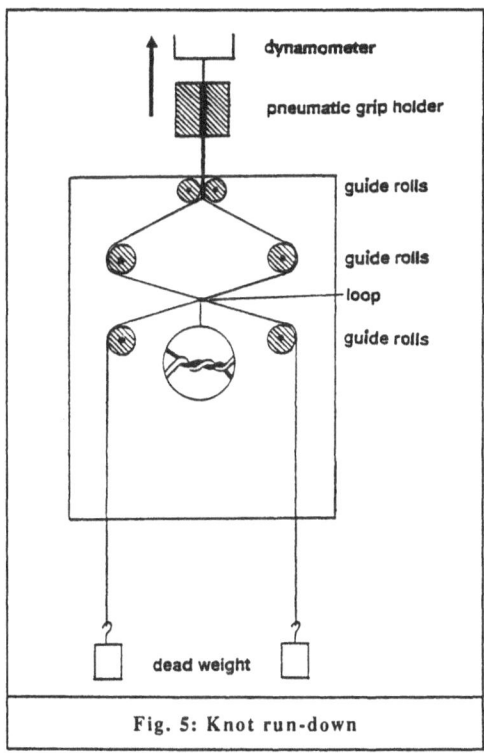

Fig. 5: Knot run-down

6. Knot snugging-down

Easy knot snugging-down, which is the displacement of a completed knot to its final position, is especially in deep operation wounds a significant point to consider.

If the surgeon needs too much force to place the knot in its final position (for example because of high yarn-yarn friction), areas of tissue are contracted leading to imperfections and necroses.

Knot snugging down is also dependent on suture stiffness, surface roughness and coating material.

Fig. 6 shows the measuring arrangement for ascertaining the needed force for knot snugging-down.

A suture with a surgeons-knot is hanged on a first fixed bolt which keeps apart the ends of the yarn coming out of the knot. A second fixed bolt represents the thumb of the surgeon which causes the knot to move.

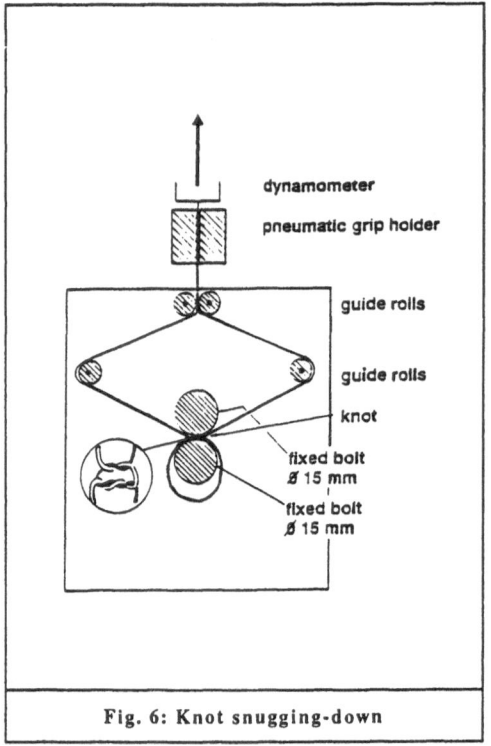

Fig. 6: Knot snugging-down

Before this measuring arrangement starts in action the knot was formed and tightened on the tensile tester with about 10% of its tensile strength at break.

At this test the characteristic value for knot snugging down, i.e. the moving of the knot, ist the plotted displacement til break minus the material elongation of the yarn at breaking strength.

On different sutures the difference in knot snugging-down behaviour of dry and wet sutures was tested, there was no significant difference to be found (wet condition means here: putting the suture for 5 minutes in physiological saline solution).

7. Knot security

Knot security, which is also called knot-holding capacity is dependent on suture stiffness, surface roughness and coating material. Knot security means, whether a knot stays under steady tension in its determined place.

To determine knot security, the similar knot-pull tensile test is not sufficient, because this test does not correspond to in vivo conditions. Knot security depends largely on that force, which pulls the knot tight. Therefore, pulling the knot tight must be done under definite conditions on a tensile tester.

A measuring device was developed, where pulling-tight the knot and testing of the knot can be made on one single device without relocating the suture.
The suture is laid around two fixed bolts (Fig. 7), a knot is formed and the knot is pulled tight with a tension of 20% of the tensile strength at break. The speed of the power traverse is 200 mm per minute. The pivoted bar is here without function.
Now the pivoted bar is fixed in the pneumatic grip holder (Fig. 8), the long ends of the suture

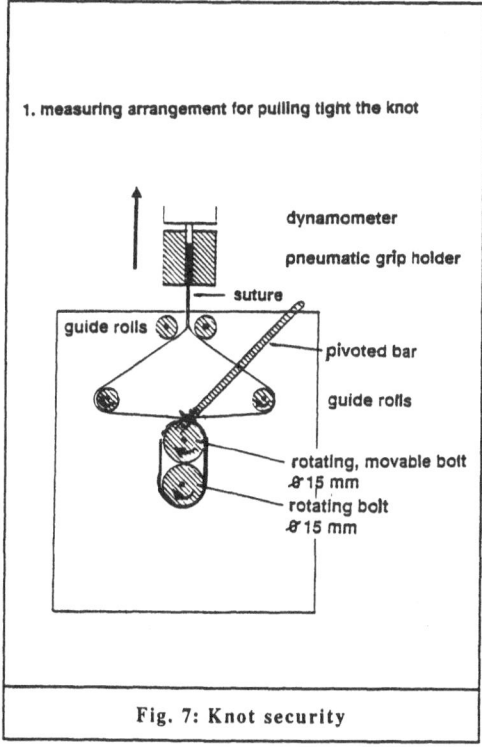

1. measuring arrangement for pulling tight the knot

dynamometer

pneumatic grip holder

suture

guide rolls

pivoted bar

guide rolls

rotating, movable bolt
$\varnothing$ 15 mm
rotating bolt
$\varnothing$ 15 mm

Fig. 7: Knot security

are cut to 10 mm, the upper bolt is disengaged from its holding and the test can be started, ascertaining that force which causes the knot to break or to slide open.

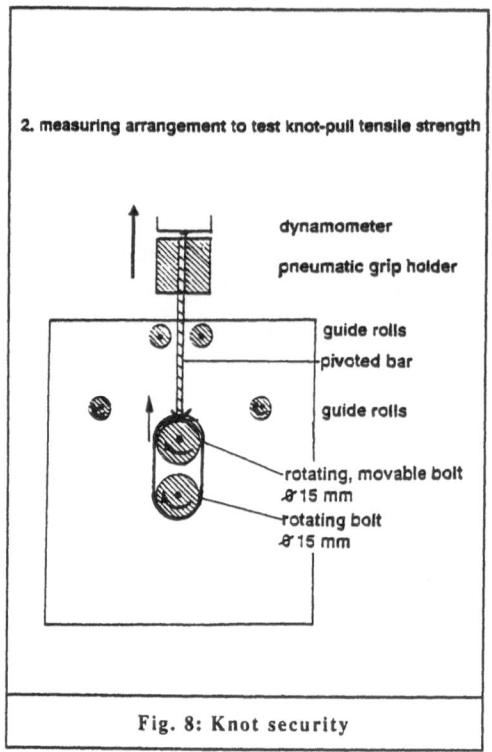

Fig. 8: Knot security

The assumption, that the movement of the upper bolt opens the knot, was taken into regard. The bolts were made rotatable to arrange the knot in the sideways space between the bolts which is the nearest aftersensation to in vivo conditions.

As a result of this measuring arrangement it was detected that a knot, which was sufficiently pulled tight, was pulled tighter by testing.

That means, that a knot is, in some cases, in vivo pulled tighter by swelling biological tissue.

8. Capillarity

Multifilament surgical suture increases the risk of bacterial infection of a wound. Braided sutures render possible tissue liquid to come out from the wound and so bacteria to get into the wound.

For testing, a surgical suture is loaded with a dead weight of 9.81 cN and sticks with its free end in a test liquid which desirably should be bovine serum.

After certain intervals of time (1, 3, 5, 10, 15, 30 min., several hours) the capillary rise of the liquid is ascertained by measuring the electric conductivity along the thread.

The capillary rise of the test liquid should be less than 5 mm within 8 hours, otherwise the material is not suitable.

In industry tests of capillarity are made with methylene-blue in deionized water. The problem arising thereby is, that water has a faster capillary rise than the methylene-blue agent. Thus looking out only for the (slower) colouring of the thread gives too good test results, therefore this test method brings no sufficient results.

9. Tissue drag

In case of tissue drag a tensile tester measures the needed strength to pull a suture through biological tissue, thus measuring surface roughness, bending stiffness and influence of coating material in one test.

A measuring arrangement used in industry is shown in Fig. 9. The angle should be 45 degree. The stitching holes should be 15 mm apart. The aorta tissue (pig) is fixed on a silicone rubber plate with pins. Power traverse speed should be 400 mm/min.

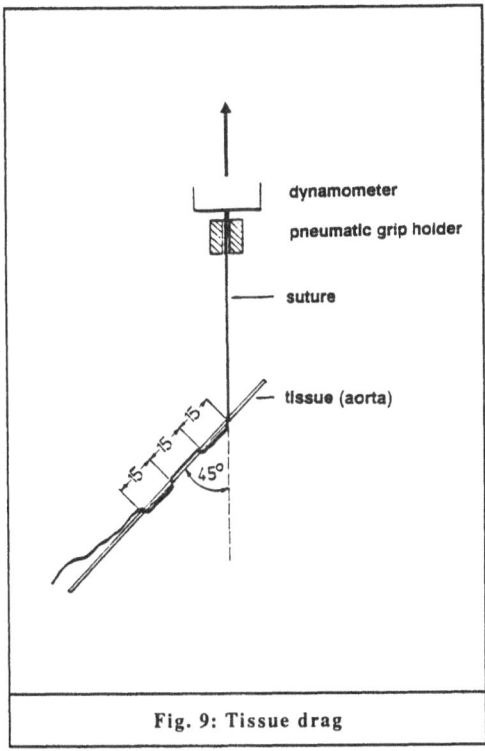

Fig. 9: Tissue drag

Some more criteria should be mentioned:

> - the needle type must be fixed for comparative tests and should be smaller than the suture, to get a good connection between suture and tissue,
> - moisture content of the aorta has a great influence on tissue drag and should be held on a constant level,
> - there are big differences in the wall thickness of aortic tissue. Therefore the test should be made with synthetic material of constant wall thickness, for example coagulated polyurethane foil.

10. Strength degradation by hydrolysis

An absorbable surgical suture should guarantee solidity for a definite length of time, otherwise the wound will break open.

Several pieces of thread were put in different basic, acid and neutral solutions and successively, in certain intervals, taken out to test the breaking strength. The test solutions should be effectively buffered, so that the derivative products do not influence pH-conditions.

All tests have to be done under sterile conditions to avoid bacterial contamination!

Fig. 10 shows the strength degradation by hydrolysis as a function of the time the threads are incubated by the test solutions.

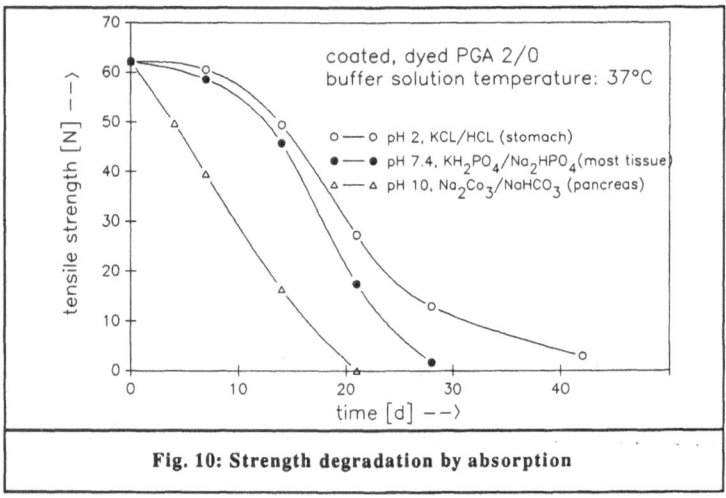

Fig. 10: Strength degradation by absorption

The long time for testing - up to 60 days - conducted to a test method where the former 37° Celsius of the test solutions are increased to 80 degree Celsius so as to reduce testing time to 1 or 2 days and bacterial contamination is neglectable.

Fig. 11 shows, that strength degradation at 80 degree happens much faster than with 37 degree test solution temperature.

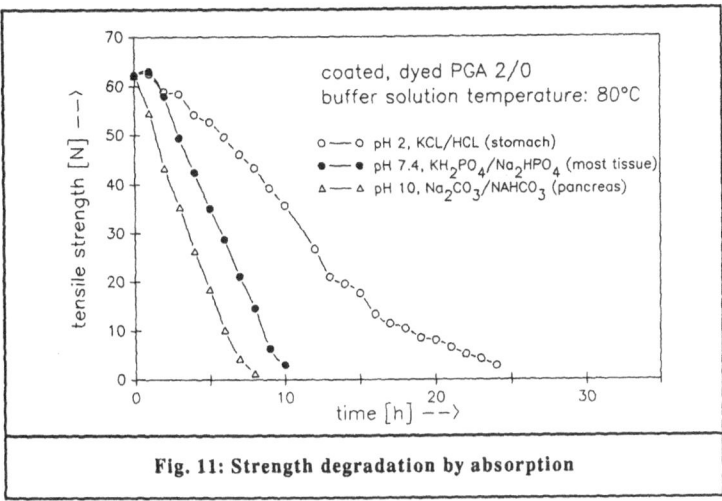

Fig. 11: Strength degradation by absorption

Because of recombination of free molecular chain ends the degradation in acid buffer solution is slower than in the other solutions.

The strength degradation of <u>uncoated</u> material would be <u>faster</u> because polyalcohol coating is more resistant against hydrolysis.

Suture material consisting of polyglycolic alcohol-polycarbonate-copolymer has more resistance against hydrolysis and basic agents, so that the degradation curves with different solutions would be more similar, the degradation slower.

We tried to find a comprehensive correlation between the 37°C test and the 80°C test. Yet every material/test-solution combination has its own correlation factor (which is composed of degradation time at 37°C versus degradation time at 80°C).

So test temperature of buffered solution should be 37°C when exact test results are needed, the 80°C test being sufficient for quality control.

11. Material behaviour under permanent stress

In regard of material behaviour under permanent stress two usual tests should be performed. The first one is the creep test, where dead weights clamped on hanging threads cause the material to creep, the elongation being ascertained.

The second one is the stress relaxation test where a certain elongation is forced upon the thread, the stress relaxation caused by material flow being ascertained in form of a hysteresis stress-strain curve.

The presented study was supported by the German ministry for Research and Technology, Bonn.

242

SELECTED DISCUSSION REMARKS:

Chu: I have two comments. The first is in regard of the capillarity. We know from similar experiments at the University of Liverpool in 1982 that the capillary reaction has two parts: the first, the intrinsic part, is dealing with the water adsorption through the fibers. The second part is the water travel along the outside of the fiber within the braided structure. And this travel is against the gravity if you keep the suture vertical. Yet in actual in vivo conditions most of the sutures are put in a horizontal position. Back to the late seventies a group of surgeons in Sweden took two test tubes, one filled up with a clear sterilized nutrient broth, the other with a bacterial broth. Then they put the suture in the middle and in a horizontal position they observed the transport of the bacteria through the suture contaminating the clear tube.
My second comment is on the last slides, showing the pH effect on the tensile strength. You retain a higher strength at pH 2 than at a near neutral pH of 7.4.
The degradation of PGA is of hydrolytical nature and depends on pH. At lower pH and higher pH you have an acid and a base catalysed hydrolysis. Therefore at neutral pH the tensile strength retention should be the highest.

Milwich: In neutral pH hydroxy$^{(-)}$ ions and hydrogen$^{(+)}$ ions are in equilibrium. In acid pH you have hydrogen$^{(+)}$ ions in excess, and that brings the possibility of a recombination of the polymer chains.

Chu: But it's a buffer medium supposed to neutralize excess acid or excess base. Do you have checked the pH of your buffer continuously?

Müller: The pH was controlled all the time. We had expected the degradation rate to be the lowest at neutral pH, but that's the result. And it should have to do with the recombination at acidic conditions.

Chu: The whole issue of the recombination of a simple organic ester under acidic conditions is no doubt. But you have a polymer and the chain movement makes the recombination difficult. After the chain is broken, the two ends particular in the amorphous phase of a polymer will move and a recombination is a pure probability. In a crystalline region the recombination may be possible, because the ends are fixed in the same position.

Müller: That may be the explanation, as you know here we have highly oriented fibers with a crystallinity of about 90 percent. But simple models do not work in this case.

THE EFFECT OF BACTERIA ON THE DEGRADATION OF ABSORBABLE SUTURES

Elser, C., Renardy, M., Planck, H.

Institut für Textil- und Verfahrenstechnik, D-7306 Denkendorf

INTRODUCTION

The presence of bacteria in surgical wounds will delay the formation of granulation tissue and thus the healing process. Sutures used for closing a wound must be able to compensate this delay by their strength. Otherwise it will cause complications that might be life - threatening for the patient. Especially in the case of absorbable sutures it is possible that the degradation is influenced - accelerated or delayed - by metabolism products of the bacteria or by bacterial enzymes. The in vivo data published in literature vary considerably, ranging from an acceleration of degradation (Dudley et al, (1)), to no influence (Thiede et al, (2)) and to a delay of the degradation of absorbable sutures (Williams, (3)) in the presence of bacteria.

These varying results might be due to the different conditions under which the experiments were made. The following experiments were carried out to develop a standardized in vitro method for measuring the effect of bacteria on the degradation of any biomaterial.

MATERIALS AND METHODS

MATERIALS

The sutures used in this study were Dexon[R] and Maxon[R] (Davis & Geck), size 2/0.

Dexon: a braided, absorbable multifilament made of polyglycolic acid. The suture is non-coated and non-stained.

$$[CH_2 - \overset{\displaystyle O}{\underset{\displaystyle O}{\overset{\|}{C}}}\ ]_n \qquad \text{polyglycolic acid}$$

Maxon: an absorbable monofilament made of polyglyconate.
Polyglyconate is a polyglycolic acid/polycarbonate triblock-copolymer.

$$H - [O - CH_2 - \overset{O}{\overset{\|}{C}}]_x - [O - R]_y - O - [\overset{O}{\overset{\|}{C}} - CH_2 - O]_x - H \qquad \text{polyglyconate}$$

The incubation liquid was a bacterial suspension based on nutrient broth inoculated with Staphylococcus aureus ATCC 6538 P (10^6/ml).

H. Planck M. Dauner M. Renardy (Eds.)
Medical Textiles for Implantation
© Springer-Verlag Berlin Heidelberg 1990

METHODS

Suture threads, 25 cm in length, were immersed in the bacterial suspension.

Incubation was performed at 37°C under agitation (120 rpm). To keep pH and the number of bacteria constant, 50% of the bacterial suspension were replaced by sterile nutrient broth every 72 hours. At the end of the immersion period sutures were removed, incubated for 2 hours in ethanol for desinfection and then dried under vacuum for 24 hours.

Controls were incubated in sterile nutrient broth and treated in the same way.

The degradation was assessed by determination of the breaking strength using a Zwick universal testing machine. The gauge length was 100 mm and the cross-head speed 200 mm/min.

RESULTS

The results are shown in table 1 and figure 1.

Bacterial Degradation

time	breaking strength							
	Dexon Staphylococcus aureus		Dexon control medium		Maxon Staphylococcus aureus		Maxon control medium	
d	%	SD	%	SD	%	SD	%	SD
0	100.0	2.1	100.0	2.1	100.0	6.6	100.0	6.6.
7	52.2	4.5	65.8	3.6	87.2	7.5	91.4	7.7
14	17.7	4.6	27.8	1.8	74.3	4.1	78.9	4.4
21	0.0	0.0	1.3	1.5	50.9	2.8	60.4	3.0
28	0.0	0.0	0.0	0.0	33.4	2.5	37.5	2.5
42	0.0	0.0	0.0	0.0	0.0	0.0	10.6	2.0

tab. 1: Degradation of absorbable sutures in vitro in the presence of Staphylococcus aureus.
SD = Standard Deviation, d = days

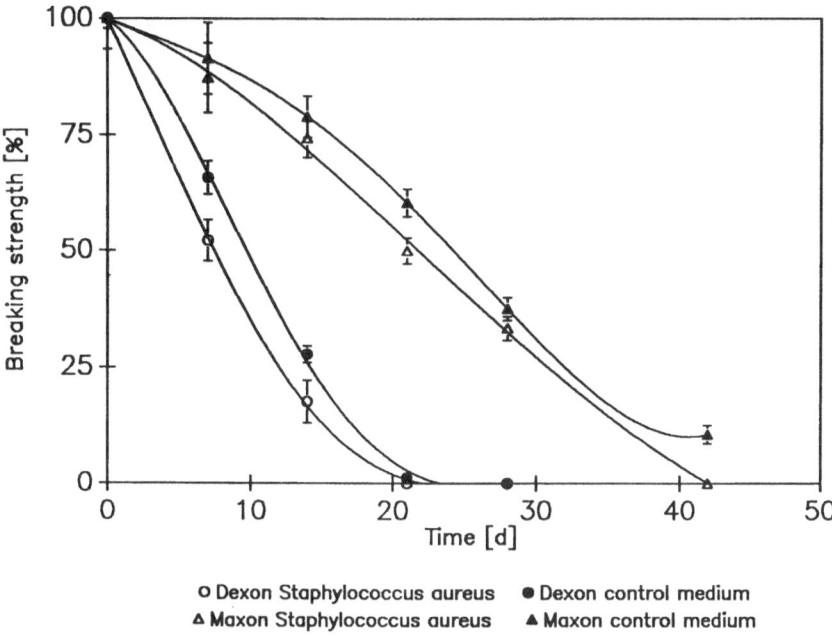

BACTERIAL DEGRADATION

○ Dexon Staphylococcus aureus   ● Dexon control medium
△ Maxon Staphylococcus aureus   ▲ Maxon control medium

fig. 1: Degradation of absorbable sutures in vitro in the presence of Staphylococcus aureus

The degradation of both Dexon and Maxon is accelerated significantly in the presence of bacteria. With Dexon 50% degradation occurred within 10 days in sterile broth respectively 8 days in the presence of bacteria. With Maxon 50% degradation was achieved within 24 days in sterile broth respectively 22 days in the presence of bacteria. Degradation was completed (i.e. no residual breaking strength measurable) within 19 respectively 21 days for Dexon in presence respectively absence of bacteria. For Maxon the time was 41 respectively 50 days.

SUMMARY

The degradation of Dexon and Maxon is accelerated in the presence of Staphylococcus aureus. This phenomenon might be due to two reasons:

1. EFFECTS OF BACTERIAL ENZYMES

Hydrolysis enzymes such as carboxypeptidase A enhance the degradation of polyglycolic acid sutures (Williams et al, (4)). Extracellularly acting enzymes produced by bacteria can have the same effect. The metabolism of Staphylococcus aureus encloses some of these enzymes: (Davis (5))

- The nuclease (phosphodiesterase) hydrolyses nucleic acids in vivo.
- The hyaluronidase supports the diffusion of bacteria into the connective tissue by hydrolysing the β 1-4 bonding of the hyaluronic acid.
- The coagulase and the staphylokinase act in the process of blood coagulation.

It is possible that all these enzymes are able to break up polymer chains of polyglycolic acid and thus enhance the degradation.

2. EFFECTS OF BACTERIAL METABOLISM PRODUCTS

Beside the enzymes Staphylococcus aureus produces some proteins that are cytotoxic for human cells:

Hemolysin destroys the membranes of erythrocytes and thus promotes hemolysis.

P-V-leukoidine causes the degradation of polymorphic leukocytes and macrophages in absence of Ca^{2+}.

Another cytotoxic protein of Staph. aureus is enterotoxin.

Again it is possible that these proteins influence the degradation of polymer chains in vivo and in vitro.

Concerning the test method we think that the described method is a valuable tool for simulating a strong in vivo-infection in vitro. We suggest it to be discussed as a standard method which would allow to compare the data of different investigators.

REFERENCES

1. Dudley, H.A.F., Kapadia, C.R.
    in: Moderne Nahtmaterialien und Nahttechniken in der Chirurgie
    Springer, 1982, 34-38

2. Thiede, A. et al
    Der Chirurg 52, 1980, 768-773

3. Williams, D.F.
    Journal of Biomedical Materials Research 14, 1980, 329-338

4. Williams, D.F., Mort, E.
    Journal of Bioengineering 1, 1977, 231-238

5. Davis, B.D., et al
    Microbiology 1980, 624-633

BIOARTIFICIAL ENDOCRINE PANCREAS

Prerequisites and Current Status

Konrad Federlin and Tobias Zekorn

Medizinische Klinik III und Poliklinik der Justus Liebig-Universität (Director: Prof. Dr. med. K. Federlin), Rodthohl 6, D-6300 Giessen (West-Germany)

ABSTRACT:

Diabetes mellitus has an incidence of 2-4% in the population of Western Europe. 20 - 25 % of patients either require insulin from the beginning (type-1-diabetes) or after years of treatment with insulin secretion stimulating drugs (secundary insulin-dependent type-2-diabetes). Although the introduction of insulin treatment remarkebly prolonged life expectancy, many patients develop severe late complications during the course of the disease. Diabetes is the most common reason for blindness and endstage renal failure leading to dialysis. Patients often develop macro- and microangiopathy and neuropathy during the course of the disease macro- and microangiopathy and neuropathy. This process can only be inhibited by an intensified insulin treatment. Successful islet transplantation would be the optimal treatment as insulin producing tissue would be replaced. Availibility of islets and immunological problems led to the idea of islet transplantation inside semipermeable artificial membranes i.e. the bioartificial pancreas (BAP). Different implantable devices are described and evaluated in this paper (BAP connected with blood circulation; macro- and microencapsulation). An introduction to problems of biocompatibility, diffusion conditions, islet survival and immuno isolation is presented together with an attempt to define required characteristics of membranes.

REASONS FOR DEATH*		%
Coma diabeticum		1.8
Artherosclerosis		76.6
coronary heart disease	53.2	
kidney	8.9	
cerebrvascular	12.8	
gangrene	1.8	
others	1.5	
Cancer		18.8
Infectious Diseases		5.4
Tuberculosis		8.8
others		5.4

*5889 diabetics from 1968 to 1968, Joslin Clinc, Boston, USA

Table 1: Reasons for death in diabetic patients

PROBLEMS OF DIABETES TREATMENT

In Western Europe, Diabetes mellitus is one of the most common diseases with an incidence of 2-4%. There are two main types of the disease: type-1-("juvenile")-diabetes mellitus (approx. 200,000 patients in West-Germany) is characterized by a manifestation mostly in young people and by a requirement of insulin injections from the beginning. Type-2-diabetes mellitus occurs mostly in elderly, obese people. Normally treatment of type-2 starts without exogeneous insulin applications, but with administration of diet (+ endogeneous insulin secretion stimulating drugs such as Glibenclamide). Later on about 15-20% of type-2-patients require insulin treatment as in type-1-diabetes.

Before the introduction of insulin treatment by

ACUTE COMPLICATIONS:
Polydipsia, Polyuria, Pruritus, Infections, Cramps, Reduction of Common Efficiency, Loss of Body Weight
Coma diabeticum

LATE COMPLICATIONS:
Microangiopathy Retinopathy, Nephropathy, Myopathy, Gastropathy, Cutaneous Ulcer
Macroangiopathy Coronary Heart Disease, Myocardial Infarction, Apoplexia, Peripheral Atherosclerosis
Neuropathy "burning feet syndrome", Pain, Crawling

Table 2: Complications in diabetic patients

H. Planck M. Dauner M. Renardy (Eds.)
Medical Textiles for Implantation
© Springer-Verlag Berlin Heidelberg 1990

	before 1922	after 1922
Life Expectancy		
age of patients:		
<10 yrs.	1.3	29.8
10 - 19 yrs.	2.7	26.7
20 - 39 yrs.	4.3	27.2
all groups	4.9 yrs	18.1 yrs
Reasons for Death (in percent)		
Coma diabeticum	63.8	1.8
Vascular Diseases	17.5	76.6

Life expectancy after occurrence of type-1-diabetes and reasons for death (A. Marble 1974)

Table 3: Life expactancy of type-1 ("juvenile") diabetic patients before and after introduction of insulin treatment

Diabetes Incidence and Organ Donors in West-Germany (1988)
200,000 type-1 diabetics
15,000 newly detected cases per yr.
1,773 Kidney-Tx
250 Heart-Tx
250 Liver-Tx
50 Pancreas-Tx
8,500 Accid.Victims
acc.to: KfH, Neu-Isenburg

Table 4: Diabetes morbidity (potential organ recipients) and number of transplantation of different organs in comparison to accident victims as a potential source for organ donors

Banting and Best [1] in 1922, life expectancy for type-1-patients was several months only [2, table 1]. The next important step forward in the history of insulin treatment was the development of intermediar insulins by Hagedorn et al. in 1936 [3].

Today, a 10-years-old diabetic child has a life expectancy which is reduced by 21 years (absence of any other risk factors as nicotin, adipositas, hypertension, alcohol) compared to his healthy friend [2]. Prior to the era of insulin treatment most diabetics died from coma diabeticum, whereas today death is most commonly caused by late complications (table 2) of the disease. In a young diabetic patient, first manifestations of late complications can be expected after 15 years of the disease.

Most commonly diabetic patients die from cardiovascular diseases (table 3). Comstam [4], Pirart [5] and others have demonstrated that the occurence of complications is related to the quality of diabetes treatment: The aim of diabetes treatment is a normalization of glucose and insulin metabolism.

First experiments to transplant insulin producing tissue were performed only with marginal success by von Minkowski in 1892 [6]. In 1966 Kelly et al.[7] performed an allogeneic pancreas transplantation that resulted in a normoglycemia for 1 week. Up to now, nearly 1400 patients were pancreas-transplanted. Unfortunately, this therapy succeeds sufficiently only in patients who suffer already from severe late complications especially nephropathy and who obtain kidney plus pancreas transplants [8].

In 1972 Ballinger et al. [9] made use of the unique anatomic situation that 1 % of a pancreas consists of small pieces of endocrine tissue and is surrounded by 99%

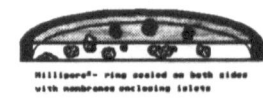

Millipore®- ring sealed on both sides with membrane enclosing islets

A-V Loop with bundles of hollow fibers surrounded by an islet containing compartment

Longitudinal section through a hollow fiber filled with islets of Langerhans

Microencapsulated islets of Langerhans

Figure 1: Different models of a Bioartificial Pancreas

of the pancreas producing exocrine digestion enzymes. Ballinger et al. were able to isolate these "islets of Langerhans" [10]; transplantation of these islets resulted in a normalization of the experimentally induced diabetes mellitus. It was demonstrated [11, 12, 13], that successful transplantation of islets reversed experimental diabetes diabetes in rodents and prevented development or progression of late complications or even partly reversed them.

A problem of human islet and pancreas transplantation in a genetically different donor-recipient system is not only rejection of the transplant but also the recurrence of the autoimmune disease [14,15] which led to type-1-diabetes by destruction of the insulin producing B-cells inside the islets. Untill today these immunological reactions prevent successful clinical application of islet transplantation. Another problem is the discrepancy between the number of diabetics and available donor organs (table 4).

For these reasons, the concept of immuno isolated islet transplantation in semipermeable artificial membranes was developed (table 5).

```
┌─────────────────────────┐
│ BIOARTIFICIAL PANCREAS  │
└─────────────────────────┘
┌────────────────────────────┐
│ Devices based on convection: │
└────────────────────────────┘
  Advantages:    - short reaction time to stimulatory signal
                 - equimolar nutrient + oxygen transport through membranes
  Disadvantages: - coagulation problems

┌───────────────────────────┐
│ Devices based on diffusion: │
└───────────────────────────┘
  Advantages:    - implantation in different compartements
                 - no anticoagulation
  Disadvantages: - nutrient,oxygen + insulin transport only by gradient
                   therefore prolonged reaction time
                 - low islet survival rates (depending on type)
                 - possible autoinhibition by insulin itself
```

Table 5: Diffusion and convection in Bioartificial Pancreas

MILLIPORE-CHAMBER (fig.1)

First experiments with islets inside a diffusion chamber were reported 1970 by Strautz [16]. During the following years different models of a Bioartificial Endocrine Pancreas (fig.1) based on diffussion or/and convection were developed.

The first experiments [16, 17, 18] were carried out with a so called Millipore chamber. Flat membranes were glued to the flat sides of a acryl ring. A disadvatage of this model was a small surface with a rather large volume. The reported results with this model were contradictory.

```
┌──────────────────────────────────────────────┐
│ CHARACTERISTICS FOR MEMBRANES FOR USE IN A BAEP │
└──────────────────────────────────────────────┘
- high diffusion rates for:  - glucose
                             - insulin
                             - glucagon
                             - other islet hormones ?
                             - products of islet metabolism (CO2,lactate etc)
                             - oxygen
                             - ionic and neutral amino acids
                             - essential fatty acids ?
                             - albumin and other carriers ?

- biocompatibility to recipient and islets  vascularized surface or no reaction
                                            toxicity to islet survival+function

- long term stability of the membrane's diffusion characteristics
                                            no biodegradibility

- simple handling to design different models and to avoid superinfection

- cut-off (and material?)which enables immuno isolation
```

MACROENCAPSULATION (fig.1)

Table 6: Reqiured characteristics of membranes for use in a Bioartificial Pancreas: A task for membran engineering

Archer et al.[19] were one of the first to use hollow fibers as a diffusion chamber. This model has a rather small volume and a larger surface. Long term transplantation success was reported where adult human [20]and neonatal mouse and hamster islets [19], human and syrian hamster insuloma tissue [20] were used. The results were not as good, where rat islet [21,22] or RIN-cells (rat insuloma cell line)[20] were used. A normalization of the diabetic state for 1 year with prevention of late complication has been reported [23]using the hollow fiber model.

MICROENCAPSULATION (fig.1)

In 1980 Lim et al. [24] published another diffusion model: single rat islets were encapsulated in small Algine-Polylysine Alginate microcapsules. They reported a reversal of the diabetic state in rats for at least 10 months [25]. Others could reproduce these results particularily if using rat [26] or porcine [27] islets.

BIOARTIFICIAL PANCREAS CONNECTED TO BLOOD CIRCULATION (fig.1)

The main disadvantage of all extravascular diffusion models is the delay of insulin response to a glucose stimulus.

The basic reasons are: A diffusion gradient must be built up before a diffusion of glucose and insulin can take place. This buildind up process results in a delayed release of insulin, depending on gradient, diffusion distance, chemical composition, charges, membrane's microstructure etc.. If a bioartificial pancreas is connected to the blood circulation like a dialysis-cartidge [28,29] certain convection forces develop and accelerate the transport of the glucose signal and insulin response. Several authors have demonstrated a complete normalization of glucose metabolism within serveral hours in using diabetic rats [29] or dogs [30]. Reach et al. [31] optimized the circulation connected model by increasing convection forces in their bioartificial pancreas. Nevertheless, the main disadvantage of all circulation models is neccessity for continous anticoagulation.

IMMUNO ISOLATION BY IMMUNO SEPARATING MEMBRANES

Membranes have to protect transplanted islets from cellular (T-lymphocytes, macrophages) and humoral specific (antibody mediated) and unspecific (IL-1 and TNF-toxicity to insulin producing B-cells) immunological destruction. To avoid cellular and specific humoral toxicity, membranes with a cut-off about 50 KD were used in most cases. Where encapsulated islets were transplanted intraperitoneally in mice with an autoimmune diabetes similar to human type-1 diabetes the islets were destroyed neither by rejection nor by the autimmune process [20]. Darqui et al. [32] demonstrated immune isolation (microencapsulated islets) in vitro by a chromium release technique. Tze et al. [29] have shown the importance of an optimized cut-off for the prevention of a penetration of cytotoxic antibodies.

Mandrup-Poulsen et al. [33] recently reported a cytotoxic effect of Interleukin 1 (17.5 kD) on insulin producing B-cells. Encapsulation of islets inside hollow fibers protects the islets from Interleukin 1-action [34] in vitro, even if a hollow fiber with a cut-off of 50 kD is used.

Further immunological experiments must prove immuno isolation of islets by artificial membranes.

BIOCOMPATIBILITY

A major condition for successful transplantation of islets in artificial membranes is a biocompatibility of the implanted membrane to the recipient and to the islet. Systematic studies of Siebers et al. [35] showed histologically that intraperitoneal implantation of hollow fibers provokes a similar foreign body reaction if the same microstructure is exposed to the recipient. Flat surfaces induce an encapsulation of the hollow fiber by connecting tissue due to foreign body reaction. Rough, cavernous surfaces, however, induce a particular ingrowth of capillaries and tissue into the surface structure. A capillarization of the membrane's surface may reduce diffusion distance and improve the function of encapsulated tissue.

Preformed membranes such as hollow fibers allow the investigation of other basic questions concerning a bioartificial pancreas:

Theodorou et al. [36] and Zekorn et al. [37] investigated in vitro conditions for diffusion of insulin through different membranes (table 5). They found remarkable differences depending mainly on the chemical structure, less on the membrane's cut-off. Recently, it was demonstrated [38,39] that an in vitro protein-coating occurs on various membranes. The result is considerate in an improvement of the insulin diffusion characteristics.

FUTURE ASPECTS

The concept of immuno isolated islet transplantation offers the possibility to transplant genetically different islet tissue (porcine islets to humans, etc.). Recently, first transplantation experiments in man were reported [40]. Nevertheless, prior to a clinical application a variety of basic questions have to be answered (table 6) in cooperation between biology, membrane engineering and medicine.

REFERENCES

1. Banting FG, Best CH (1922) J. Lab. Clin. Invest. 7:251
2. Marble A (1974) Horm. Metab. Res. Suppl. 4:153
3. Hagedorn HC, Norman-Jensen B, Krakup NB, Wodstrup I (1936) J. Amer. Med. Ass. 106:177
4. Constam GR (1977) Med. Klin. 72:695
5. Lauvaux JP, Pirart J (1974) Diabetologia 10:383
6. von Minkowski O (1892) Klin. Wochenschr. 29:90
7. Kelly WD, Lillehei RC, Merkel FK, Idezuki Y, Goetz FC (1967) Surgery 61:827

8. Sutherland DER, Moudry KC, Fryd DS (1989) Diabetes 38:46

9. Ballinger WF, Lacy PE (1972) Surgery 72:175

10. Langerhans P (1869) Inaugural-Dissertation, Med. Fakultät der Friedrich Wilhelm-Universität, Lange, Berlin

11. Federlin K, Bretzel RG, Schmidtchen U (1976) Horm. Metab. Res.8:404

12. Hoffmann L, Mandel TE, Carter WM, Koulmanda M, Martin FIR, Campbell DG, McMillan N (1983) Metabolism 32:451

13. Orloff MJ, Macedo A, Greenleaf GE, Girard B (1988) Transplantation 45:307

14. Woehrle M, Markman JF, Silvers WK, Barker CF, Naji A (1986) Surgery 100:334

15. Sibley RK, Sutherland DER, Goetz F, Michael AF (1985) Lab. Invest. 53:132

16. Strautz RL (1970) Diabetologia 6:306

17. Gates RJ, Lazarus NR (1977) Lancet 17:1257

18. Helmke K, Freitag F, Schneider R, Laube H, Badenhop T, Federlin K (1982) in: Islet-Pancreas-Transplantation. K Federlin, E Pfeiffer, S Raptis (eds.) Thieme-Stratton, New York:45

19. Archer J, Kaye R, Mutter G (1980) J. Surg. Res. 28:77

20. Altman JJ (1988) J. Diabetic Complications 2(2):68

21. Zekorn T, Filip L, Mauer K, Doppl W, Bretzel RG, Federlin K (1989) Diabetes 38. Suppl. 1:298

22. Zekorn T, Siebers U, Filip L, Mauer K, Schmitt U, Bretzel RG, Federlin K (1989) Transpl. Proc. 21,1:2748

23. Altman JJ, Houlbert D, Callard P, McMillan P, Soloman BA, Galletti P (1986) Diabetes 35:625

24. Lim F, Sun A (1980) Science 210:908

25. Sun AM, O'Shea GM, Goosen MFA (1984) Appl. Biochem. Biotech. 10:87

26. Reach G, Darqui S, Chicheportiche D (1989) in"Workshop on Methods in Islet Transplantation Research" Bad Nauheim/Giessen June 22-24. accepted for publication in Horm. Metab. Res.

27. Calafiore R, Calcinaro F, Basta G, Falorni A, Brunetti P (1989) in"Workshop on Methods in Islet Transplantation Research" Bad Nauheim/Giessen June 22-24, accepted for publication in Horm. Metab. Res.

28. Chick WL, Like AA, Lauris V (1975) Science 187:847

29. Tze WJ, Wong FC, Chen LM, O Young S (1976) Nature 264:466

30. Tze WJ, Tai J, Wong FC, Davis HR (1980) Diabetologia 19:541

31. Reach G, Jaffrin MY, Desjeux JF (1984) Diabetes 33:752

32. Darqui S, Reach G (1985) Diabetologia 28:776

33. Mandrup-Poulsen T, Bendtzen K, Nielsen JH, Bendixen G, Nerup J (1985) Allergy 40:424

34. Zekorn T (1989), unpublished data

35. Siebers U, Zekorn T, Sturm R, Bretzel RG, Planck H, Renardy M, Trauter J, Zschocke P, Federlin K (1989) in 3rd. International ITV Conference on Biomaterials. Medical Textiles for Implantation. Stuttgart, June 14-16, accepted for publication, Springer-Verlag Heidelberg 1990

36. Theodorou NA, Howell SL (1979) Transplantation 27:350

37. Zekorn T, Komp U, Bretzel RG, Federlin K (1987) Horm. Metab. Res. 19:87

38. Renardy M, Zschocke P, Planck H. Trauter J, Zekorn T, Siebers U, Federlin K (1989) in 3rd. International ITV-Conference on Biomaterials. Medical Textiles for Implantation. Stuttgart, June 14-16, accepted for publication, Springer-Verlag, Heidelberg 1990

39. Zekorn T, Siebers U, Schmitt U, Bretzel RG, Planck H, Renardy M, Trauter J, Zschocke P, Federlin K (1989) in 3rd. International ITV-Conference on Biomaterials. Medical Textiles for Implantation. Stuttgart, June 14-16, accepted for publication. Springer-Verlag, Heidelberg 1990

40. Wu ZG, Shi ZQ, Lu ZH, Yang H, Shi FY, Zheng XR, Sun A (1989) 35th Annual Meeting of the ASAIO, Dallas (USA)

IMPROVEMENT OF INSULIN DIFFUSION BY PROTEIN COATING
OF MEMBRANES FOR USE IN A BIOARTIFICIAL PANCREAS

T. Zekorn, U. Siebers, R.G. Bretzel, M. Renardy,[*] H. Planck[*] and K. Federlin

Medizinische Klinik III und Poliklinik der Justus Liebig-Universität, Rodthohl 6,
D-6300 Giessen and (*)Institut für Textil- und Verfahrenstechnik, D-7000 Denkendorf (West-Germany)

ABSTRACT:

Encapsulation of islets of Langerhans in artificial membranes may solve
immunological problems of islet allo- and xenotransplantation in diabetes
mellitus. With this respect insulin diffusion characteristics of membranes have
to be examined in vitro. In this study, the impact of a preincubation of
membranes for insulin diffusibility was evaluated. Preincubation in 10 % fetal
calf serum (but not protein-free medium) particular of Amicon XM 50 hollow
fibers resulted in a long term improvement of insulin diffusibility in vitro.

INTRODUCTION:

Successful transplantation of islets of Langerhans should be an ideal treatment
of type-1 (insulin dependent) diabetes mellitus. Transplantation of islets
enclosed in artificial, semipermeable membranes might solve problems of
autoimmunity and may prevent islet allo- and xenograft rejection [1,2]. Poor
success rates of transplantation experiments described [3,4] might have been
caused by insufficient diffusion characteristics of the membranes [5] for insulin,
oxygene etc.. In vitro diffusion kinetics of insulin for distinct membranes were
described by our group and others [6,7]. In these experiments we studied the
influence of protein coating on in vitro insulin diffusibility of artificial,
semipermeable membranes.

MATERIALS AND METHODS:

Islets of Langerhans were isolated from about 260 grs. weighting
male LEWIS/Han rats (Zentralinstitut für Versuchstierzucht GmbH,
Hannover[Germany]) by a Collagenase-digestion and hand-picking technique
after in vivo arterial perfusion-distension with neutral red solution [8].

After intensified washing XM 50 hollow fibers (Amicon Corporation, Danvers,
MA. [USA], 1 cm length, 1 mm i.d., cut-off approx. 50,000 Daltons,
Polyvinylpolyacrylic copolymere) were preincubated (free-floating tissue culture)

H. Planck M. Dauner M. Renardy (Eds.)
Medical Textiles for Implantation
© Springer-Verlag Berlin Heidelberg 1990

either in 5 ml of RPMI 1640 culture medium (Gibco Europe GmbH, Karlsruhe, Germany) with or without 10 % heat inactivated fetal calf serum (FCS, Gibco GmbH) at 37 C (95 % humidity, 5 % CO_2, 95% air) for two or 6 days. Antibiotics were added to the medium (Penicillin 100 U/ml, Streptomycin 100 ug/ml, Amphotericin B 25 ug/ml, Gentamycin 50 ug/ml). The medium was changed every alternate day.

Each hollow fiber was seeded with 100 freshly isolated islets and sealed on both sides. In control experiments non-preincubated XM 50 hollow fibers were seeded with the same amount of islets.

For insulin diffusion tests, the devices were equilibrated by incubation for 45 mins. at baseline conditions in 1.5 ml Krebs- Ringer-buffer (KRB) containing 2.75 mmol/l glucose. Afterwards they were exposed to a glucose challenge of 16.5 mmol/l (KRB, pH 7.4, 1 % bovine albumin [RIA-grade BSA, SIGMA, Heidelberg, Germany], 37 C shaking waterbath). 100ul-Batches of incubation medium were taken at 15, 30, 45, 90 and 135 mins., immediately frozen and stored at -20 C. Insulin content was determined by radio immunoassay (Fa. Serono, Freiburg, West-Germany) with rat insulin as standard (NOVO, Copen-hagen, Danmark).

In another series of experiments XM 50 hollow fibers were preincubated by the method described above up to 14 days in RPMI 1640 + 10 % FCS. After seeding with 100 freshly isolated islets glucose challenge was performed as described.

STATISTICAL ANALYSIS:

All experiments were performed with duplicate samples. The level of significance compared to controls was regarded as $p = 0.05$. Statistical analysis was performed by two-way analysis of variance with repeated measures.

RESULTS:

There were no significant differences between the different groups (fig. 1) up to a stimulatory challenge of 30 mins.. From the 45. min. there were significant differences between HF incubated in FCS for 2 or 6 days compared to non-pretreated HF. The same was seen in comparison to HF preincubated for 2 or 6 days in RPMI without FCS. No statistically significant differences were found between HF preincubated in RPMI+FCS for 2 versus 6 days. After the 90th minute 2-days-RPMI+FCS-preincubated HF showed a significantly higher insulin diffusibility compared to HF incubated in RPMI without FCS for 2 or 6 days.

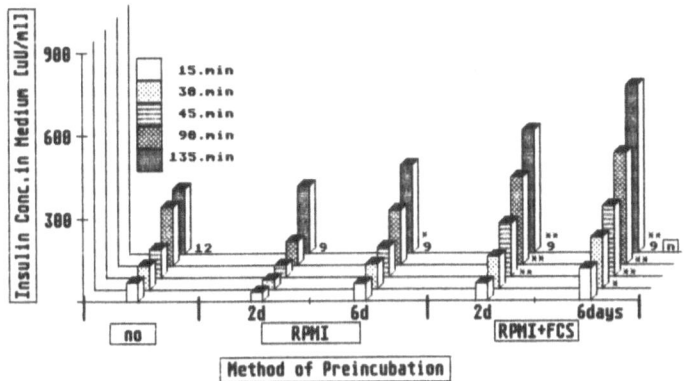

Fig. 1: Glucose (16.5 mmol/l) stimulated insulin release of freshly isolated
islets enclosed in hollow fibres preincubated in medium RPMI 1640
or RPMI 1640 + 10 % FCS for different days.
100 islets per test, values given as mean; significances *: p<0.05;
**: p<0.02; n = number of experiments:

In the second series of experiments using hollow fibers preincubated in RPMI
with 10% fetal calf serum for prolonged periods up to 14 days no statistically
significant difference of the glucose stimulated insulin response was found at
any time (fig. 2). But, in all experiments using pretreated HF the insulin
response differed significantly after the 30. min. compared to non-pretreated HF.

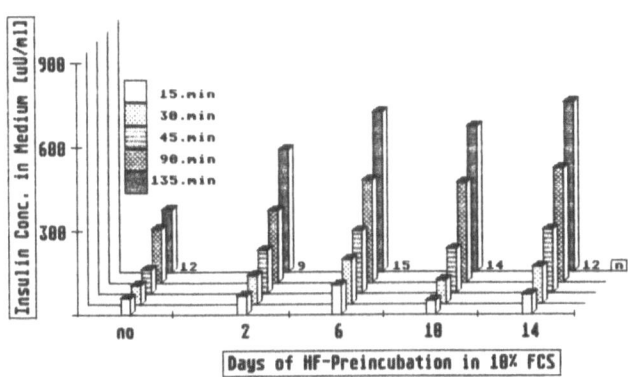

Fig. 2: Glucose (16.5 mmol/l) stimulated insulin release of freshly isolated
islets enclosed in hollow fibres (HF) preincubated in RPMI 1640
+ 10 % FCS for a prolonged period up to 14 days.
Values given as mean; significance after 30 mins. in all experiments
compared to non-treated HF : p < 0.05; after 45 mins. p < 0.02.
n = number of experiments.

DISCUSSION:

In order to examine the effect of protein-coating on the insulin diffusibility of semipermeable membranes Amicon XM 50 hollow fibers were incubated for several days in medium with or without serum proteins.

In our study we demonstrated in a static diffusion model an adequate insulin release of encapsulated islets to a glucose challenge. The amount of insulin was dependend on the type of pretreatment of the XM 50 membrane used in a hollow fiber device. Deconditioning alone or preincubation in serum free medium resulted in a poor insulin release compared to preincubation in the same medium but containing 10% serum. From the data presented we speculate that a protein coating of the membrane's surface is responsible for the improvement of the insulin diffusion characteristics shown (at least for 14 days).

Since the studies of Baier [9] it is well known that artificial surfaces are coated by serum proteins after contact with circulating blood within a very short time. This coating may change the behaviour of membranes (hemocompatibility, etc). Wizemann [10] described an adsorption of insulin by a polysulfone membrane. Although the membrane's cut-off was approx. tenfold higher than the mol. weight of insulin he could not observe any penetration of the peptid hormone. Our group [5,6] found remarkable differences in insulin diffusibility of different dialysis or ultrafiltration membranes tested.

In further studies proteins bound to membranes should be identified by specific methods. Preliminary results from studies with different membranes (polyetheretherketone- and polysulfone-HF) showed an improvement of insulin diffusion by pre treatment with serum proteins (data not shown).

A coating of artificial surfaces takes place within several moments [9]. However, the aim of our study was to test long term stability of diffusion characteristics of HF in vitro. This could be demonstrated in culture medium containing 10% Serum up to 2 weeks.

From our results we conclude that protein coating has a beneficial effect on insulin diffusibility of artificial semi permeable membranes to be used in a bioartificial pancreas. Further studies have to prove direct evidence of protein coating including the impact on the diffusibility of other substances as substrates, glucagon, somatostatin, oxygene in vitro and its importance for transplantation experiments.

ACKNOWLEDGEMENTS:

The skilled technical assistance of Iris Hollerith, Barbara Preiss and Uta Röhm is gratefully acknowledged. This study was supported by a grant of the German minister for research and technology (FKZ 07024806).

REFERENCES:

1. Darquy S, Reach G (1985) Diabetologia 28: 776
2. Altman JJ, Penfornis A, Boillot J, Maletti M (1988) Trans. Am. Soc. Artif. Intern. Organs 34: 247
3. Theodorou NA, Vrbova H, Tyhurst M, Howell SL (1980) Diabetologia 19: 313
4. Helmke K , Freitag F, Schneider R , Laube H , Badenhop T, Federlin K (1982) Horm. Metab. Res. Suppl. 12: 45
5. Zekorn T, Siebers U, Filip L, Mauer K, Schmitt U, Bretzel RG, Federlin K (1989) Transplant. Proc. 21,1: 2748
6. Zekorn T, Komp U, Bretzel RG, Federlin K (1987) Horm. Metabol. Res. 19,2: 87
7. Theodorou NA, Howell SL (1979) Transplantation 27,5: 350
8. Bretzel RG (1984) Pflaum-Verlag, München, 18
9. Baier RE (1977) New York Academy of Sciences, New York: 17
10. Wizemann V, Velcovsky HG, Bleyl H, Brüning S, Schütterle G (1985) Contr. Nephrol. 46: 61

SELECTED DISCUSSION REMARKS:

Von Recum: Are there only advantages of the protein coating or could you also imagine disadvantages?

Zekorn: These results are very preliminary, but it is possible, that other proteins - may be antibodies, too - have a better chance to diffuse through the membrane after protein coating and so immunoisolation may be affected.

POLYMERIC MEMBRANES FOR USE IN A BIOARTIFICIAL
DIFFUSION PANCREAS: INSULIN DIFFUSION IN VITRO

Renardy, M., Zschocke, P., Planck, H., Trauter, J.,

Zekorn, T.*, Siebers, U.*, Federlin, K.*

Institut für Textil- und Verfahrenstechnik, D-7306 Denkendorf

* Medizinische Klinik III und Poliklinik der Universität, D-6300 Giessen

SUMMARY

Ultrafiltration membranes with a molecular cut-off > 67 kD were developed by phase inversion technique from polysulfone (PSU), polyetheretherketone (PEEK) and polyvinylidenefluoride (PVDF) for use in a biohybrid pancreas. Unexpectedly the membranes showed a very low insulin diffusion in vitro. A protein coating of the membranes by preincubation in cell culture medium with 10% fetal calf serum for 6 days resulted in a significant improvement of the insulin diffusion.

INTRODUCTION

It is known, that the treatment of diabetics with exogenous insulin therapy is not able to prevent the secondary complications of the disease. These complications can only be prevented by a more precise blood glucose regulation which requires a more precise insulin delivery. Today it is widely accepted that the transplantation of islets of Langerhans is the optimal treatment of insulin dependent diabetes (1). To protect islets from the immune rejection of the host, they should be included in semipermeable membranes.

METHODS

Ultrafiltration membranes were developed by phase inversion technique from polysulfone (PSU), polyetheretherketone (PEEK) and polyvinylidenefluoride (PVDF). The membranes had a molecular cut-off > 67 kD.

The insulin diffusion was measured using a test chamber made from stainless steel (fig. 1). Two membranes with a membrane area of 1.23 cm^2 each were fixed in the test chamber, creating a chamber volume of 200 µl. After filling the chamber with 200 µl of Tris-buffer pH 7.4 with 0.25% bovine serum albumin (BSA) and 0.1% sodium-azide, containing 61.6 mU insulin, the test

H. Planck M. Dauner M. Renardy (Eds.)
Medical Textiles for Implantation
© Springer-Verlag Berlin Heidelberg 1990

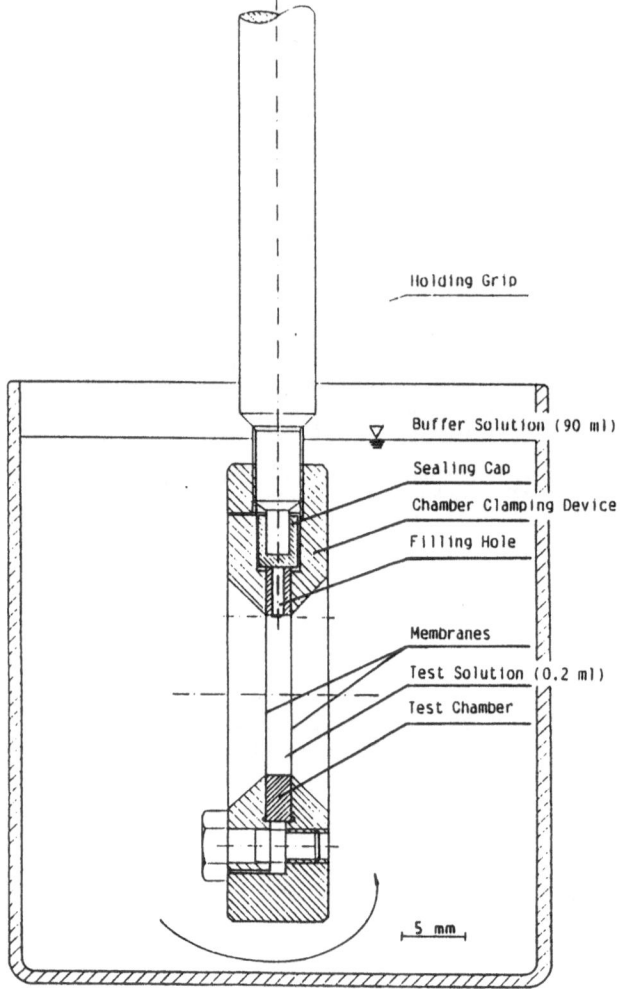

Fig.: 1: Test chamber for the evaluation of membrane permeability

Test conditions:

Membrane area: 2 x 1.23 cm^2

Chamber volume: 200 μl

Buffer volume: 90 ml

Test solution: Tris-buffer pH 7.4 + 0.25% BSA + 0.1% sodium-azide

Insulin assay: Radioimmunoassay

chamber was fixed in a beaker containing 90 ml of insulin-free Tris-buffer (pH 7.4 + 0.25% BSA + 0.1% sodium-azide). Insulin diffusion was measured over 24 hours at 37°C under agitation. Under these conditions hexameric insulin with a molecular weight of 35 kD is present in the solution (2).

The results of insulin diffusion after different time intervalls are expressed in %, with 100% meaning complete distribution of the insulin in the whole volume (682.9 μU/ml).

Insulin diffusion was measured with untreated membranes and with membranes preincubated in cell culture medium with 10% fetal calf serum for 6 days.

RESULTS

Unexpectedly the untreated membranes showed a very low insulin diffusion in vitro. After 3 hours no insulin diffusion was detectable with the PSU-membrane (fig. 2) and the PEEK-membrane (fig. 3), with the PVDF-membrane 26% of the insulin had permeated within 3 hours (fig. 4). After 24 hours the results were: 19% for the PSU-membrane, 39% for the PEEK-membrane and 57% for the PVDF-membrane. The insulin that had not permeated could not be found inside the membrane compartment.

The preincubation of the membranes in cell culture medium with 10% fetal calf serum for 6 days resulted in a significant increase of the insulin diffusion rates. With preincubated

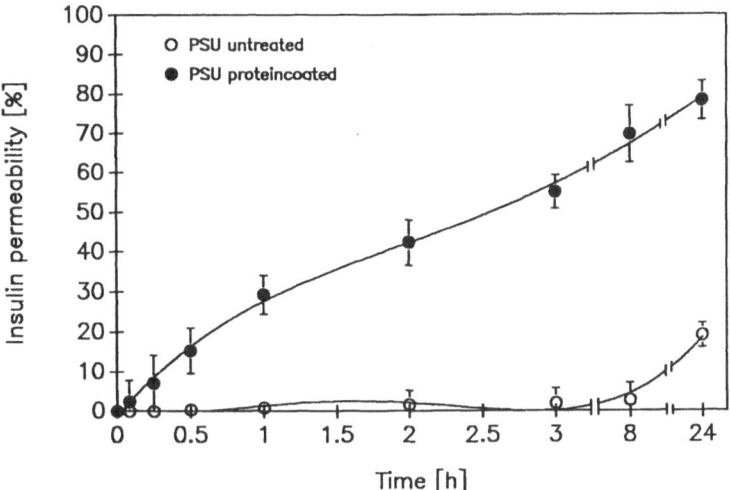

Fig. 2: Insulin permeability of a PSU-membrane, molecular cut-off > 67 kD, untreated and after preincubation in 10% fetal calf serum for 6 days

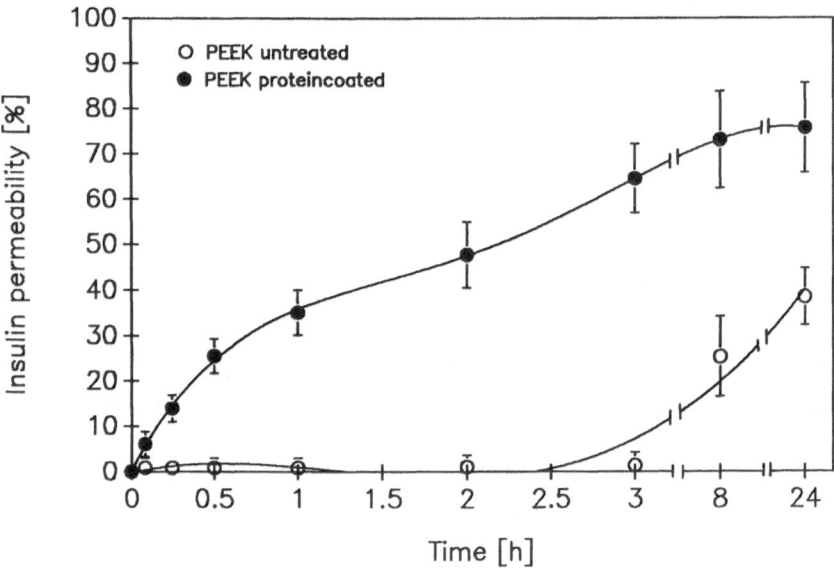

Fig. 3: Insulin permeability of a PEEK-membrane, molecular cut-off > 67 kD, untreated and after preincubation in 10% fetal calf serum for 6 days

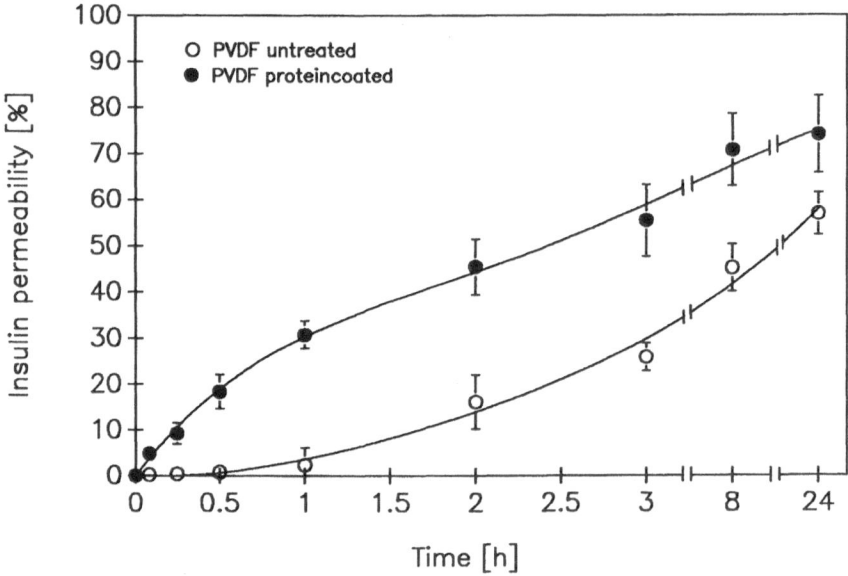

Fig. 4: Insulin permeability of a PVDF-membrane , molecular cut-off > 67 kD, untreated and after preincubation in 10% fetal calf serum for 6 days

membranes the lag-phase at the beginning of the experiment seen with untreated PSU- and PEEK-membranes was missing. Within 3 hours 55% of the insulin had permeated through preincubated PSU-membranes, 65% through preincubated PEEK-membranes and 59% through preincubated PVDF-membranes. After 24 hours 76-80% of the insulin had permeated through all preincubated membranes (fig. 2-4).

DISCUSSION

With untreated membranes a difference in insulin diffusion can be seen depending on the polymeric nature of the membrane (PVDF > PEEK > PSU, fig. 5), which cannot be explained by differences in the molecular cut-off.

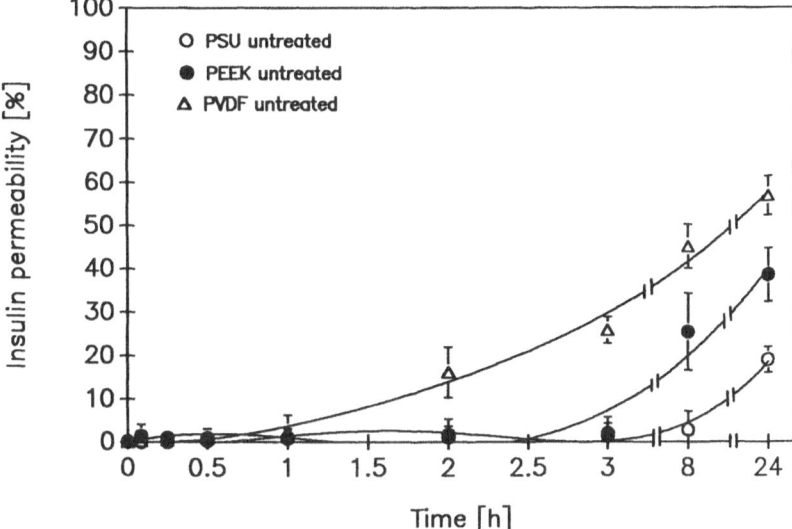

Fig. 5: Insulin permeability of untreated membranes of PSU, PEEK and PVDF, molecular cut-off > 67 kD

The fact that less than 100% of the insulin is detectable suggests that there is a considerable adsorption of insulin to the membrane polymer, especially during the first 3 hours of the experiment. The amount of adsorbed insulin depends on the nature of the polymer.

A preincubation of the membranes in 10% fetal calf serum seems to diminish the amount of adsorbed insulin. This fact can be explained by the hypothesis that all protein binding sites on the polymer surface are occupied by serum proteins before insulin contacts the surface. Furthermore this protein coating seems to "mask" the surface of the polymers leading to comparable diffusion rates for these three polymers after preincubation (fig. 6).

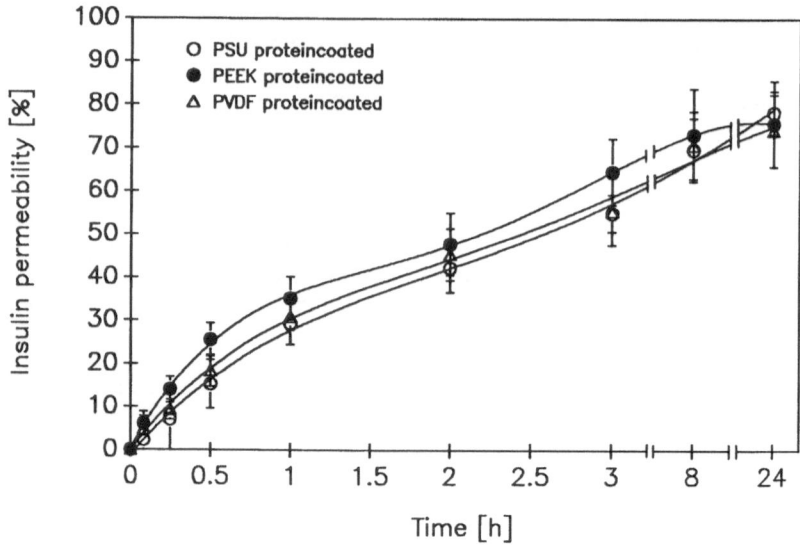

Fig. 6: Insulin permeability of a PSU- PEEK- and PVDF-membrane , molecular cut-off > 67 kD, after preincubation in 10% fetal calf serum for 6 days

Further work has to be done to evaluate the influence of polymer chemistry and polymer charge on the adsorption of proteins and the consequence on insulin permeability.

This work was supported by the German Ministry of Research and Technology.

REFERENCES

[1] Zekorn, T. et al.
Bioartificial Pancreas: The Use of Different Hollow Fibers as a Diffusion Chamber
Transplantation Proceedings, Feb. 1989
[2] Obermeier, R.: Personal Communication

MORPHOLOGICAL AND FUNCTIONAL STUDIES
ON IMPLANTED DIFFUSION MEMBRANES

U.Siebers. T.Zekorn. RG.Bretzel. R.Sturm. H.Planck*.
M.Renardy*, J.Trauter*. P.Zschocke* and K.Federlin.

Med.Klinik III der Universität. Rodthohl 6. D-6300 Giessen
und ITV Denkendorf. Körschtalstr.26, D-7306 Denkendorf (*)

INTRODUCTION:

Immunoisolated transplantation of islets of Langerhans inside preformed
membranes may be a suitable therapy of insulin dependent diabetes mellitus
(1.2). In vitro. morphological and functional survival of islets inside the device
similar to that of free floating islets has been demonstrated previously (3).
Nevertheless. for a transplantation therapy, behaviour of the artificial membranes
in vivo must be investigated. In this study foreign body reaction towards
different membranes was examined histologically and insulin diffusibility of
explanted membranes was determined.

MATERIALS AND METHODS:

1. Different types of membranes (Amicon XM 50 HF, Grace Inc.. USA; PEEK
10a. PEEK 4.1, Polysulfone HF. ITV Denkendorf) were implanted into the
peritoneal cavity of non-diabetic male Lewis rats (Savo GmbH. Kißlegg. FRG).

H. Planck M. Dauner M. Renardy (Eds.)
Medical Textiles for Implantation
© Springer-Verlag Berlin Heidelberg 1990

After 21 days the membranes were retrieved, fixed, paraffin embedded, cut into 5 um thin sections and stained H.E. or Masson-Goldner for histological examination. The following parameters were examined:

a) influence of different polymers

b) influence of different surface structures

2. Amicon XM 50 (Grace Inc. USA) and PEEK 10A capillary membranes (ITV Denkendorf) were sterilized by gamma irradiation and implanted intraperitoneally into non-diabetic Lewis rats. After 21 days the membranes were retrieved and seeded with 100 islets each. Islets had been isolated from Lewis rats by a technique described before (4) using neutral red injection, collagenase digestion and purification by hand- picking. Then, a glucose stimulation test was performed (16,7 mmol/l glucose in Krebs-Ringer-Buffer, pH 7.4). Insulin release after 90 minutes of stimulation was determined by radioimmuno assay (RIA-gnost, Behring, Marburg, FRG) using a rat insulin standard (Novo Ind., Denmark). The amount of insulin was compared to that released by islets in not pretreated capillary membranes. Islets in free floating culture served as controls.

The degree of foreign body reaction was judged by histological examination of 12 representative sections per membrane. Grading was done according to the following scheme:

$$1 = \text{no reaction}$$
$$2 = \text{monolayer}$$
$$3 = \text{« 50 um}$$
$$4 = 50 - 100 \text{ um}$$
$$5 = \text{» 100 um}$$

Statistical analysis was done by Mann-Whitney-Wilcoxon test.

RESULTS:

Polysulfone, Polyetheretherketone and Polyvinylacrylic-Copolymere all provoked only a mild foreign body reaction. The degree of connective tissue formation did not differ significantly between the three polymers tested.

Different structures of the exposed membrane-surface caused a completely different type of foreign body reaction. A smooth, homogenous, "closed" surface structure without lightmicroscopically visible pores led to the formation of a fibrous capsule , whereas an "open" cavernous surface structure allowed the inspreading of blood vessels and connective tissue . This phenomenon is illustrated in figure 1.

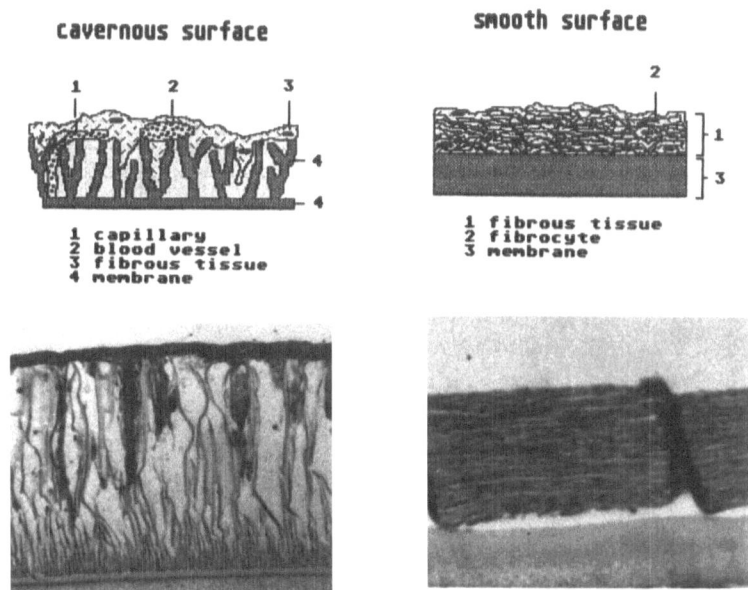

Fig.1.: Morphological reaction towards an implant a) with a smooth surface (on the right) or b) with a cavernous surface (on the left). Schematic graph (above) and corresponding histological section (below).

The results of the ex vivo stimulation are shown in figure 2. Insulin release of explanted XM 50 capillary membranes was about 50% of controls in free floating culture (552 ± 60 uU/ml vs. 1089 ± 152 uU/ml). Untreated XM 50 membranes only released 20.7% of controls (225 ± 56 vs. 1089 ± 152 uU/ml ; p« 0.002). Explanted PEEK 10a membranes even showed an insulin release of 64.2% of controls (699 ± 173 uU/ml vs. 1089 ± 152 uU/ml). This difference between free floating islets and islets inside explanted PEEK 10a membranes is not statistically significant. Untreated PEEK 10a membranes released 43.9% of controls (478 ± 84 uU/ml vs. 1089 ± 152 uU/ml ; p « 0.05).

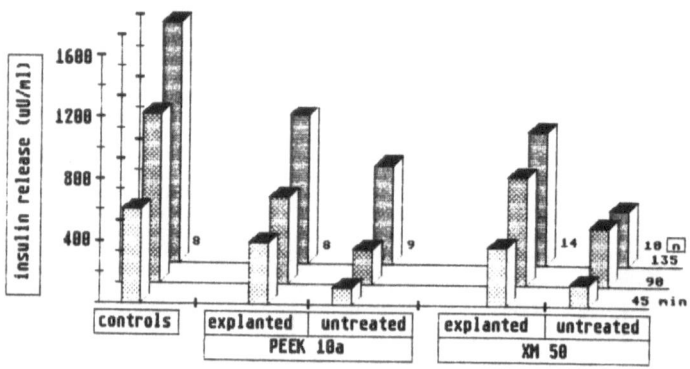

Fig.2.: Insulin release of explanted XM50 and PEEK 10a membranes in comparison to not pretreated ones and free islet controls. Mean value of n experiments.

The degree of foreign body reaction is shown in the following table (table 1):

Table 1: Histomorphometrical results after intraperitoneal implantation into rats for 21 days. Grading according to the scheme described above. Values are given as percentage of the sections judged.

GRADE	1	2	3	4	5
XM 50	–	50.0%	44.2%	5.8%	–
PEEK 10a	–	54.0%	38.0%	8.0%	–

DISCUSSION:

Foreign body reaction towards an artificial implant is an overall problem in biotechnology. Any artificial material that remains inside the organism induces a response of the host, but the extent and the type of this reaction varies according to the material itself, the host's reactivity (5), the implantation site and other factors.

In our experiments intensity and type of connective tissue reaction towards implanted artificial membranes mostly depended on the structure of the surface. Membranes consisting of the same polymer but showing a different ultrastructural design of the exposed surface produced quite a different histomorphological reaction, whereas membranes with similar surface structure showed a similar reaction of the tissue whatever the chemical composition was.

These findings are supported by others (6,7) reporting a milder tissue reaction towards implants with a textured, porous surface compared with chemically identical smooth surface controls. For the use in a bioartificial pancreas, a capillarisation of the outer compartements of the membrane may be an important factor minimizing the diffusion distance.

2. Intense foreign body reaction towards the artificial membranes used was supposed to be responsible for the failure of early transplantation experiments with the millipore chamber (8,9,10), perhaps because of a sticking of the pores. Even a coating of the membrane with collagenase (11) did not prevent fibrosis.

Unexpectedly, both membranes tested in this study, showed better insulin diffusibility after an implantation period of 21 days compared with untreated ones. Previous in vitro studies have already proven the beneficial effect of a protein coating on insulin diffusibility (12). Probably some kind of protein coating takes place in vivo as well. As the average thickness of the tissue layer formed upon the membrane's outer surface was less than 50 um in our experiments, this additional diffusion distance did not inhibit significantly the diffusibility for insulin .
Changes in the kinetics of insulin permeation are likely but could not be detected in our incubation system. We could not establish a correlation between the thickness of the fibrous tissue layer and insulin diffusibility in this study.

Ex vivo experiments of Theodorou et al. (13) support our results. Comparing insulin diffusibility of explanted and fresh membranes with J-125 labeled insulin they found no significant difference. Moreover, they did not find a correlation between the degree of tissue reaction and transplantation success. The relatively bad diffusion characteristics of untreated membranes may be due to an unspecific binding of insulin molecules to the membrane's polymer. If these binding sites are saturated by a previous protein coating, either in vitro or in vivo, insulin permeation is significantly improved. A sticking of the pores by tissue formation was not observed in our experiments.

Acknowledgements: The authors want to thank Iris Hollerith and Uta Röhm for skillful technical assistance. This study was supported by a grant of the German Ministry for Research and Technology (FKZ 07024806).

References :

1 Altman JJ, Penfornis A, Boillot J, Maletti M (1988) ASAIO Transactions 34: 247

2 Altman JJ (1988) J Diab Compl 2: 68

3 Zekorn T, Siebers U, Bretzel RG, Federlin K (1989) Int J Artif Organs (in press)

4 Lacy PE, Kostianovsky M (1967) Diabetes 16: 35

5 Imai Y, Watanabe A (1988) Artif Organs 12: 451

6 Whalen R (1988) ASAIO Transactions 34: 887

7 Taylor SR, Gibbons DF (1983) J Biomed Mater Res 17:205

8 Maratos E, Taub RN, Bramis J (1976) Mount Sinai J Med 43: 415

9 Kemp CB, Scharp DW, Knight MJ (1973) Surg Forum 24: 297

10 Buschard K (1975) Horm Metab Res 7: 441

11 Jolley WB, Hinshaw DB, Call TW, Alvord LS (1977) Transpl Proc 9: 363

12 Zekorn T, Siebers U, Bretzel RG, Planck H, Renardy M, Trauter J, Zschocke P, Federlin K (1989) Proc 3rd Int ITV Conference, Springer, Heidelberg

13 Theodorou NA, Vrobva H, Tyhurst M, Howell SL (1980) Diabetologia 18: 313

FIRST EXPERIENCES ON EPITHELIALISATION
OF AN EXPANDED PTFE TRACHEAL PROSTHESIS

H.J. Gerhardt[1], O. Kaschke[1], M. Wenzel[2], F. Böhm[3] and K. Haake[1]

Humboldt- Universität Berlin, Charité, Schumannstr. 20/21
DDR-1040 Berlin

1) HNO-Klinik, 2) Anatomie, 3) Dermatologie

Abstract: A critical evaluation of the literature concerning experiments on the development of an alloplastic tracheal prosthesis shows that infection and progressive obstruction of the prosthesis are among others important causes for failures.
For our experiments we used at first GORE-TEX Ⓡ vascular prostheses, reinforced and FEP ringed, internodal distance 30 μ. The aim of our experiments was to achieve complete ingrowth of fibrous tissue to prevent infection of the hollow spaces of the prosthesis later on and then to line the inner surface with respiratory epithelium to prevent obstruction of the prosthesis by an excess of granulation tissue.

The first step was the implantation of the prosthesis into the sternocleidomastoid muscle under sterile conditions in dogs. Scanning electron microscopic examination revealed, that the outer reinforcing layer (0,004 mm, 10 μ pores) seems to be an obstacle for rapid ingrowth of fibrous tissue into the wall of the prosthesis (0,67 mm). Therefore we tried to create perforations trough the wall but without obstructing the adjacent pores. We succeded by using an excimer laser (wave-length 193 nm). The laser performed channels had a diameter of 120-200 μ. A pretreatment of Gore-Tex reinforced prostheses in this way seems to provide better conditions for rapid ingrowth of fibrous tissue together with vessels.
The next series of experiments was performed with Gore-Tex vascular prosthesis, 30 μ internodal distance, FEP ringed, but without a reinforcing outer layer. One week after implantation the luminal surface of the prosthesis showed a thin homogenous fibrin layer. After two weeks this layer was already replaced by fibroblasts. Those prostheses then have been seeded one week after implantation

---

Supported by the Main Topic Research Artificial Organs and Biomaterials M 28 of the GDR

H. Planck M. Dauner M. Renardy (Eds.)
Medical Textiles for Implantation
© Springer-Verlag Berlin Heidelberg 1990

with autologous isolated respiratory cells. Three weeks later the
inner surface of the prosthesis was completely lined with cells
showing many microvilli and various stages of differentiation.
Therefore this schedule seems to be a reasonable one for replacement of tracheal defects with prostheses prepared in this way.

Surgical Concept for Application of an Alloplastic Trachea

Our surgical concept comprises the following steps:

1. The porous material of the prosthesis at first is implanted under
   sterile conditions into a suitable muscle to achieve complete
   ingrowth of connective tissue into the wall filling all the pores
   to prevent infection of hollow spaces later on.
   Of special interest was to find out the period necessary for the
   connective tissue to reach the inner surface.

2. To prevent progressive obstruction of the prosthesis a rapid and
   complete epithelialisation of the inner surface is of utmost
   importance. We tried to realize this by seeding autologous respiratory epithelial cells into the prosthesis when the ingrowing
   connective tissue has provided a suitable layer for adhesion of
   the epithelial cells on the inner surface.

3. The third step then is to transplant the in this way prepared
   prosthesis into the tracheal defect together with the surrounding
   muscle and its preserved vascular supply like a flap.

Ingrowth of Connective Tissue into Gore-Tex ® Vascular Prostheses

Material and methods: For our experiments we used at first Gore-Tex ®
vascular prostheses, reinforced and FEP ringed, internodal distance
30 μ. The experiments have been performed on dogs (ASTI). After pre-
medication with atropine and morphine general anaesthesia was
achieved by intravenous injection of 400 mg Pentobarbital, as well
as 5 milligrams Polamivet ® (Levomethadonhydrochloride, Hoechst,
München) and 0,2 mg Combelen ® (Methylparaben, Bayer, Leverkusen)
i.m. Spontaneous breathing was maitained. From a midline incision
both of the sternocleidomastoid muscles were exposed, incised and
then 3 cm long pieces of a Gore-Tex ® vascular prosthesis, closed
at both ends by Silastic ® sheeting 500-1, glued on with Silastic ®
Medical Adhesive Silicon Type A, Dow Corning, implanted into the

belly of the muscle. The wound then has been closed in layers.
Postoperatively 0,6 mill. units of Retacillin Ⓡ (Jenapharm) has
been administered.

In a first series the implants were harvested after 2,4 and 6 weeks,
immediately opened and rinsed with sterile physiological saline and
then prepared for scanning electron microscopic examination.

Results: Two weeks after implantation a thin fibrinous membrane at
the inner surface of the prosthesis could be seen with adherent
blood cells. After 4 weeks on the now more compact fibrin layer
fibroblasts in reticular structure could be detected. Macroscopically
cordlike excess granulations on the inner surface were to be seen.
After 6 weeks the layer of fibroblasts was markedly thickened. Below
a thin layer of fibrin was interposed toward the PTFE structures. The
lumen of the prosthesis in part was already obliterated. Fig. 1-3.
Scanning electron microscopic examination of the reinforced Gore-Tex
vascular prosthesis alone aroused suspicion, that the outer reinforcing layer (4 µ wide, pores of about 10 µ) could be an obstacle for
rapid ingrowth of fibrous tissue into the 0,67 mm wide wall of the
prosthesis. Above all the necessary ingrowth of arterioles with a
greater diameter is completely hampered. Fig. 4-5.

Tissue Ingrowth after Additional Laser Perforation

One way to overcome this problem could be to create additional pores
of 100-200 µ through the wall. Preconditions for a positive result
are the avoidance of changes of the biophysical properties of the
material by this procedure, for instance by the effect of extreme
high temperature, and on the other hand to keep open the pores adjacent to the channels.

Method: We succeeded by using an excimer laser with the wave-length
of 193 nm (Lasertype LPX 205 i CC, Fa. Lambda-Physik, Göttingen). The
density of the energy required was 6 J/cm^2. 200 Pulses per perforation were necessary.

By this procedure we achieved sharply defined perforating channels of
100-200 µ in diameter without a marginal wall or signs of deformation
of the adjacent pores. Fig. 6.

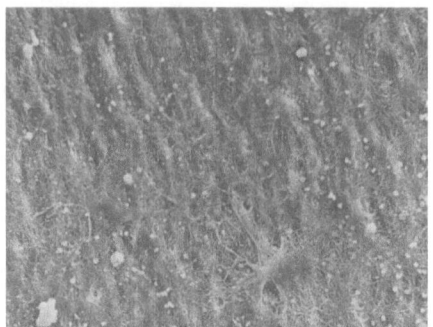

Fig. 1: Gore–Tex R Vascular Prosthesis, reinforced, 30 μ internodal
distance, 2 weeks after implantation.
200 x.

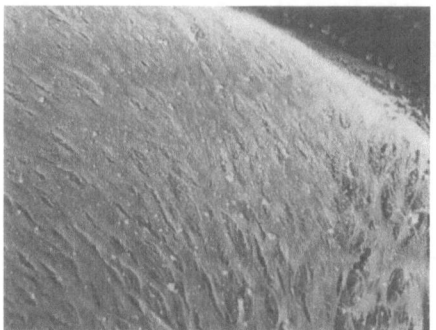

Fig. 2: Same material, four weeks
after implantation. 160 x.

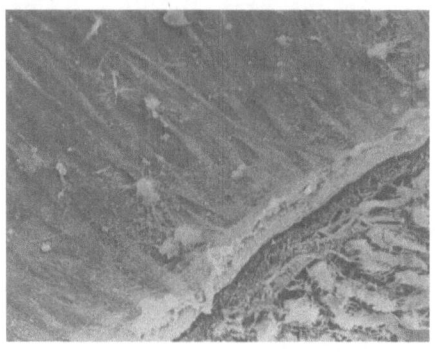

Fig. 3: Same material, six weeks
after implantation. 370 x.

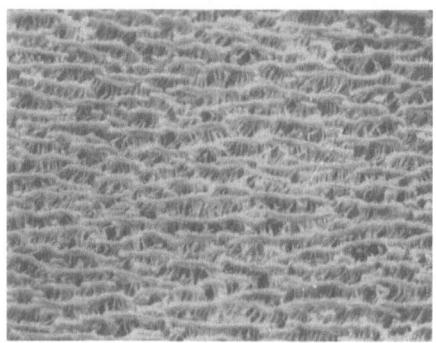

Fig. 4: Same material. Inner surface,
open pores. 190 x.

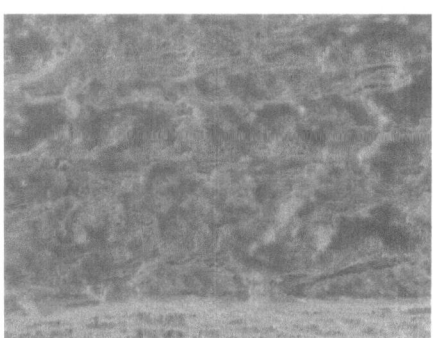

Fig. 5: Same material. Outer surface
reinforced, pores less than 10 μ in
diameter. 190 x.

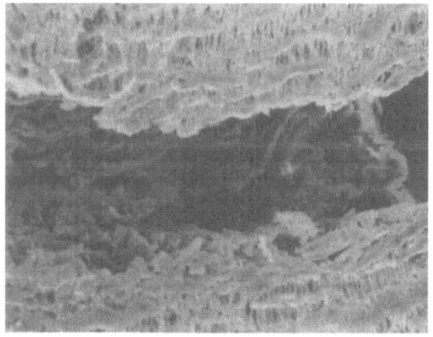

Fig. 6: Laser performed channel
through the wall of the prosthesis.
190 x.

In a second series of experiments on dogs we tried and to define the effect of those channels on the course of tissue ingrowth into the Gore-Tex prosthesis. The implantation was done in the same way as described before.

Results: After one week already a homogenous fibrin layer on the inner surface could be seen. At the outer opening of a laser channel ingrowing cell structures were visible. On the other hand on the outer reinforced surface surrounding the opening of the channel only a few adherent cell structures could be detected. Fig. 7-8.

After two weeks cell structures in the laser channels were clearly visible as well as ingrowth of connective tissue in lateral direction towards the adjacent parts of the wall of the prosthesis. On the inner surface a thin reticular layer of fibroblasts could be seen. Fig. 9-10.

After three weeks the inner surface was lined by compact and cord-like connective tissue. At the luminal openings of the laser channels clearly could be seen that connective tissue growing through the channels spreaded over the adjacent inner surface. Fig. 11-12.

Summarizing these experiences we would like to state that to perforate systematically a reinforced Gore-Tex prosthesis leads to more rapid ingrowth of connective tissue and perhaps a better blood supply later on of the epithelium on the inner surface by larger arterioles having the chance to develop at the channel sites.

Tissue Ingrowth into Non-Reinforced Vascular Prostheses

Another way to improve the ingrowth of connective tissue into the wall of the prosthesis, of course, is to use non-reinforced ones. We are grateful to the Gore Company for having provided us recently with this material, too. This enabled us to perform further experiments with this material in a third series.

Results: As expected the ingrowth of connective tissue into a non-reinforced prosthesis occured much faster.

After one week already a thin but homogenous fibrin layer was established at the inner surface, probably appropriate to adhesion of seeded epithelial cells. Fig. 13.

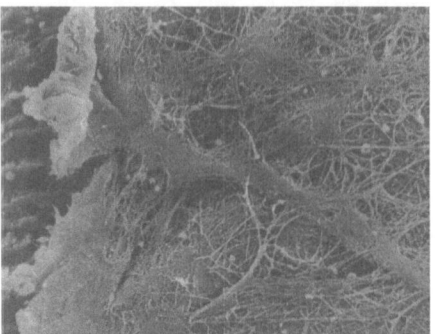

Fig. 7: Laser perforated Gore-Tex R
Vascular Prosthesis, 30 μ internodal
distance, one week after implantation. 490 x.

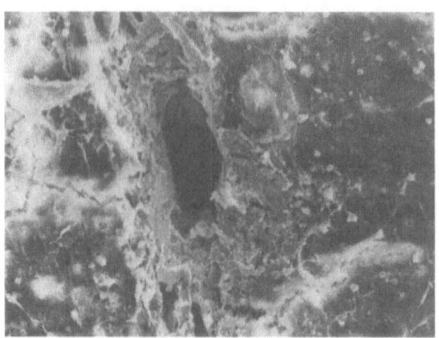

Fig. 8: Same material. Outer channel
opening with ingrowing cell structures. 205 x.

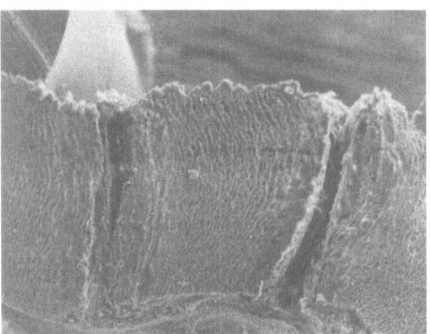

Fig. 9: Same material. Ingrowth of
connective tissue from the channels
in lateral direction two weeks after
implantation. 60 x.

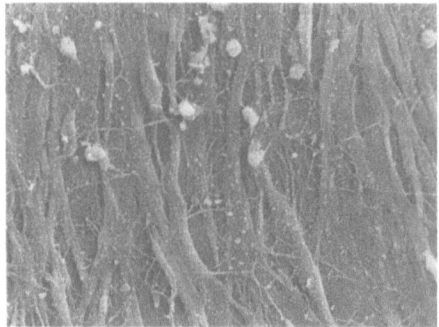

Fig. 10: Same material. Two weeks
after implantation thin reticular
layer of fibroblasts on inner surface. 560 x.

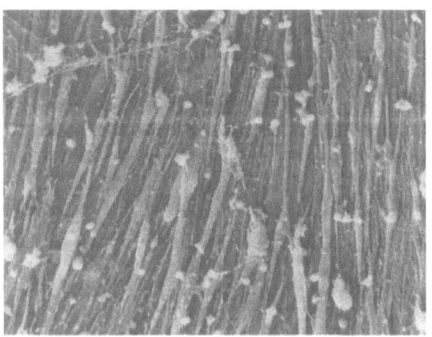

Fig. 11: Same material. Three week
after implantation connective tissue
on the inner surface. 500 x.

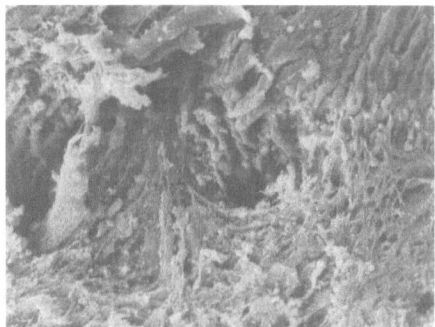

Fig. 12: Same material. Connective
tissue growing through a channel,
spreading over adjacent inner surface. 400 x.

After 2 weeks a layer of fibroblasts was clearly visible. Macro-scopically the luminal surface was completely covered with tissue showing in part cordlike thickening. Fig. 14.
These findings led to the decision to continue our experiments on epithelialisation of the prosthesis only with this material.

Experiments on Epithelialisation of the Prosthesis with Isolated Respiratory Epithelial Cells

To prevent threatening progressive obstruction of the prosthesis by excess of granulations a rapid and complete epithelialisation of the inner surface must be achieved. In order to overcome this problem we developed a method enabling us to seed autologous respi-ratory epithelial cells into a previously into muscle implanted prosthesis.

Material and method: About 2 cm^2 of tracheal mucosa were taken from the dog involved in this experiment and then the trachea has been closed again by Prolene R sutures.
Blood has been rinsed out with Dulbecco-buffer (NaCl 8,0 g, Na_2HPO_4 1,2 g, K_2HPO_4 0,2 g, aqua ad 1000 ml, pH 7,2) and addi-tional antibiotics (50 IE/ml penicilline G, 50 µg/ml streptomy-cine and 25 IE/ml nystatine). Then the specimen was immediately incubated in the same fresh solution at 4° C for one hour.

After keeping it in 0,02 % EDTA-buffer for 10 min at room tempe-rature, the specimen was kept in 0,25 % trypsin-buffer solution (1:250, Difco, Chicago, Ill.) at a temperature of 4° C for 16 hours.

After that time the mucosal cells could be carefully separated from the submucosa by forceps and scalpel and isolated by pipetting. The vitality of the isolated cells amounted to more then 95 % in the trypan exclusion test.
The schedule of this experimental series was as follows:
1. Implantation of non-reinforced and by silastic sheeting at the ends closed pieces, 2 cm long, of Gore-Tex vascular prosthesis, FEP ringed, 30 µ internodal distance, in muscle.
2. At the sixth day taking of tracheal mucosa from the same ani-mal. Processing of the epithelial cells as described above.

280

Fig. 13: Gore-Tex R Vascular Prosthesis, non-reinforced, 30 μ internodal distance, one week after implantation. 250 x.

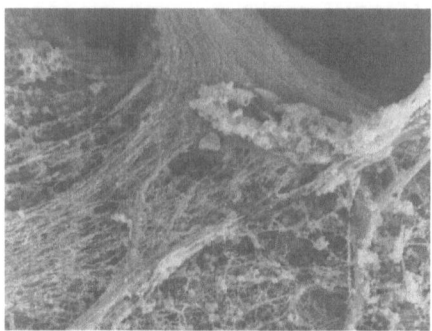

Fig. 14: Same material. Two weeks after implantation. Layer of fibroblasts, cordlike thickening. 345 x.

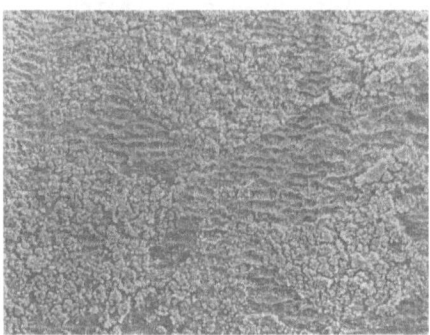

Fig. 15: Same material. One week after implantation seeded with autologous epithelial cells. Examination 30 days later. 90 x.

Fig. 16: Vascular prosthesis, non-reinforced, with Silastic R sheeting 500-5 as inlay and Silastic tube for injection.

Fig. 17: Vascular prosthesis as shown in Fig. 16. Examination 20 days after cell seeding. Nearly complete epithelialisation. 400 x.

Fig. 18: Section of Fig. 17, higher magnification (2100 x). Cells with many microvilli. No completely differentiated ciliated cells.

3. At the seventh day injection of 0,5 ml suspension containing
   isolated epithelial cells from about 0,25 cm^2 tracheal mucosa
   into the prosthesis.
4. Harvesting of the samples after 20 and 30 days, respectively,
   for examination.

Results: Samples from this series taken 30 days after seeding of
epithelial cells showed in low magnification (90 x) in part a good
but not complete epithelialisation of the inner surface. In well
epithelialised areas the cells were in various stages of differentiation, most of them displaying microvilli as described by Bottema,
too. We could not find completely differentiated ciliated cells.
Bare areas showed a dense layer of fibroblasts but no signs of excess
granulations and therefore no tendency to obstruction of the prosthesis. Fig. 15.

To achieve a more even seeding of the injected cells we modified the
method as follows:
An adequate piece of Silastic Ⓡ sheeting 500-5 was inserted into the
prosthesis so that it lays against the luminal wall without putting
much pressure on it. Additionally a thin Silastic Ⓡ tube, inner diameter 0,3 mm, outer diameter 0,65 mm, was inserted through the wall
so that its ends lay inside the cleft between the inner prosthetic
wall and the Silastic sheeting. At the outer side these tubes were
fixed by Silastic adhesive against one of the reinforcing rings.Fig.16.
The implants then were closed as usual. We believe that by injecting
the epithelial cell suspension into the cleft between the luminal
wall and the Silastic sheeting after cutting the tube outside to open
it, more cells in a more even way by the effect of cohesion will
adhere to the fibrin layer inside of the prosthesis.
The results of an additional series seems to confirm this. The
samples harvested 20 days after cell seeding showed nearly complete
epithelialisation of the luminal wall. Some small clefts in the epithelial layer were only artefacts that followed from the processing
procedure for scanning electron microscopy. Fig. 17.
The cells again showed many microvilli and various stages of differentiation, but no completely differentiated ciliated cells. Fig. 18.

<u>Conclusions:</u> According to a formulated wound management model (2) establishing of a tissue layer at the luminal surface of a prosthesis starts with sweeping in of fibrin and fibroblasts from the implantation layer through the pores into the lumen. Fibrin forms a more and more dense network with enclosed or adhering granulocytes, macrophages and erythrocytes. The reinforcing layer of the Gore-Tex vascular prosthesis obviously delays the formation of this fibrin layer which is more rapidly established in laser perforated or non-reinforced material. A similar difference in the course of formation of a fibroblast layer could be detected.

Fibroblasts are able to survive even under poor conditions. The vitality of higher differentiated epithelial cells depend on a sufficient vascular supply. A laser perforated wall may provide better preconditions for ingrowth of larger arterioles feeding the capillary network below the epithelial layer later on. A reasonable schedule for preparing an epithelialised prosthesis seems to be cell seeding one week after implantation into muscle. Another three weeks later the prosthesis seems to be ready for replacing tracheal defects. We have to test this by continuing our experiments.

<u>References:</u>

1 Bottema JR, Havinga P, Kalicharan D, Karrenbeld A, Wildevuur ChHR, Molenaar I (1982) In: Bottema JR (ed) The microporous tracheal prosthesis, chapter 8. Proefschrif, Groningen
2 Hunt TK, Winkle jr Hv (1979) In: Hunt TK, Dunphy JE (ed) Fundamentals of wound management. Normal repair 8-21. Appleton-Cent-Crofts, New York

SELECTED DISCUSSION REMARKS:

von Recum: At some other applications we saw this morning problems of dead space and accumulation of serum fluid. I wonder whether that interferes with your seeding techniques on your intra-muscular implants?

Gerhardt: Of course, when we explanted these specimens, they were filled up with serum. But obviously it didn't interfere with the proliferation and spreading of the cells. So I believe, before transplantation the serum can be removed.

Turnquist: I wonder how much time you have to act in case of severe damage of the trachea? It sounds, that this preparation would require a certain time.

Gerhardt: The transplantation of trachea will not be a matter of emergency. When you have a patient after an accident tracheatomy is the first step and can overcome the immediate dangerous problems.

Nakamura: What is about the air leakage through the small pores produced by the laser?

Gerhardt: During 2 or 3 weeks i.m. implantation these pores are closed completely by connective tissue. So when you transplant this trachea after 4 weeks, it is completely closed.

Mestres: We have a poor vascular supply in the trachea area. Obviously the poor healing and later on infection of the suture lines cause very serious problems.

Gerhardt: Concerning the vascular supply we intend to transplant the prepared trachea together with the surrounding muscle, preserving the feeding arteries and venes to the sterno-cleidomastoid muscle. With regard to the suture line, perhaps it would be good to have a little cuff for the anastomotic site.

PROLIFERATION AND DIFFERENTIATION OF
HUMAN CILIATED EPITHELIA ON PTFE-PROTHESES

F. Böhm[1], M. Wenzel[2] and H.J. Gerhardt[3]

From the Department of [1]Dermatology, [2]Anatomy and
[3]Otolaryngology, Humboldt University, Berlin, GDR

For several years there have been efforts to develop an artificial trachea. Our efforts aimed at the development of a suitable
transplant for a complex substitution of destroyed tracheal segments. It was our aim to line expanded polytetrafluorethylene
(ePTFE) with human mucosal cells. Therefore we made experiments
on the growth of human mucosal cells on ePTFE. We were especially
interested in the proliferation kinetics of the cells on a relatively coarse net structure and in the differentiation ability of
mucosal cells under such conditions.

We succeeded in former experiments in the separation, cultivation and differentiation of human ciliated epithelia out of
ethmoidal cells (1). Cell layers could be kept in the open in-vitro
system in collagen-coated plastic dishes up to three weeks. A monolayer was already visible after 3-4 days of cultivation, first
regressive changes were visible on 7th to 10th day. The cells had
an extremely high proliferation and differentiation speed. We were
not able to influence differention and further increase of proliferation by reduction of Ca-concentration in the growth medium
(3). In this respect, there is a substantial difference between
mucosal cells and epidermal cells (2). We report here on the
essentially modified growth of human ciliated epithelia on ePTFE
and demonstrate the results by means of scanning electron
microscopy.

Supported by the Main Topic Research Artificial Organs and
Biomaterials M28 of the GDR

H. Planck M. Dauner M. Renardy (Eds.)
Medical Textiles for Implantation
© Springer-Verlag Berlin Heidelberg 1990

MATERIAL AND METHODS

The procedure for cell preparation, cultivation and scanning electron microscopy was already described (1). We chose the following procedures:

Cell preparation: Mucous membrane was taken from human ethmoidal sinuses during transethmoidal hypophysis surgery (approx. 2-3 sqcm). Blood was rinsed out with Dulbecco buffer (NaCl 8.0 g, Na_2HPO_4 1.2 g, K_2HPO_4 0.2 g aqua ad 1000 ml, pH 7.2) with an addition of antibiotics (50 IE/ml penicillin G, 50/ug/ml streptomycin and 25 IE/ml nystatin). The specimen was immediately incubated in the same fresh solution at 4°C for one hour. After placing the specimen in 0.02% EDTA buffer for 10 min. at room temperature, it was kept in 0.25% trypsin buffer solution (1:250, Difco, Chicago, III) at a temperature of 4°C for 16 h. Subsequently the mucosal cells could be carefully separated from the mucosa with forceps and scalpel and then isolated by pipetting. The vitality of the isolated cells amounted to more than 95% in the trypan blue exclusion test.

Cultivation: The mucosal cells were kept in quantities of approx. $1x10^6$ ciliated cells/ml in the following growth medium: Parker 199, pH 7.0 (Staatliches Institut für Immunpräparate und Nährmedien, Berlin), with hydrocortisone (0.4 mg/ml), L-glutamine (2 mM), HEPES (20 mM), human transferrin (5 mg/ml, Sigma Chemical Co., St. Louis, Mo.) insulin (20/ug/ml), human epidermal growth factor (10^{-9}M, Firma Serva, Heidelberg, BRD), penicillin G (50 IE/ml), streptomycin (50/ug/ml) nystatin (25 IE/ml), fetal calf serume 20% and sodium hydrogene carbonate 0.35 g/l. The cell suspension (1 ml each) was seeded at a temperature of 34°C on specimens of ePTFE (4-5 sqmm, internodal distances about 30/um) (Fa. W.L. Gore & Ass., Inc.) prepared in advance with collagen type IV (20/ug/ml, Sigma Chemical Co., St. Louis, Mo.) and cultivated at a temperature of 34°C in a CO_2-incubator (5% CO_2 in air, 95% water saturation). The medium was changed one day after seeding and then every 3-4 days.

Scanning electron microscopy: The medium was removed for the
SEM and the cell preparations were washed twice with physiological sodium chloride. After that they were fixed with 2.5%
glutardialdehyde in 0.15 M cacodylate buffer, pH 7.4 at 22°C for
2 h. Then they were again fixed, this time with 1% osmium
tetroxide in the same buffer at 22°C for one hour. Dehydration
followed by an increasing sequence of alcohol and critical point
dehydration. The last step was platinium sputtering of the
preparations. The preparations were observed and photographed
in a Tesla scanning electron microscope Type BS 3000 (CSSR).

RESULTS

SEM-pictures made one day after seeding show an initial
adhesion of the cells on the ePTFE surface, the rate being
reduced to approx. 70% compared with cultures in plastic dishes.
Above all, the morphology of the adhering cells is essentially
different (always compared with cells on plane conditioned surfaces, s. (1)). Obviously, the mucosal cells adhere ball-shaped
to the threads of the plastic skeleton and keep this shape
even for days (Fig. 1 and 2). In the beginning, the surface of
the cells appears relatively smooth but increasingly becomes
filiform after hours and days.

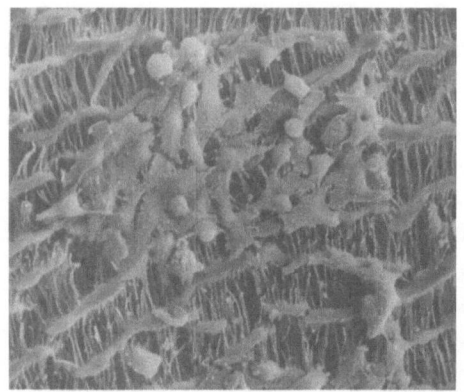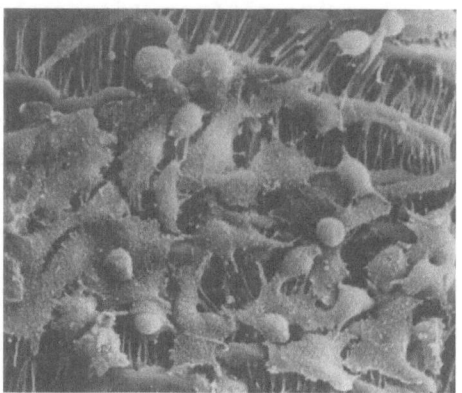

Fig. 1 and 2. Mucosal cells from ethmoidal sinuses one day after seeding on ePTFE (SEM, x 810, x 1260).

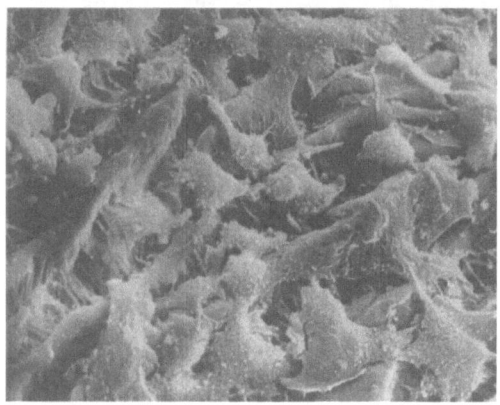

Fig.3. 3-day-old mucosal cell cultures (SEM, x 1800).

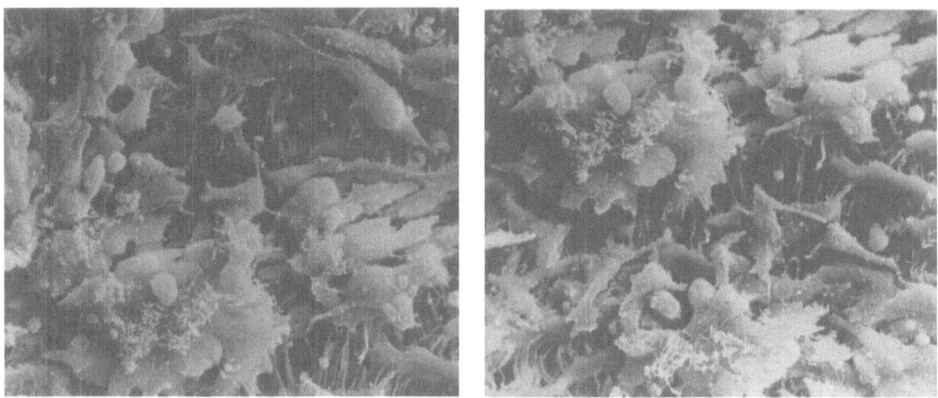

Fig. 4 and 5. Ciliated epithelia of 5-day-old mucosal cell cultures (SEM, x 1240, x 1280).

Fig. 6. Mucosal cells with kinocilia 7 days after seeding on ePTFE.
Amongst them are cells with structures similar to microvilli (SEM,
x 2200).

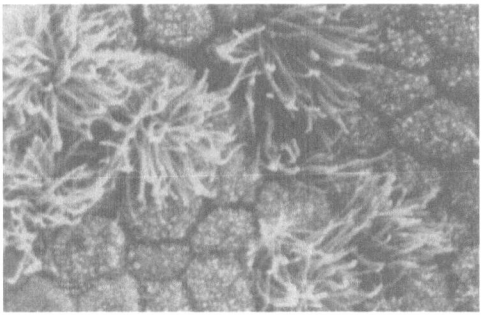

Fig. 7. 5-day-old mucosal cell culture on a Petri dish (SEM, x 5000).

Simultaneously, the cells form slim appendices of cytoplasm like pseudopodia which they push along the plastic threads. Within 24 to 36 h of cultivation all processes of initial spreading appear to be three-dimensional whereas they appear two-dimensional on limited plane areas. Most cells change on the second day of cultivation. After getting in contact with other cells, the cell bodies become bigger and loose their ball-shape. This is the first step to a monolayer. Differentiation processes towards ciliated or goblet cells are not visible within the first three days. The cells form a submonolayer or, in some cases, a monolayer with punctiform vertical growth (Fig. 3).

5 days after seeding of the mucosal cells first kinociliated cell clusters are visible (Fig. 4 and 5). However, the bumpy and irregular character of the culture remains. Many differentiated cells are visible and an increased thick-ening of the cell layer can be observed. The coexistence of ball-shaped and smooth cells, flattened, starlike and ciliated plain cells is maintained up to the 8th to 10th day. Only afterwards ciliated cells, goblet cells and polygonal plain mucosal cells dominate in the culture (Fig. 6). An orderly cell formation as on even culture dishes (Fig. 7) are not visible on ePTFE. The proliferation speed of the cells on ePTFE is reduced to 50 to 70% compared with the standard culture. Differentiation does not seem to be reduced. High prismatic cell layers as typical for a mucosa in vivo were not observed.

Nevertheless proliferation and differentiation are still faster than that of human epidermal cells (4, 5).

In the further process of cultivation mostly cell clusters are formed and not so often larger cell areas.

Even after 14 days of cultivation the synthetic material was not completely lined with cells. Regressive changes as can be observed in standard culture already on the 10th day were not observed on ePTFE till the end of the experiments after 3 weeks.

DISCUSSION

Obviously, the type of the synthetic material and its space structure have essential impact on the proliferation kinetics, the morphology and the differentiation process of a mucosal culture. The cells show completely different reactions within the three-dimensional net of plastic threads than on the plain surface of culture dishes.

The reduced initial adhesion rate results from the fact that many cells drop first into the net structure of the ePTFE. On the other hand adhesion of the cells to the relatively thin plastic threads is more different.

Obviously even surfaces offer better conditions for the spreading of the cells than net structures. In the case of a standard cell culture, the cells are spreading directly to mono- and polylayers. But on ePTFE first a net structure of cells slowly develops followed by area proliferation. Only subsequently the cells differentiate. This seems to be the reason why it takes more time to form a functioning mucous membrane.

With regard to an integrated tracheal prothesis ingrowth of connective tissue into the porous prothetic wall as a first step before seeding of epithelial cells on the inner surface seems to be the most promising way to go on.

SUMMARY

The growth of human ciliated epithelia on expanded polytetra-fluorethylene was examined. Compared with the standard culture, differences are observed in adhesion, spreading and differentiation of the cells. The development of areas of kinociliated layers takes more time. The cells first form net structures which slowly develop to layers. The impact of the results on the development of an integrated tracheal prothesis is being discussed.

REFERENCES

1 Böhm F, Wenzel M, Gerhard HJ. Acta Otolaryngol (Stockh) 107: in press.
2 Böhm F (1986). Dermatol Monatsschr 172:14.
3 Hennigs H, Holbrook KA, Juspa SH. (1983) J invest Dermatol 81:50
4 Marcelo CL, Tomich J. (1983) J invest Dermatol 81:64.
5 Wilkinson DJ. (1987) J invest Dermatol 88:198.

Address:
F. Böhm, Department of Dermatology, Humboldt University, Schumannstraße 20/21, Berlin 1040, G.D.R.

CONTROLLED MICROPOROSITY:
A KEY DESIGN PRINCIPLE FOR ARTIFICIAL SKIN.

Hinrichs, W.L.J.*, Kuit, J.*, Feijen, J.*, Lommen, E.J.C.M.P. **, Wildevuur, Ch.R.H.**

* University of Twente, Department of Chemical Technology, P.O. Box 217, 7500 AE Enschede, The Netherlands.

** University Hospital of Groningen, Department of Experimental Cardiopulmonary Surgery, P.O. Box 30001, 9700 RB Groningen, The Netherlands.

Summary

An ideal artificial skin should meet two opposing requirements. It should be sufficiently occlusive to prevent rapid wound dehydration and bacterial invasion, but it should also be permeable enough to allow passage of wound exudate.

These opposing requirements have been met using microporous polyurethane structures. A porous bilayer artificial skin was fabricated using immersion precipitation techniques. The top layer (thickness: 10-20 µm) contains micropores (pore size: < 0.7 µm) which are interconnected so that fluid drainage can take place. However, the size of the micropores is such that the evaporative water loss is reduced and the penetration of bacteria is prevented. To this top layer, a bottom layer (thickness: 100 µm) with a fibrillar sponge-like structure (pore size: 50-100 µm) is attached. The polymer matrix of this layer also contains interconnecting micropores and forms a continuum with the micropores in the top layer. The design of this layer guarantees a firm adherence of the artificial skin to the wound surface.

In vitro and in vivo tests revealed that the opposing requirements mentioned above are indeed met. Treatment of full thickness skin defects on the backs of guinea pigs with the polyurethane artificial skin showed that the porous structure prevents penetration of bacteria, limits the evaporative water loss but allows an adequate wound drainage. Furthermore, the polyurethane artificial skin firmly adhered to the wound surface. The polyurethane artificial skin did not need regular replacement but could be left on the wound surface until the wound was fully healed.

Future clinical trials will indicate to which wound types the artificial skin can be applied successfully.

H. Planck M. Dauner M. Renardy (Eds.)
Medical Textiles for Implantation
© Springer-Verlag Berlin Heidelberg 1990

Introduction

The healing of skin defects can be impaired when the wound becomes dehydrated and/or infected. Therefore, an often used design for a wound dressing is based on solid structures, for example Opsite[®1,2] and Spandre[®1]. The occlusive nature of solid structures prevents extensive water loss and bacterial penetration. However, in most cases an effective drainage of the wound is not achieved because the transport of fluids is restricted to evaporation of water through the dressing. A poor wound drainage may lead to accumulation of wound exudate under the dressing which can cause tissue maceration, detachment of the dressing from the wound surface and infections. Consequently, these dressing have to be replaced regularly, thereby increasing the risks of infecting the wound[3,4,5]. Therefore, much attention has been paid to the development of new materials and design concepts to increase the water vapour permeability (ie. Omiderm[® 1,6]). However, despite these efforts, in most cases, the passage of water vapour through the dressing does not keep pace with the production of wound exudate when applied on highly exuding wound types, like burns[7].

One approach to solve the problem of wound drainage is the application of macroporous structures or hydrogels, for example Coldex[®1] and Surfasoft[®1]. Some of these types of dressings, like Duoderm[®1], are linked to an occlusive top layer. Fluid build-up beneath the macroporous dressings without the occlusive top layer is prevented because these materials are highly absorbent and fluids are allowed to pass through the dressing. However, bacteria are free to penetrate the wound and also wound dehydration might take place. An effective wound drainage is achieved at the early stages after application of macroporous structures or hydrogels which are supplied with an occlusive top layer. However, when the absorbent layer is saturated, fluid build-up beneath the dressing takes place which leads to leakage of exudate at the edges or blister formation. Like solid structures, these dressings need frequent changing with the additional risk of bacterial contamination.

The aim of this study was to combine the design principles of the types of wound coverings mentioned to create an artificial skin which meets the opposing requirements of being permeable enough to assure an effective wound drainage but being occlusive enough to prevent dehydration and bacterial invasion.

A polyurethane artificial skin consisting of two porous layers was developed. The top layer is supplied with micropores smaller than 0.7 μm whereas the bottom layer has a sponge-like structure containing pores with a size in the range of 50-200 μm.

The fabrication procedures and the physical characteristics of this artificial skin will be presented. Biobrane[®1,8], a commercially available biosynthetic skin substitute was used as reference material throughout this study. We have choosen Biobrane[®] because the design principles of this product show some similarities with our artificial skin. Biobrane[®] is a bilayered membrane which consists of a knitted nylon mesh to which a thin silicon rubber toplayer is mechanically bonded. Collagenous peptides obtained from porcine skin are chemically coupled to both layers. The silicon toplayer is supplied with pinholes to allow wound drainage.

Methods and Materials

Preparation of the artificial skin

The artificial skin was fabricated according to a modified technique described by Gogolewski et al[8,9]. First the macroporous bottom layer was prepared. A 8.6% (w/v) solution of Biomer[®] (Ethicon, Sommerville, N.J., USA), a polyetherurethane urea, in N,N-dimethylacetamide (DMAc) was prepared. Sodium citrate (Janssen, Beerse, Belgium) with a particle size of 63-106 μm (80 g/100ml), obtained by recrystallization from a water/methanol-mixture and fractionated on particle size with Tamson test sieves, was added to this solution under vigorous stirring. The resultant suspension was heated to 50 °C and then degassed. A layer of this suspension was cast on a glass plate using a doctor blade which was adjusted to a height of 350 μm. Then the glass plate was immersed in an ethanol/water-mixture (6:1(v/v)) at room temperature after which the polymer coagulated. After two minutes the coagulated product was exposed to air until nearly dry. Subsequently, the microporous top layer was applied, now using a 11.5% (w/v) solution of Biomer[®] without salt particles. This solution was also heated to 50 °C and degassed before use. A layer of this solution was cast on top of the first layer using the doctor blade with the same height adjustment. The two layers were first immersed into ethanol at room temperature for ten minutes and thereafter into a waterbath also at room temperature for further coagulation, extraction of solvent residues and dissolution of the salt particles. After thouroughly washing with water and ethanol the polyurethane artificial skin was dried and sterilized by ethylene oxide or gamma irradiation.

The microstructure of the artificial skin was examined using a Jeol JSM-35CF scanning electron microscope.

In vitro experiments

Bacteriology

To investigate whether bacteria are able to pass the polyurethane artificial skin or Biobrane®, circular discs of the membranes (n=30) were placed on agar plates. Subsequently, the center of each disc was inoculated with 0.2 ml of an overnight culture of Pseudomonas aeruginosa. After 24 hours of incubation at 37 °C, the agar plates were checked for bacterial contamination using standard procedures[11].

Water vapour permeability

A gravimetric method described in reference 12 was used to determine the water vapour permeability through the polyurethane artificial skin and Biobrane®. Briefly, a glass cup, filled with distilled water, was covered with the material (n=5) and heated to 37 °C using an electrical heater. Water vapour permeability measurements were carried out in an environmental chamber in which the temperature and the relative humidity were kept constant at 27 °C and 45 % respectively. The water vapour permeability was calculated from the water loss determined by weighing the glass cup at different time intervals. These data were compared with the evaporative water loss of an uncovered glass cup.

Drainage capacity.

To determine the drainage capacity of the polyurethane artificial skin and Biobrane®, one end of a hollow glass tube was sealed with the membrane (n=20). Subsequently, the tube was filled with serum. The amount of serum passing through the material was measured as a function of time. During the experiment, the hydrostatic pressure was kept constant by adding serum to the tube. The specific drainage resistance R was calculated using the formula R=P/Q*A, where P represents the hydrostatic pressure (height of the column and the specific gravity expressed in cm H_2O), Q the flow and A the membrane surface area.

In vivo experiments

Implantation

Full thickness skin defects created towards the panniculus carnosus of male guinea pigs (600-700 g) were used as a wound model[11]. Excisions (6x8 cm) were performed under sterile conditions after the animals were anaestisized with Fluothane® and a 2:1 mixture of N_2O/O_2 gas. The skin defects were closed with the polyurethane artificial skin (n=8), Biobrane® (n=8) or left uncovered (n=8).

Water vapour permeability

The animals were housed in small cages to limit their movements. The cages were kept at a constant temperature of 27 °C and a relative humidity of 50 %, while air currents were avoided as much as possible. Water vapour permeability determinations were performed using a measuring probe of a ServoMed® Evaporimeter Ep1B[13] at a distance of 1 cm from the wound surface.

Wound drainage

The drainage capacity of the materials was determined qualitatively by visual wound inspection.

Results and discussion

Preparation of the artificial skin

An artificial skin has to be made from an elastic and flexible material to be applicable at irregular or flexing wound surfaces. Furthermore the material must exhibit good biocompatibility, must be sterilizable, and may neither provoke any toxic, allergic nor antigenic tissue reactions [3,4]. Polyurethanes, which are already used in several biomedical devices such as pacemaker leads, catheters, and the artificial heart [14] do meet these properties. Another consideration for the use polyurethanes in the fabrication of artificial skin are the good handling properties, for example its suitability for the immersion precipitation techniques used in this study[15].

The newly developed polyurethane artificial skin can be described as a highly elastic bilayered membrane with a relatively thick (100 µm) macroporous bottom layer with a high porosity (>80%) to which a relatively thin (10-20 µm) microporous top layer is attached (see figure 1). The bottom layer is fabricated with, mainly spherically shaped, macropores (pore size: 50-100 µm) which are highly interconnected. The high porosity and the high degree of interconnectivity of the macropores result in a fibrillar sponge-like structure. Furthermore in the polymer matrix a second type of pores can be observed. These pores have a size smaller than 5 µm and are also interconnected. In the top layer only one type of pores are present. Also these pores, which have a size of 0.2-0.7 µm and a density of $0.2/\mu m^2$ at the surface, are interconnected and form a continuum with the pores of the bottom layer.

The polyurethane artificial skin is prepared using a two step immersion precipitation procedure. With immersion precipitation[16], a polymer solution is brought into contact with a non-solvent. By the exchange of the solvent by the non-solvent, the mixture is rendered thermodynamically instable. The Gibb's free energy of the system can be lowered by phase-separation at which polymer-rich and polymer-poor domains are formed. The polymer matrix is obtained after solidification of the polymer-rich phase, while the pores originate from the polymer-poor domains. With this technique, the process parameters such as the type of polymer, solvent and non-solvent as well as polymer concentration and temperature determine the final pore structure. With the preparation of the bottom layer an extra porosity inducing element is introduced with the addition of the sodium citrate particles to the polymer solution. During the coagulation of the polymer, by immersing the suspension into an ethanol/water-mixture, the salt particles are entrapped in the polymer matrix. At a later stage of the procedure, pores arise by dissolving the salt particles in the extraction bath. The size and the density of these pores are determined by the size of the salt particles and the polymer/salt-ratio. Hence, the two types of pores originate from two different mechanisms. The presence of the salt particles will have some influence on the diffusional processes during the phase-separation and therefore on the pore structure originated from this process. But, if required, the size and the density of the two types of pores can be adjusted at will more or less independently by changing the process parameters mentioned earlier.

The top layer is prepared using a polymer solution without salt particles. Therefore only one type of pores are present in this layer. These pores are interconnected with the pores of the bottom layer and form channels throughout the whole structure.

The two layers are firmly linked together, which is explained as follows. The bottom

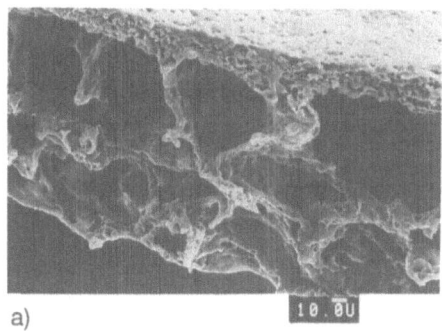

a)

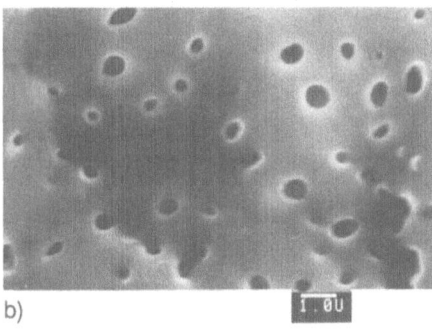

b)

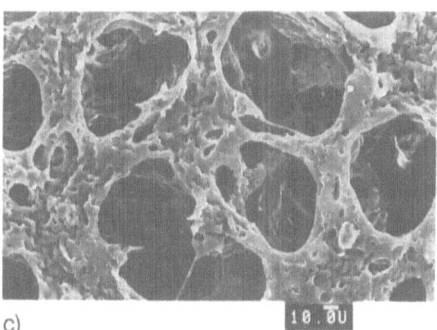

c)

Figure 1. Scanning electron micrographs of the polyurethane artificial skin.

    a) Cross section.

    b) Top layer.

    c) Bottom layer.

layer is exposed rather shortly to the ethanol/water-mixture. Because of this and the fact that DMAc has a low volatility as compared to ethanol/water, there will be some solvent left in the coagulated film after the drying step. As a consequence, when the second layer is cast on top of the bottom layer, the two layers are fused to a certain extent and the polymer of the bottom layer near the boundary will be partly redissolved before coagulation of the polymer in the ethanol bath takes place.

Available artificial skin substitutes need frequent replacement, especially when applied at highly exuding wounds. However, an ideal artificial skin should be designed such that once applied, the dressing can be left on the wound bed until the damaged skin is healed. Among other things, such a material should have occlusive characteristics to prevent bacterial invasion and limit the evaporative water loss, but it should also be permeable enough to allow wound drainage.

By means of in vitro and in vivo experiments we have investigated whether the polyurethane artificial skin does meet these requirements.

In vitro experiments

The in vitro bacteriologic test did show that the microporous top layer of the polyurethane artificial skin can successfully prevent infections. Only in one case out of thirty was bacterial contamination of the agar plates found. Bacterial penetration in this particular case was problably due to the accidental use of an imperfect sample with a pore or pores larger than 0.7 µm on the top layer. In testing Biobrane®, eleven out of thirty discs were found to be permeable to Pseudomonas aeruginosa. This high incidence of bacterial penetration can be ascribed to the pinholes present in the toplayer. Apparently, these pinholes are large enough to allow the passage of bacteria. In nineteen samples, no bacterial penetration had occurred. Most probably, the pinholes are unevenly distributed over the surface. As a consequence, some samples were not provided with pinholes through which the bacteria could migrate.

The in vitro evaporimetry test revealed that the water vapour permeability of the polyurethane artificial skin and Biobrane® was reduced to respectively 23.0 % (89.3 $g \cdot m^{-2} \cdot h^{-1}$) and 10.6 % (41.0 $g \cdot m^{-2} \cdot h^{-1}$) as compared to an open water surface (386.2 $g \cdot m^{-2} \cdot h^{-1}$). From these data, it can be concluded that the polyurethane artificial skin limits the evaporative water loss substantially, although Biobrane® is more effective in this respect. In vitro, the polyurethane artificial skin showed a very high drainage capacity as compared to Biobrane®. The polyurethane artificial skin yielded reproducible values for the specific drainage resistance with a mean of 1.08 $cm \cdot H_2O \cdot ml^{-1} \cdot cm^{-2} \cdot sec$ whereas measurements using Biobrane® were less reproducible, yielding a specific drainage resistance with a mean of 3.54 $cm \cdot H_2O \cdot ml^{-1} \cdot cm^{-2} \cdot sec$. The high drainage capacity of the polyurethane artificial skin can be ascribed to the micropores which form channels in the top layer through which fluids can flow. The high reproducibility of the measurements indicate a homogeneous distribution of the micropores. The pinholes in the Biobrane® top layer also provide a possibility for drainage. However, the density is much lower which results in a higher specific drainage resistance. Furthermore, the low reproducability of the measurements indicate an uneven distribution of these pinholes.

In vivo experiments

The polyurethane artificial skin as wel as Biobrane® were used to close full thickness skin defects created on the backs of guinea pigs. Immediately after application, the evaporative water loss, as compared to the open wound, was limited to 40 % when using the polyurethane artificial skin and 18 % when using Biobrane®. Thus both materials are less effective in reducing the evaporative water loss in vivo than in vitro. However, in time the reduction of the evaporation rate became more pronounced and after 7-8 days of application, the water vapour permeability of normal guinea pig skin was reached. This increase of the evaporative water loss reduction in time and the differences with the in vitro experiments can be explained by the fact that both materials allow wound drainage and that the production of wound exudate reduces in time. The channels formed by the micropores in the polyurethane artificial skin provide capillary forces resulting in the transport of wound exudate to the outer surface of the dressing where it will evaporate more easily. This effect will give rise to a higher evaporative loss than might be expected from the in vitro experiments. Hence, when the production of wound exudate reduces, also the evaporative water loss reduces. Furthermore, when the transport of fluids subsides, the micropores are obstructed by crust formation resulting in increased occlusive characteristics of the artificial skin. Experiments using Biobrane® also showed that the initial evaporative water loss in vivo was increased as compared to in vitro . Furthermore , like the polyurethane artificial skin, Biobrane® showed a decrease in the evaporation rate in time. However, probably due to the poor drainage capacity of this material, these effects are less pronounced. Both materials showed a good initial adherence to the wound bed. After three days of application, the development of small fluid filled pockets between Biobrane® and the wound surface could be observed. These fluid filled pockets became infected and were enlarged in time until the material detached from the wound surface. This occurred between the 5th and the 15th day after application. In contrast, using the polyurethane artificial skin, wound exudate was noted to pass the material, while fluid build-up beneath the dressing and infections could not be observed. Furthermore, the dressing remained fully adherent to the wound bed during the entire time of wound healing (7 weeks) without the need of suturing. Noteworthy was the observation that the polyurethane artificial skin was very firmly adhered to the wound surface. Wound closure using only the microporous top layer resulted in a dramatic decrease of the adherence forces (data not shown). This observation indicates that the macroporous bottom layer makes an important contribution to these forces. The improved adherence of the bilayer artificial skin to the wound surface can be explained by the following considerations: a good initial adherence is achieved by the capillary forces exerted by the micropores in both layers. At a later stage the adherence forces are taken over by the deposition of fibrin. Since the bottom layer has a highly porous structure, there is a large contact surface with many anchoring points available for fibrin deposition. Thereafter the ingrowth of wound tissue will lead to an increased adherence.

In conclusion, the in vitro tests demonstrate that the polyurethane artificial skin is impermeable to bacteria and substantially limit the evaporative water loss but allows the passage of fluids. Treatment of full thickness skin defects with the bilayer structure confirmed that these opposing requirements for an ideal artificial skin are successfully met in an in vivo situation: neither infection, wound dehydration nor accumulation of wound exudate under the polyurethane artificial skin occurred. Furthermore, the dressing remained firmly adhered to the wound bed. Therefore, the polyurethane artificial skin did not need frequent replacement, but could be left on the wound surface until the skin was fully regenerated.

The applicability of this artificial skin to different wound types will be evaluated in future clinical trials.

References

1 Brochures of commercial available wound dressings and skin substitutes: Biobrane®, Coldex®, Duoderm®, Omiderm®, Opsite®, Spandre®, Surfasoft®.

2 May SR (1984) In: Wise DL (ed) Burn wound coverings. Vol 2. CRC Press, Inc. Boca Raton

3 Park GB (1978) Biomat Med Dev Art Org 6(1):1

4 Quinn KJ, Courtney JM, Evans JH, Gaylor JDS, Reid WH (1985) Biomaterials 6:369

5 Townsend PLG (1976) Burns 2:82

6 Behar D, Juszynski M, Ben Hur N, Eldad A, Tuchman Y, Sterenberg N, Rudensky B (1986) J Biomed Mater Res 20:731

7 Cristofoli C, Lorenzini M, Furlan S (1986) Burns 12:587

8 Woodroof EA (1984) In: Wise DL (ed) Burn wound coverings. Vol 2. CRC Press, Inc. Boca Raton

9 Gogolewski S, Pennings AJ (1982) Macromol Chem Rapid Commun 3:839

10 Gogolewski S, Pennings AJ (1983) Makromol Chem Rapid Commun 4:675

11 Lommen EJCMP (1988) Artificial skin. Dissertation, Groningen

12 Annual book of ASTM standarts, American society for testing material (1980) Philadelphia

13 Nilsson G (1977) On the measurement of evaporative water loss. Dissertation, Linkoping

14 Lelah MD, Cooper SL (eds) (1986) Polyurethanes in medicine. CRC Press, Inc. Boca Raton

15 Koenhen DM, Mulder MHV, Smolders CA (1977) J Appl Polym Sci 21:199

16 Kesting RE (ed) (1972) Synthetic polymer membranes. Mc Graw Hill, New York

SELECTED DISCUSSION REMARKS:

Jerusalem: How can the device be sterilized?

Hinrichs: It can be sterilized by gamma irradiation and as well by EtO-gas.

Becker: Your method of preparing the sample seems to be not so easy. I would imagine that you would have a lot of artifacts or empty spaces in the bottom-layer because of the very thin top-layer of about 10 μm. How reproducible is your method?

Hinrichs: The important step is the drying step. If you dry the bottom-layer too far you get too strong fusion and then you get large holes in the top-layer. If the bottom-layer is too wet when you apply the top-layer, you get a precipitation of the polymer before you coagulate it in the bath and that also creates irregularities in the top-layer of the material. Therefore every batch needs to be tested with SEM.

Becker: There is a foambased polyurethane-sheet in GDR for artificial skin (SYSpur-derm (R), VEK Synthesewerk). On one side it has pore sizes of 100 - 200 μm and on the other side about 1 - 2 μm. It's a soft foam and therefore the pores are open and connected. The thickness of SYSpur-derm is about 500 μm, so your consumption of materials is 5 times lower. May be it would be a good idea to evaluate your material in comparison to SYSpur-derm.

von Recum: I'd like to ask Prof. Becker whether your porosity of 1 - 2 μm is as effective as a bacterial filter as the material presented here having pores of somewhat like 0.5 μm.

Gogolewski: We have tested a similar material based on patents from Groningen and compared it as well with the material from GDR. Unfortunately you don't have this kind of antibacterial protection with the material coming from GDR as with that developed in Groningen. Still it is very critical that you do not exceed 0.7 μm pores at the protective layer as far as the antibacterial protection is concerned. Otherwise you have quite fast infection.

Gerhardt: Why not combine these very good properties for covering a wound with your material with cell-seeding using keratinocytes?

Hinrichs: What we did here was a very superficial wound. At deeper skin defects you have ingrowth of epidermal cells from the wound edges and you need cellseeding therefore. In the moment we are working on burns and we are combining a mesh graft technique and microskin graft technique in combination with this artificial skin.

A NEW KIND OF COLLAGEN MEMBRANE TO
BE USED AS LONG TERM SKIN SUBSTITUTE

B. Hafemann, B. Sauren, R. Hettich

Clinic of Burns, Plastic- and Reconstructive Surgery
Klinikum of the RWTH Aachen, Pauwelsstrasse, 5100 Aachen, F.R.G.

Abstract: A new kind of collagenous membrane has been tested under
different aspects for its suitability to improve wound conditions
and to protect the wound especially in cases of deep skin defects.
Preliminary experiments on rats demonstrated, that the membranes exhibit a considerable haemostyptic effect and rapid adherance.

They are getting vascularized and integrated into a newly formed
connective tissue. Thus they remain in place and form an adequate
deep dermal layer with reasonable mechanical stability, which can be
covered with different kinds of permanent skin or epidermal grafts.
Wound contraction is significantly reduced, before and after application of the definitive cover.

Introduction: Since long it has been acknowledged, that the most
effective way to treat deep extensive thermal injuries is to excise
the necrotic tissue and - since any open wound is prone to be infected - to permanently cover these areas with autogenic skin as early
as possible. In cases of extremely large defects, though, even techniques to expand the available autogenic donor skin soon reach their
limits. By mesh- (1) or patch-grafting (2) one can achieve an effective recipient-site/donor-site autograft surface area ratio of up to
4 : 1 (3). Even with the chinese method of intermingled skin grafting, one cannot expect satisfying results if the ratio lies much
beyond 20 : 1 (4, 5). Modern cell culture techniques achieve an ex-
pansion of autogenic epidermis by a factor of several thousands (6,
7), but since this needs about 3 - 4 weeks, the surgeon is confron-
ted with the difficult task to provide an effective temporary cover
for the yet open wound areas, until e. g. donor-sites can be used
for another time.

To deal with this problem many different materials have been tested

H. Planck M. Dauner M. Renardy (Eds.)
Medical Textiles for Implantation
© Springer-Verlag Berlin Heidelberg 1990

for their suitability (8, 9, 10). Through years of clinical experience a number of features which are necessary for skin substitutes
have been worked out (Tab. 1).

Tab. 1:   NECESSARY PROPERTIES OF SKIN SUBSTITUTES
           - Absence of antigenicity
           - Haemostyptic effect
           - Tissue compatibility
           - Absence of local and systemic toxicity
           - Water vapor transmission similar to normal skin
           - Impermeability to exogenous microorganisms
           - Rapid and sustained adherence to wound surface
           - Inner surface structure that permits ingrowth of
             fibrovascular tissue
           - Prevention of wound contraction
           - Flexibility and pliability to permit conformation
             to irregular wound surface
           - Elasticity to permit motion of underlying body tissue
           - Resistance to linear and shear stresses
           - Prevention of proliferation of wound surface flora
             and reduction of bacterial density of wound
           - Tensile strength to resist fragmentation and retention
             of membrane fragments when removed
           - Biodegradability (important for "permanently" implanted
             membranes)
           - Low cost
           - Indefinite shelf life
           - Minimal storage requirements

Until today there is no single dressing - biological or synthetic -
that satisfies all these requirements and the quest for an ideal
skin substitute is still going on. Several different lines of research have been pursued. In more recent years there seems to be a
tendency to construct two-layered grafts (11, 12), thus copying in
different ways the morphology of normal skin, of which the epidermis
controlls the evaporative loss of fluids and serves as a barrier
against exogenous microorganisms, while the predominant feature of
the dermis is to provide mechanical stability.

As a dermal substitute, collagen is an often preferred material
especially because of its low antigenicity, haemostyptic effect,
not to forget its easy availability and the fact, that collagen is

the major component of natural dermis. Collagen can be produced
in a variety of physical forms like gels, sponges, films etc. Most
of them have been tested as skin substitutes either alone or in combination with other materials, but it has to be considered, that the
three-dimensional texture of the intercellular matrix has a tremendous influence on the behaviour of the cells, by which it is colo-
nized (13). So we think, that to serve as a long term dermal substitute, collagen must be used in a form as native as possible.

This applies for the collagenous membranes which were subject of our
investigation.

We tested the membranes under different aspects for the suitability
to improve wound conditions especially in cases of deep and extensive skin defects. We report here our results about a first set of
transplantation experiments on rats.

Materials and Methods: The membranes are of animal origin and are
purified essentially by a procedure, which has been published by
Oliver et al. (14). They are delivered to us in dried form, either
native or crosslinked with hexamethylenediisocyanate (HMDIC). We
analyzed them by standard histological methods and scanning electron
microscopy.

Grafting experiments were done using the histoincompatible, highly
inbred rat strains DA as recipient and Lewis as donor (15). Animals
were purchased from the "Zentralinstitut für Versuchstierzucht"
(Hannover, F.R.G.).

Rectangular pieces of the back skin measuring 7 x 5 cm were excised
down to the muscle fascia (Fig. 1). These defects were covered by
same sized pieces of membrane, which were sutured to the wound margins. To control loss of fluids by evaporation, the membranes were
covered by a polyurethane membrane (OpSite^R). After different times
the wounds were definitively covered with thin and thick split
thickness skin grafts as well as chinese intermingled skin grafts
which are used in our clinic expecially for the treatment of extremely extensive deep burns.

At certain time intervalls grafts were evaluated by planimetry, pho-
tographic documentation, and standard histological analysis of biopsies.

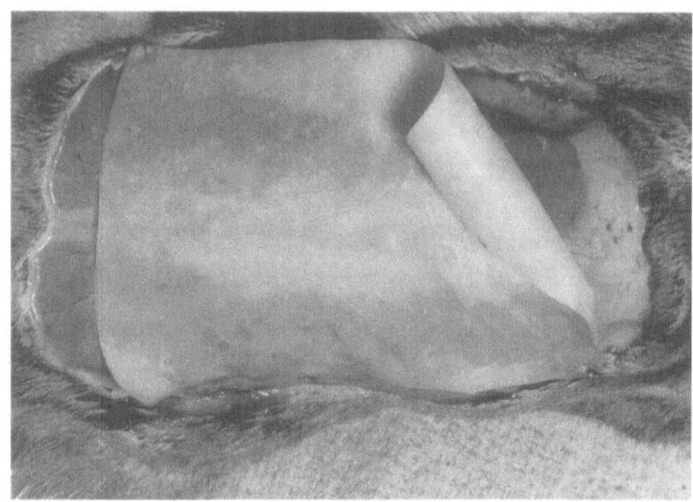

Fig. 1: Membrane placed on the muscle fascia before suturing it into place

Results: After rehydration the thickness of the membranes varies between 0.5 and 1.5 mm approximately. The preliminary characterization by standard histological techniques and electron microscopy shows, that long collagenous fibre-bundles of various sizes running in random directions are the major structural elements of the membranes (Fig. 2). Elastin seems also to be present in abundance (Fig. 3).

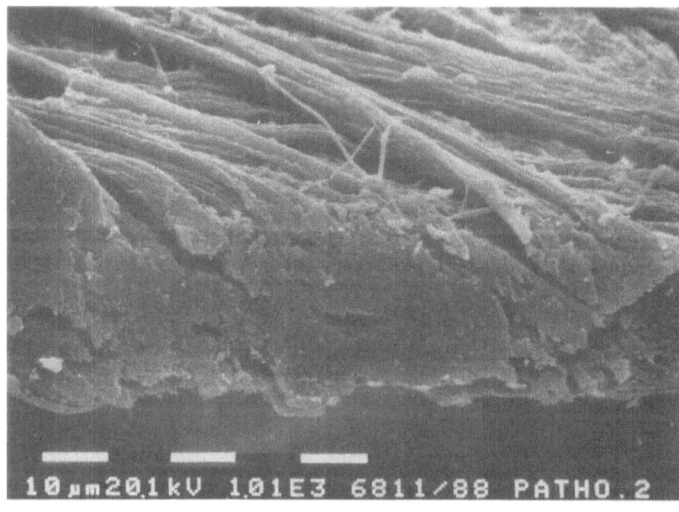

Fig. 2: Membrane as it appears under the scanning electron microscope

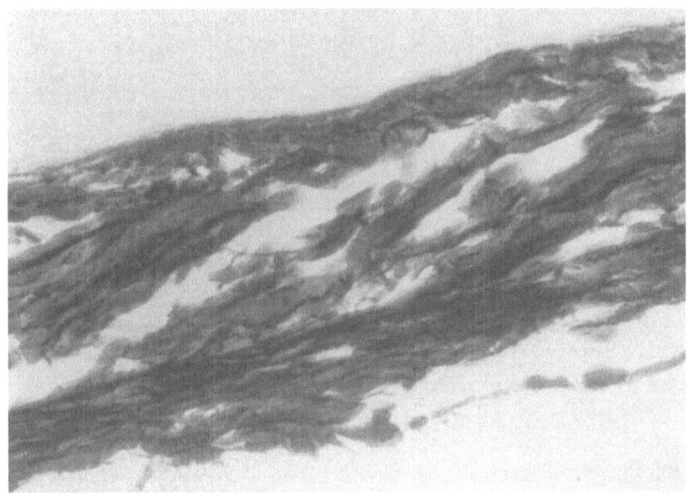

Fig. 3: Thin section of the collagenous membrane showing elastic
        structures in abundance (E. v. G., orig. magn., 250x)

The elastic features resemble very much those of normal skin.

After grafting together with OpSite[R] the membranes adhered well to
the wound and after a few days it became impossible to pull them off
without causing severe bleedings. 2 - 3 weeks post transplantation
the OpSite[R] cover and some fibrinous exudates were removed (Fig. 4).
The wound area had a clean transparent appearance showing many small
blood vessels (Fig. 5). The previously white colour of the membrane

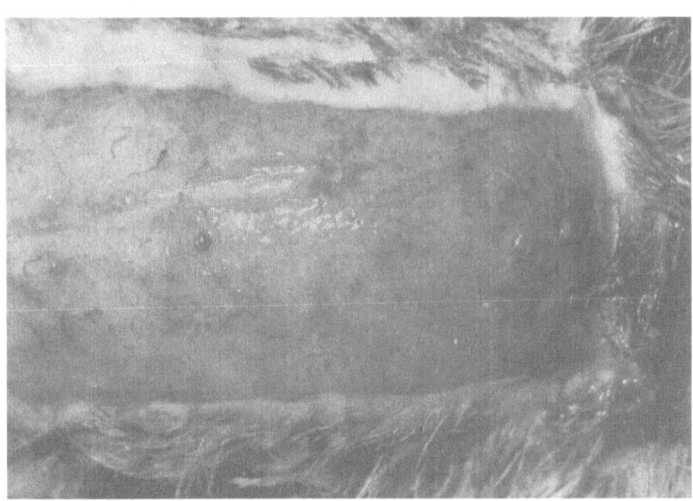

Fig. 4: HMDIC-crosslinked membrane - grafted wound 25 days post
        transplantation

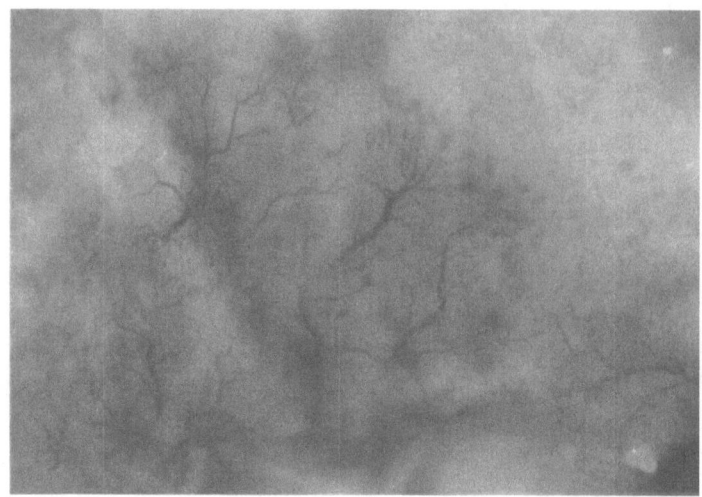

Fig. 5: Close-up view of the wound in Fig. 4 showing many small blood vessels

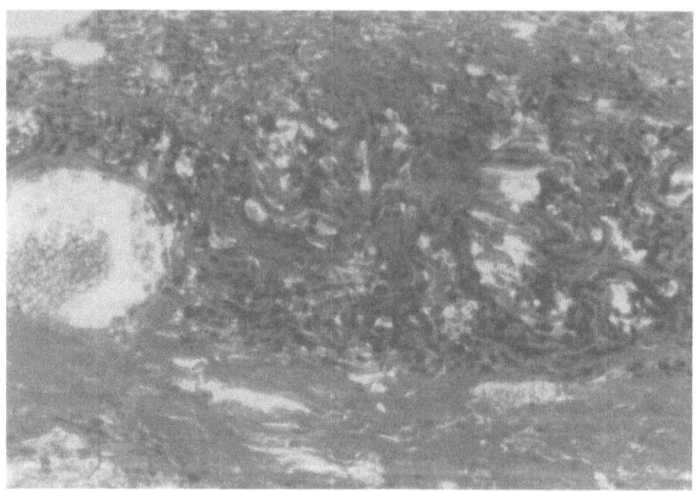

Fig. 6: Histological appearance of the membrane 25 days post transplantation showing the elastic fibres an several blood vessels (E. v. G., orig. magn.; 250x)

had changed to light red. Though very well vascularized the wound did hardly show any bleeding. The histological evaluation confirmed, that membrane structures were totally integrated into the wound surface (Fig. 6). As far as it could be judged by palpating and pulling, the integrated membrane exhibits considerable mechanical stability.

A comparision between different wound coverings (Fig. 7) showed, that defects, which had been coverd with OpSite^R alone were almost

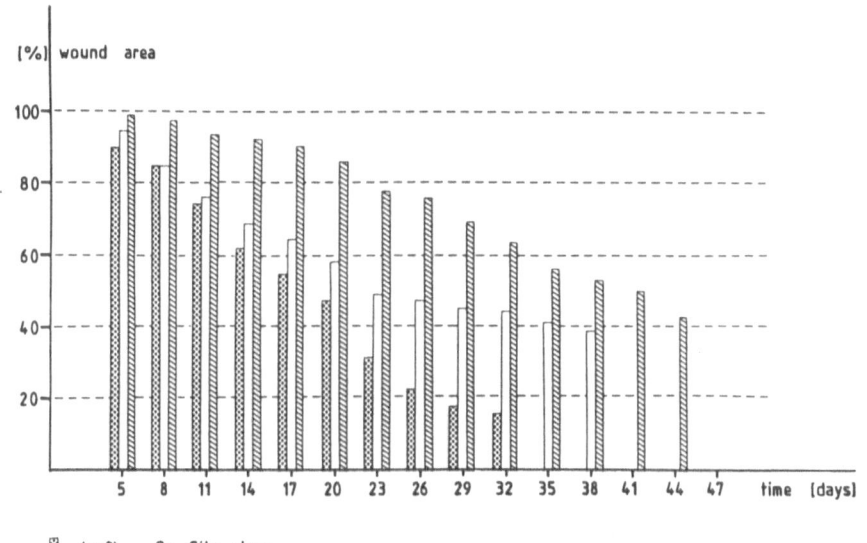

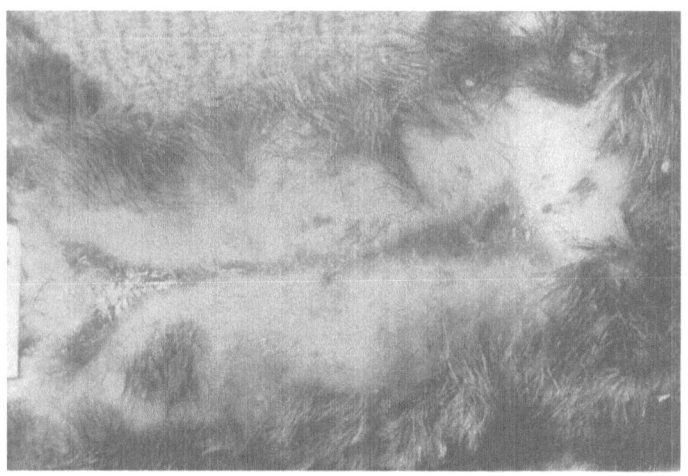

- 🔲 (n=8) Op-Site alone

- ▯ (n=8) membrane not crosslinked

- 🔲 (n=9) membrane crosslinked with HMDIC

Fig. 7: Wound contraction with different kinds of coverings

Fig. 8: Same kind of defect as shown in Fig. 1, 25 days post operation which had been covered with OpSite^R alone.

completely closed about 4 weeks post operation by wound contraction
(Fig. 8). This process was considerably delayed with a non-cross-
linked membrane. Membrane crosslinked with HMDIC showed even better
results.

To study the suitability of the integrated collagen membrane as a
wound bed for permanent closings like those which are commonly used
in our burn unit, we grafted thick and thin autogenic split thickness skin and also chinese intermingled skin.

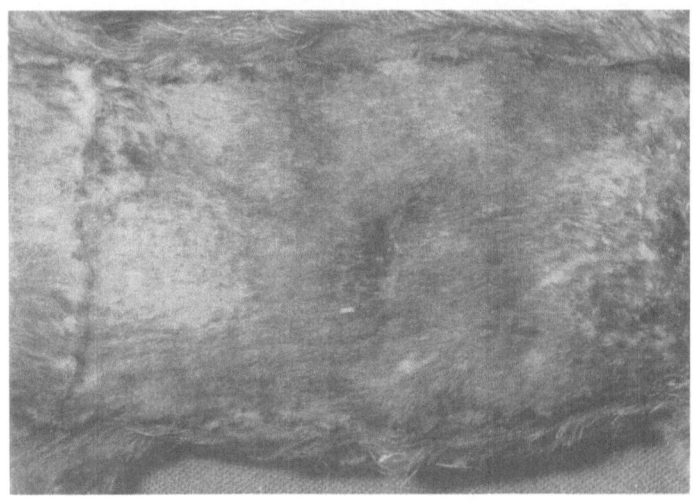

Fig. 9: Autogenic thick split thickness skin graft 10 days after
transplantation onto an integrated membrane

Fig. 10: Chinese intermingled skin graft 10 days after transplantation onto an integrated membrane

The take rate was not at all different from that on freshly excised wounds (Figs. 9 and 10).

The histologic evaluation about four weeks after grafting the inte-grated HMDIC-crosslinked membrane with split thickness skin (Fig.11) shows elastic fibres in abundance and remnants of collagen fibres between the skin graft and the muscle fascia i. e. the primary wound. After definitive grafting the wound contraction slowed down to cease completely after about two weeks.

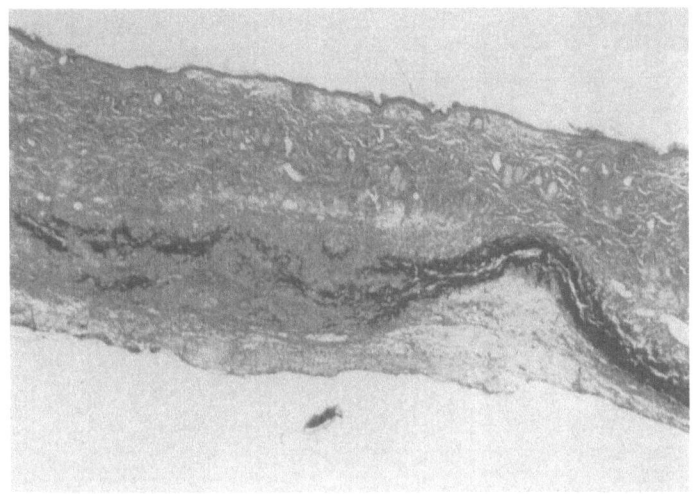

Fig. 11: Histological section through an autogenic thick split thick-ness graft approximately 4 weeks after transplantation onto an integrated HMDIC-crosslinked membrane (E. v. G., orig. magn., 100x)

Figures 12 and 13 give an impression of the stable condition of the grafts after about 4 months.

The situation did not change much even after more than one year. The histological section (Fig. 14) shows an almost normal skin. Collage-nous and elastic fibres of membrane origin could not be identified anymore.

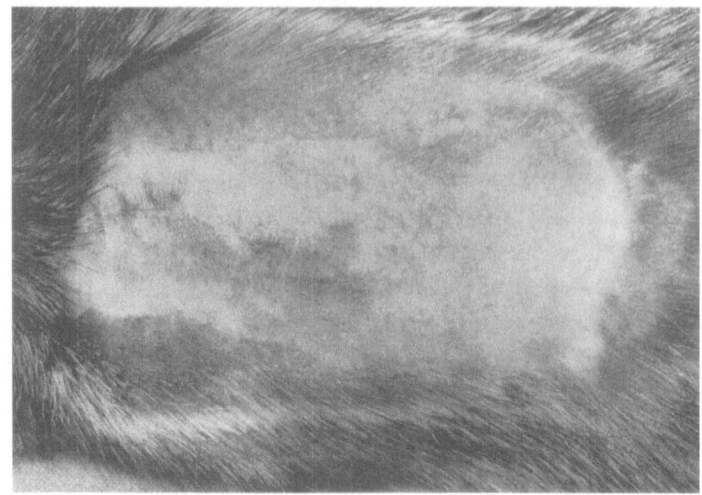

Fig. 12: Autogenic thick split thickness skin graft 140 days post
         transplantation

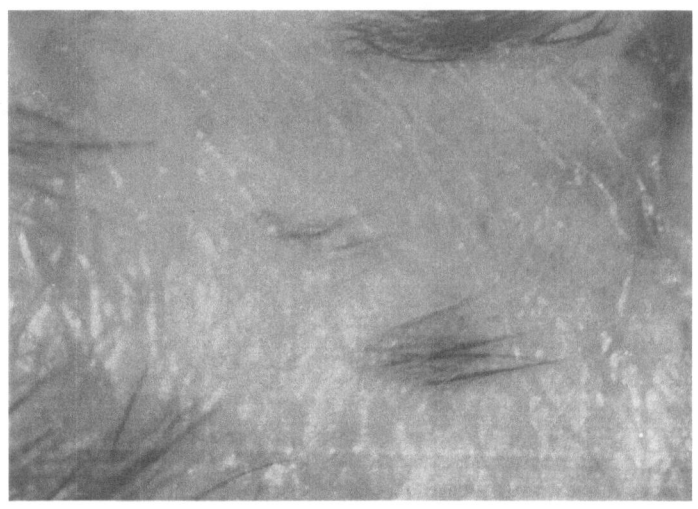

Fig. 13: Close-up look of an intermingled skin graft 150 days post
         transplantation. Autogenic islets can be identified by
         newly grown hairs.

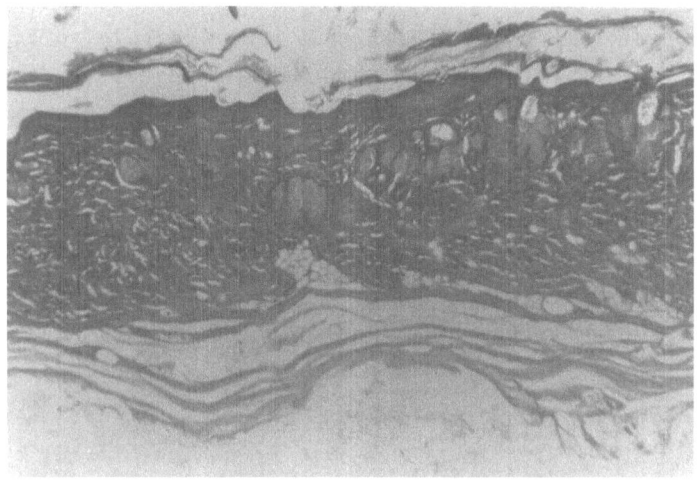

Fig. 14: Histological section through an autogenic thick split thickness skin graft about 15 months post transplantation

Discussion: Through early debridement of deep burn wounds followed
by the application of a permanent cover the absorption of burn
toxins, loss of body fluids and danger of infection by exogenous
bacteria, resulting in numerous, often fatal complications can be
prevented.

To cover extremely large defects requires the repeated harvesting at
the same donor sites for autogenic split thickness skin. Since it
takes at least 14 days for the skin e. g. on the hairy scalp to re-
cover and 3 - 4 weeks in other areas, in large burns open wounds
sometimes have to be protected by some kind of temporary cover.

So far with materials available for this purpose, besides pain for
the patient, this frequently leads to wound sepsis and overshooting
granulation tissue. This again can often result in terrible contrac-
tions, even if the wound has been surgically revised before grafting it permanently.

Especially for this problem the grafting of epidermal sheets alone
does not offer an adequate solution, since it is not very effective
to prevent wound contraction.

As to the results of our preliminary experiments, the collagenous
membranes in combination with a layer which controls evaporation
turned out to have a number of promising features under the aspects
of wound protection and wound improvement. The good adherence of the

material to the wound reduces fluid accumulations which are known to facilitate bacterial growth. Though we were not able to quantify it, we got the impression, that the membrane also exhibited a reasonable haemostyptic effect. This agrees with the often described function of collagen in the process of blood coagulation (16, 17, 18). Normal granulation tissue is characterized by a very good blood supply which favours a vascularization of skin grafts. It is on the other hand not very resistant to mechanical stress, which often results in loss of transplants due to extensive bleeding. This seems to be avoidable through the use of collagen membranes like those we tested.

After the integration of the membrane the wound is not only well vascularized but also enforced against mechanical stress by the collagen-elastin matrix of the membrane, which in the months after the application of the permanent cover appears to be slowly absorbed or replaced by autogenic tissue.

If these preliminary results can be confirmed through further experiments and clinical experience, these collagenous membranes could turn out to be of tremendous help for the surgeon not only to save the lives of severely burned patients, but also to improve the reconstructive results.

References:
1 Tanner JC, Vandeput J, Olley JF (1964) Plast Reconstr Surg 34:287
2 Gabarro P (1943) Br Med J 1:723
3 Peeters R, Hubens A (1988) Burns 14:239
4 Yang CC, Shih TS, Chu TA et al (1980) Burns 6:141
5 Yang CC, Shih TS, XU WS (1982) Plast Reconstr Surg 70: 238
6 Green H, Kehinde O, Thomas J (1979) Proc Natl Acad Sc USA 76:5665
7 Gallico GG III, O'Connor NE, Compton CC et al (1984) N Engl J Med
   311:448
8 Davies JWL (1984) Burns 10:94
9 Davies JWL (1984) Burns 10:104
10 Alsbjörn B (1984) Scan J Plast Reconstr Surg 18:127
11 Burke JF, Yannas IV, Quinby WC et al (1981) Ann Surg 194:413
12 Bell E, Ehrlich PH, Sher S et al (1981) Plast Reconstr Surg
   67:386
13 Gospodarowicz D, Vlodawsky J, Greenburg G et al (1979)
   In: Sato GH, Roos R (eds) CHS Conferences in Cell Proliferation,
   561, Cold Spring Harbor, New York
14 Oliver RF, Hulme MJ, Mudie A et al (1975) Nature 258:537
15 Arenas O, Wagener K, Kunz H (1981) J Immunogenet 8:307
16 Gordon JL (1979) Nature 278:13
17 Wilner GD, Nossel HL, Leroy EC (1968) J Clin Invest 47:2608
18 Nieviarowski S, Stuart RK, Thomas DP (1966) Proc Soc Exp Biol Med
   123:196

SELECTED DISCUSSION REMARKS:

Jerusalem: What is the time until elastic fibers occur?

Sauren: We have taken histological sections about 6 to 7 weeks after transplantation and we could find only few elastic fibers. If you remember the last slide, it was taken after one year, and you cannot find any structures any more. So occurance of elastic fibers must be in between 7 weeks and one year.

THE GORE-TEX SURGICAL MEMBRANE FOR TEMPORARY SKIN CLOSURE AFTER COMPLICATED CARDIAC OPERATIONS

C.A.Mestres, J.L.Pomar, C.Barriuso, S.Ninot
and J.Mulet

Cardiovascular Surgery. Hospital Clínico
y Provincial. University of Barcelona.
Barcelona. Spain.

ABSTRACT

Between April 1986 and March 1989, 12 patients, 7 male and 5 female
with a mean age of 50.0 years underwent temporary skin closure after
complicated cardiac operations with the Gore-Tex Surgical Membrane
and the indications to delay primary closure of the sternume were
myocardial edema, uncontrollable hemorrhage and additional cannulations needed for mechanical circulatory support. Seven patients
(58%) finally underwent sternal closure. Five patients survived
and left the hospital (42%). No superficial nor deep infection developed in survivors. The analysis of 3 explanted membranes showed
no bacterial growth.

Although open-heart surgery is routinely and safely performed world-
wide with generally speaking low mortality and reasonably low morbidity, several and potentially lethal intraoperative complications
can sometimes develop. Myocardial edema and massive intraoperative
bleeding can preclude routine closure of the sternum and delay it
for a number of hours or days. Several methods of skin coverage
have been tested, however the experience is scanty[1,2].

Following our previous experience in this field[3], we have been using
recently the polytetrafluoroethylene Gore-Tex* Surgical Membrane
(GSM) in patients whose clinical condition precluded conventional
primary closure of the chest after cardiac operations.

MATERIAL AND METHODS.

From April 1986 until March 1989, The GSM has been used at our Institution to cover the skin when primary sternal and skin closure
was not feasible because of extreme myocardial edema and distention

*W.L.Gore&Ass., Flasgtaff, Arizona, USA

H. Planck M. Dauner M. Renardy (Eds.)
Medical Textiles for Implantation
© Springer-Verlag Berlin Heidelberg 1990

or uncontrolable bleeding developed after cardiac operations. Twelve
patients were included in the experience, being 7 males and 5 fema-
les, ages ranging from 7 to 70 years and a mean age of 50.0 ± 20.1
years. Surgery was indicated because of coronary artery disease
in 6 patients, valve disease in 3 and congenital heart disease,
aortic dissection and cardiocutaneous fistula en 1 each (Table 1).
Operations performed are summarized in Table 2. Aortic cross-clam-
ping time averaged 66.72 ± 17.63 minutes (range 40-96) in the 11
patients who underwent surgery under cardioplegic arrest. Cardiopulmonary bypass time averaged 140.66 ± 38.42 minutes (range 87-240).

Table 1

Preoperative Diagnosis
_____

- Ischemic heart disease	6
- Prosthetic valve dysfunction	2
- Calcified aortic stenosis	1
- Type A acute aortic dissection	1
- Discrete subaortic stenosis	1
- Chronic chest wall fistula (post TOF)	1

TOF=Tetralogy of Fallot

Table 2

Operations performed
_____

- Coronary artery bypass grafting	6
- Prosthetic valve replacement (MVR)	2
- Native valve replacement (AVR)	1
- Repair of aortic dissection	1
- Resection of subaortic stenosis	1
- Right ventricular outflow tract patch	1

MVR=Mitral valve replacement; AVR=Aortic valve replacement

A total of 18 pieces of GSM were implanted in the 12 patients. The
size of the GSM was 12.0 x 12.0 cm and its thickness 0.1 mm. Initially in the experience, the GSM was anchored to the skin edges by
using 3/0 Prolene suture. In the last cases, a running 5/0 Gore-
Tex suture has been used for such purpose.

RESULTS.
One patient died in the operating room from low cardiac output syn-
drome under mechanical circulatory assistance condition. An intra-
aortic balloon (IABP) and a left ventricular assist device (LVAD)
were implanted, however he died from profound refractory cardiogenic
shock. Four patients were extremely ill at the time of leaving the

operating room and died in the ICU at 3,3,3 and 12 hours postoperatively with the chest still open.

**Seven patients (58%) later underwent removal of the GSM and sternal
closure at a mean of 2.28 ± 1.25 days after the operation.** Two additional patients ultimately died 3 weeks later because of multiorgan
failure and pulmonary embolus. Thus, the 7 patients who died accounted for a 58% mortality rate and the 5 survivors represented a 42%
survival rate in such a series of very ill patients.

No superficial nor deep infection developed in the 7 patients who
underwent delayed sternal closure, including the 5 survivors and
2 other patients who died. Absence of infection was assessed by
mediastinal cultures of tissues taken at sternal closure. Analysis
of the retrieved GSM in 3 patients showed no bacterial growth.

DISCUSSION.

Myocardial edema and increased cardiac size secondary to prolonged
periods of cardiopulmonary bypass, uncontrollable bleeding after
post-bypass coagulopathy and additional cannulations for mechanical
circulatory assistance have been the three indications for delayed
sternal closure in our experience. How to cover the skin defect
when there is no possibility to approximate the sternum and soft
tissues has been a subject of interest for surgeons. The need to
cover the mediastinum to avoid deep infection has led surgeons to
use several materials in such critical situations[1]. Our experience
with the GSM has been encouraging because this membrane allowed
us to adequately close the surgical wound until cardiac size returned to normal or mediastinal hemorrhage stopped. Subsequent sternal
closure was later completed. The GSM has sealed the wound and represented a barrier against mediastinal infection, a potentially lethal
complication after major cardiac surgery. In fact, cultures of mediastinal tissues taken at the time of secondary closure of the
sternum in the 7 patients to returned to the operating room at a
mean of 2.28 days after the initial operation showed no evidence
of bacterial growth and no superficial nor deep infection developed
in any case.

The problem of non-conventional cannulations has also been managed by using the GSM. Two of our patients required mechanical support of their failing hearts. One female patient underwent transaortic implantation of a IABP and one male patient underwent placement of a LVAD by cannulating the left atrium and the ascending aorta. The chest was left open because of cardiac edema and the extra space required for the large-bore cannulas. No cardiac compression by cannulas was noted.

Summarizing, we should conclude that the GSM has shown a good performance at the time of covering the skin defect in cases of patients in critical condition after complicated cardiac operations, who required to delay the primary closure of the sternum. this performance should be assessed in terms of: 1) Avoid cardiac compression, 2) Allow for mediastinal packing for uncontrollable hemorrhage and 3) Adequately isolate the mediastinum.

REFERENCES.

1 Gielchinsky I, Parsonnet V, Krishnan B, Silidker M, Abel RM (1981) Ann Thorac Surg 31:273
2 Josa M, Khuri S, Braunwald NS, VanCisin MF, Spencer MP, Evans DA, Barsamian E (1986) J Thorac Cardiovasc Surg 91:589
3 Mestres CA, Ninot S, Palliso FJ, Barriuso C, Abad C, Morales MA, Pomar JL, Mulet J (1987) Presented before the XVIII World Congress of the International Society for Cardiovascular Surgery. Sydney (Australia)
4 Ugorji CC, Turner SA, McGee MG (1980) Cardiovasc Dis Bull Texas Heart Inst 7:307

**PATHOLOGICAL ANALYSIS OF THE EXPLANTED GORE
-TEX SURGICAL MEMBRANE AFTER SURGICAL
IMPLANTATION**

C.A.Mestres and J.L.Pomar

Cardiovascular Surgery. Hospital Clínico
y Provincial. University of Barcelona.
Barcelona. Spain.

ABSTRACT

The polytetrafluoroethylene Gore-Tex Surgical Membrane has been
used at our Institution for pericardial closure, temporary skin
closure after cardiac operations and to close the retroperitoneum
over aortic vascular prostheses. Five Gore-Tex membranes have been
available for examination. Light microscopy showed that neither
tissue ingrowth nor bacterial growth have been developed in any
of the samples studied. In addition, only a minimal epicardial
reaction in the area beneath the membrane was seen in one of the
cases of pericardial closure. This material seems to act as an
adequate substitute for biological barriers such as the pericardium
and the skin.

The Gore-Tex Surgical Membrane* is a very thin sheet of expanded
polytetrafluoroethylene that has initially been designed as a pericardial substitute in an attempt to reduce or minimize adhesions
formation to provide a safe and efficient plane of dissection which
is of utmost interest when dealing with a cardiac reoperation.
In addition, we have been using it also as a substitute of other
biological barriers such as the skin and the retroperitoneum. In
the last few years, some information about clinical experiences
with the GSM has been available, however very few information has
been collected about GSM explants[1-3]. We have been able to collect
some GSM explants and histological analysis has been performed
to ascertain if there has been tissue penetration or bacterial
colonization.

*W.L.Gore&Ass., Flagstaff, Arizona, USA

H. Planck M. Dauner M. Renardy (Eds.)
Medical Textiles for Implantation
© Springer-Verlag Berlin Heidelberg 1990

MATERIAL AND METHODS.

From April 1985 until March 1989, the GSm has been used at our
Institution for pericardial closure and temporary skin closure
after cardiac operations. In addition, few patients underwent covering of aortic vascular prostheses. Pericardial closure has been
performed 34 times in 33 patients, temporary skin closure in 12
patients and closure of the retroperitoneum on 6 patients. Five
membranes have been available for examination, 2 from the pericardial closure group (Group A) and 3 from the skin closure group
(Group B). No retrievals from the retroperitoneum are available
up to now.

The samples in both groups were fixed in 10% formalin. Sections
were taken through the samples and including suture lines and brown
clot-like material along surfaces. According to standard techniques
hematoxilin&eosin, Milligan's trichrome and fibrin stains were
performed[4]. Light microscopy studies were performed under 2.5x,
25x, 50x and 100x magnification.

RESULTS.

Group A (Pericardium).
Two membranes were available for examination at 13 days and 3
months postoperatively. The patients were initially operated on
for mitral valvuloplasty and aortic valve replacement.

In both cases, the GSM was essentially clean along both surfaces,
with a few fibrous tissue tags present. A few erythrocytes and
fibrin-protein matrix were also present along both surfaces. A
dark brown material was present due to the breakdown of the erythrocytes. No calcium was noted. There was no evidence of infection
and no tissue attachment to the GSM was noted.

Group B (Skin).
Three membranes were retrieved at 4, 2 and 4 days respectively
after implantation for temporary skin closure. The patients were
initially operated upon for cardiocutaneous fistula, aortic valve
replacement and coronary artery bypass grafting. Light microscopy
revealed a few erythrocytes and elements associated with the break-

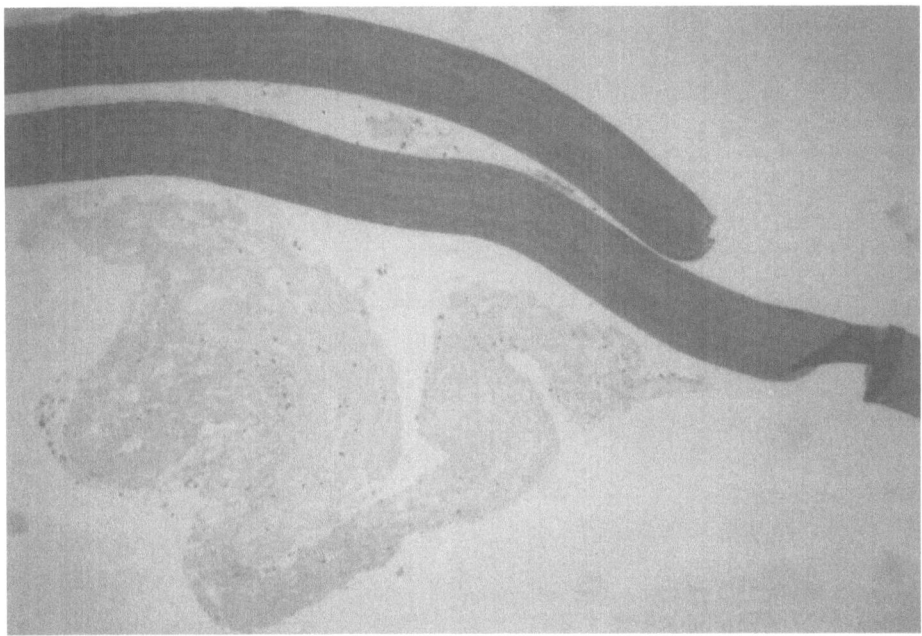

Fig.1   Group A GSM clean of tissue, not adhered. Milligan's trichrome stain.
         25x magnification.

down of erythrocytes, along the inner surface. No inflammatory
reaction was present. A ver small amount of calcium associated
with the breakdown of thrombus was noted in one case. No evidence
of infection within the GSM or its surfaces was noted.

DISCUSSION.

The GSM has been used by us for three different purposes: pericardial[5], skin and retroperitoneum substitution. Up to now, results
seem to be acceptable in terms of good surgical handling, absence
of complications and good performance as assessed by examination
of retrievals. In addition, no inflammatory epicardial reaction
was noted beneath the GSM in the pericardial group.

The need for pericardial substitutes has grown since the seventies
and also the use of synthetic materials for temporary skin closure,
although in smaller series. Regarding the GSM, a few published
material is available and also a few clinical retrievals came for
examination. In addition, very few membranes have been histologically studied. Those studies showed similar findings to the data

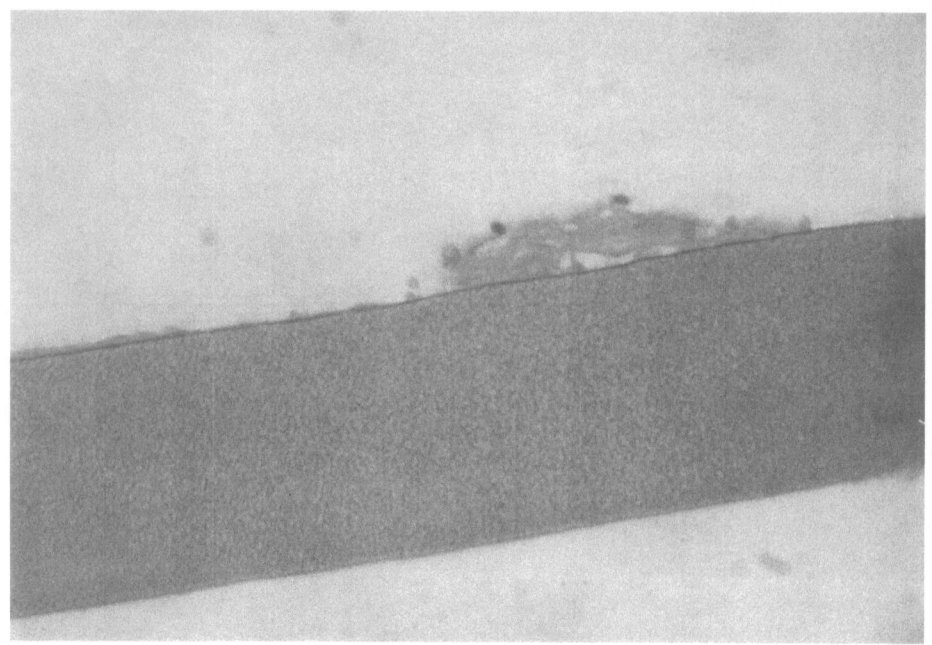

Fig.2   Group B GSM. Normal appearance of GSM. A few erythrocytes and leukocytes
along one surface. Milligan's trichrome stain. 100x magnification.

presented by us herein supporting the fact that no tissue incorporation has been noted[2,3]. On the other hand, the absence of signs
of infection strongly support the fact that the GSM seems to be
an adequate substitute for biological barriers. Our previous experience with other materials such as polyurethane for similar purposes was initially satisfactory, however the very severe inflammatory epicardial reaction found in two reoperations has led us to
consider the GSM as a better alternative for pericardial closure[5,6].

The main difference between the pericardial and skin groups is
that the expected implantation time is longer in the pericardial
group. In the skin group, what is for us of utmost importance is
that the GSM acts as a protecting barrier by isolating the mediastinum. In the pericardial group, as reported by others and confirmed by us, the GSM prevents adhesion formation and limits tissue
attachment. Other findings have been the maintenance of long-term
flexibility, original thickness and size[7,8].

We finally should conclude that the GSM has shown a good performance as a pericardial and skin substitute by limiting tissue incor-

poration and because no infection developed in any case or position.

REFERENCES.

1 Heydorn WH, Daniel JS, Wade CE (1987) J Thorac Cardiovasc Surg 94:291
2 Minale C, Hollweg G, Nikol S, Mittermayer C, Messmer BJ (1987) Thorac Cardiovasc Surgeon 35:312
3 Harada Y, Imai Y, Kurosawa H, Hoshino S, Nakano K (1988) J Thorac Cardiovasc
  Surg 96:811
4 Manual of Histological Staining Methods of the Armed Forces Institute of Pathology (1960)
5 Mestres CA, Cugat E, Ninot S, Gómez JF, Pomar JL (1986) Thorac Cardiovasc Surgeon 34:137
6 Mestres CA, Rives A, Cugat E, Ninot S, Alcaraz A, Pomar JL (1986) Polyurethanes in Biomedical Engineering II. Elsevier
7 Opie JC, Larrieu AJ, Cornell IS (1987) Ann Thorac Surg 43:383
8 Revuelta JM, García-Rinaldi R, Val F, Crego R, Durán CMG (1985) J Thorac Cardiovasc Surg 89:451

SELECTED DISCUSSION REMARKS:

Becker: Do you know the diameter of the pores of the membrane?

Mestres: 1 μm.

BIOABSORBABLE NON-WOVEN FABRIC FOR SURGERY

Tatsuo NAKAMURA, Y. Shimizu, S. Watanabe, K. Shiraki,
S.-H. Hyon, M. Suzuki[*], T. Shimamoto[*], Y. Ikada

Research Center for Medical Polymers & Biomaterials,
Experimental Surgery and Chemistry of Biomedical Polymers
Kyoto University, 53 Kawahara-cho, Shogoin Sakyo-ku, Kyoto 606, Japan

ABSTRACT: Bioabsorbable non-woven fabrics were fabricated from polyglycolide(PGA) and poly-L-lactide(P-L-LA). Their degradation in vitro and in vivo was evaluated. Non-woven fabric made from PGA showed rapid degradation; retention of breaking strength is about 50% at 10 days, almost zero at 21day. The rate of degradation of non-woven fabrics made from P-L-LA was slow and in vivo the mechanical strength of the implanted samples of P-L-LA seemed to be increased as far as 4 weeks after implantation perhaps this is due to ingrowthed connective tissue. Both of non-woven fabrics showed neither strong reaction nor rejection.

INTRODUCTION: Bioabsorbable synthesized polymers have been widely used as suture materials for surgery. Most commonly used are -aliphatic polyesters such as polyglycolide (PGA), polylactide (PLA) or their copolymers, for instance Polyglactin 910. As for surgical suture materials, since their initial function disappears after wound healing, the ideal material for such sutures would be bioabsorbable suture which has an adequate absorption rate in accordance with both wound healing and the situation in which it is used.

On the other hand, surgical meshes or fabrics are used for many purposes, which can be classified into two gross categories ; 1) permanent reconstruction material e.g., for artificial vessels, and 2) temporary scaffold (framework) for new tissue generation. As Chu[1] pointed out, non-absorbable materials have one common disadvantage - if infection occurs they must be removed. Since in the living body connective tissue easily grows into implanted textile or knitted fabric, removal of the implanted material is often difficult and dangerous.

In the case of anastomosis or suture of fragile tissue, pledgets or buttresses have been used, particularly in cardiovascular surgery, where they have become the standard mode. These materials are Teflon or Dacron and have often been reported to cause infection or consequent aneurysm, necessitating subsequent removal[2].

We have produced non-woven fabric made from PGA, prepared pledgets from it, and have reported their in vivo and clinical superiority to conventional non-absorbable pledgets for pulmonary surgery[3]. Recently, pediatric cardiovascular

H. Planck M. Dauner M. Renardy (Eds.)
Medical Textiles for Implantation
© Springer-Verlag Berlin Heidelberg 1990

surgeons' have reported on the clinical application of the bioabsorbable sutures, Dexon and Vicryl[4] , claiming good results. The most suitable absorption rate of pledgets or bolsters (buttresses) for cardiovascular surgery, however, has not yet been determined.

In this study we prepared non-woven poly-L-lactide (P-L-LA) fabric, which is considered to be degraded more slowly than PGA, and evaluated its mechanical characteristics and compared its absorption behavior in vivo and in vitro.

MATERIAL: Materials evaluated in this study were two types of non-woven fabric. One was made from PGA, another P-L-LA. Using polyfilament yarns, tubular-knit meshes were fabricated, and non-woven fabrics were made with the scratching method.

PGA non-woven fabric is 0.5 mm thick, weight 240 g/m^2, and the PGA fiber is 32 - 39 d.

P-L-LA non-woven fabric is 0.45 mm thick, 203 g/m^2, and the P-L-LA fiber is 32 - 39 d.

RESULTS AND DISCUSSION: Samples (4 x 1.5 cm) were grasped longitudinally at both sides 1 cm wide, then their breaking strengths were determined. Samples with a 5-mm incision made transversely in the middle were also tested (Fig. 1).

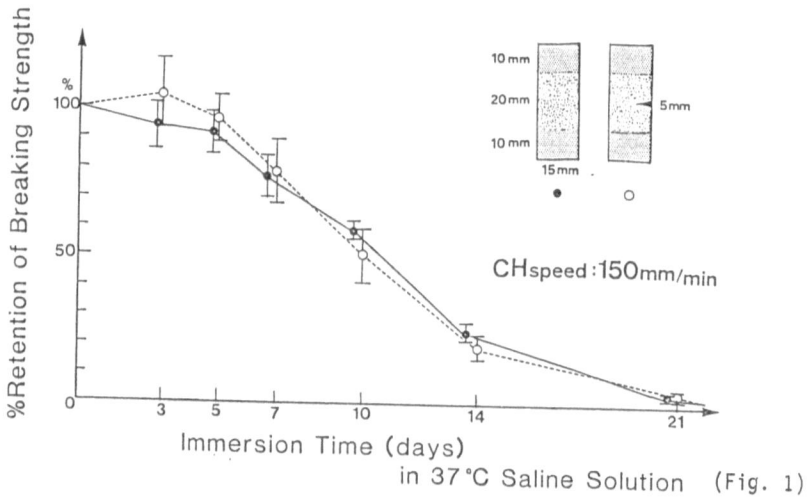

CHANGE OF TENSILE BREAKING STRENGTH
OF PGA NON-WOVEN FABRIC (in vitro)

in 37 °C Saline Solution   (Fig. 1)

Tear strength testing with double stainless steal wire was also carried out (Fig. 2).

BREAKING STRENGTH OF PGA–PLEDGET (in vitro)

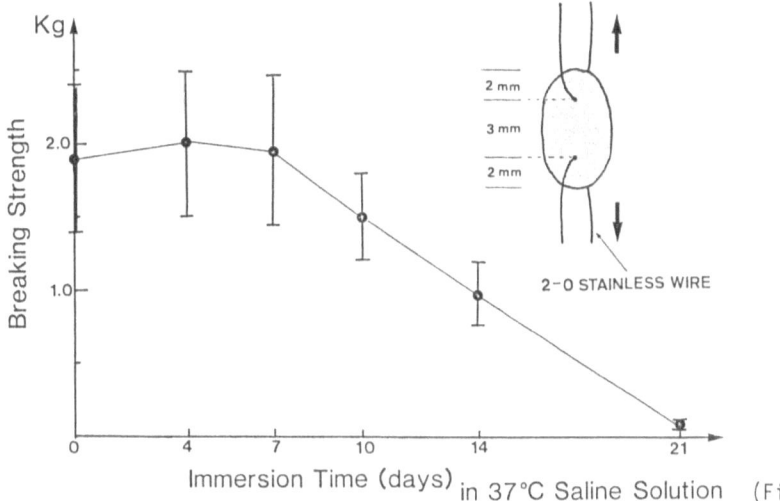

Immersion Time (days) in 37°C Saline Solution (Fig. 2)

The strength, as pledgets against suture tearing force was maintained for one weeks. At 3 weeks, it became almost zero.

As fibers of non-woven fabrics become entangled, non-woven fabric is strong against tearing, as shown in Figs. 1 and 2. This mechanical property is adequate for pledgets or buttresses.

Fig 3 indicates the mechanical strength change of P-L-LA non-woven fabric implanted subcutaneously in the rabbits. As far as 8 week the tensile strength of implanted fabrics seems to be increased after implantation. As it is difficult to remove the ingrowthed tissue from the fabrics, measurement of the samples were carried out. Since the tensile strength of P-L-LA fabric was maintained as far as 3 months. Therefore the extirpated sample became stronger against tearing due to the ingrowthed connective tissue. That may be the reason of the increasing the tensile strength.

Chu and Williams[5] reported that degradation of ..-aliphatic polyester in vivo is different from that in vitro, although they noted that hydrolysis by water molecules is believed to be the major mechanism causing polyglycolide sutures to degrade during the initial stage. Furthermore, it is difficult to examine the degradation of the implanted fabric, because tissue grows easily into the intima (space between fibers) of the fabric and isolation of the fabric from newly grown tissue is almost impossible. Hence we carried out the tests in vitro as described above. In vivo histological reactions of the tissue and degradation of P-L-LA non-woven fabric, which was slower than that of PGA, will be discussed in the session. We are currently researching the application of these P-L-LA pledgets for cardiovascular surgery, in which longer duration and security is required.

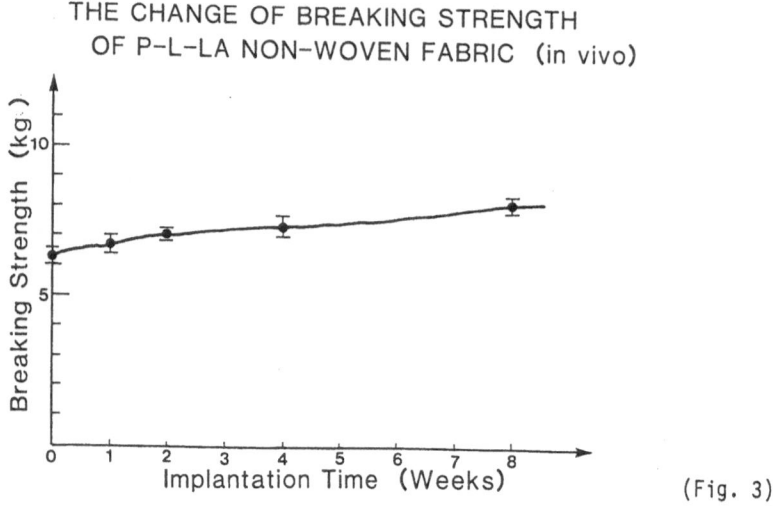

(Fig. 3)

ACKNOWLEDGMENT The assistance of JMS Co. in measuring the sample mechanical properties is acknowledged.
*Medical Development Unit, GUNZE Co., Ikurashinmachi, Ayabe-shi, Kyoto 623, JAPAN.

REFERENCES

1. Chu, C.C. and Welch, L., J. Biomed. Mater. Res., 19, 903-916, 1985
2. Looser, K.G. et. al., J. Thorac Cardiovasc Surg. 72, 280-281, 1976
3. Nakamura T. et. al., Transactions 3rd World Biomaterials Congress, 277, 1988
4. Kanazawa, H., J.J.A.T.S. 35, 799-803, 1987
5. Chu C.C. and Williams, D.F., J. Biomed. Mater. Res. 17, 1029-1040, 1983

SELECTED DISCUSSION REMARKS:

Becker: Your fibers are coloured. Is there any reason for dying and does it affect the degradation?

Nakamura: In surgery it is more easy to find again dyed fibers. The dye is non-toxic and it is absorbable. The influence of the dye on the material was not checked, but ....

Gogolewski: These dyes are proved by FDA and they are the same as used for Vicryl- and Polyglactin-sutures. And the dye doesn't affect the degradation time.

Planck: What is the specific strength of the yarn?

Gogolewski: 0.3 - 0.6 GPa tensile strength.

WATER VAPOR AND OXYGEN PERMEABILITIES
OF POLYETHERURETHANES

Mutlu Karakelle and Robert A. Taller

Becton Dickinson Polymer Research
P.O. Box 1285, Dayton, OH 45401, USA

ABSTRACT

Effects of soft segment content, hydrophilicity of polyether macroglycol soft segment and presence of polydimethylsiloxane co-soft segment on oxygen and water vapor permeabilities of polyether-urethanes were investigated. Polyetherurethanes were synthesized using a bulk synthesis procedure and solution cast as thin films. Oxygen and water vapor transport measurements were performed at 38 °C, 100% relative humidity (RH) and 4 mm Hg oxygen gradient, and 40 °C and 80% RH test gradient, respectively. The soft segment content and structure had profound influences on oxygen and water vapor permeabilities of the polyetherurethanes. Water vapor and oxygen permeabilities of polyetherurethanes increased with increasing soft segment content. Polytetramethylene oxide based polyether-urethanes (PTMO-PEU) had high oxygen permeabilities and polyethylene oxide based polyetherurethanes (PEO-PEU) had high water vapor permeabilities. Addition of silicone co-soft segment increased water vapor permeability and slightly reduced oxygen permeability of PTMO-PEU. Its effect on PEO-PEU was the reverse of that of the PTMO-PEU systems.

INTRODUCTION

Polyetherurethanes have been widely used for biomedical applications because of their excellent biocompatibility and unique physical properties [1]. Textile backed polyetherurethane films have good biaxial flexibility and dimensional stability upon hydration. Oxygen and water vapor permeabilities of polyetherurethanes are of interest to the biomedical community for tissue ingrowth cuffs, vascular prostheses as well as barrier and breathable film applications. Polyetherurethanes are segmented block copolymers consisting of hard and soft segments, therefore, their properties are strongly dependent

H. Planck M. Dauner M. Renardy (Eds.)
Medical Textiles for Implantation
© Springer-Verlag Berlin Heidelberg 1990

on their structure [2]. The soft segment content and structure have profound influences on the properties of polyetherurethanes [3]. A wide range of physical properties can be obtained by altering the soft segment content and the polyether macroglycol composition.

A number of studies have been reported on water vapor and oxygen permeabilities of polyetherurethanes. Schneider et al. reported soft segment structure - water vapor permeability relationship of polyurethanes [4]. PEO based polyurethane was found to have the highest water vapor permeability. McBride et al. studied diffusion of oxygen, hydrogen and carbon dioxide gases through PTMO and polytetramethylene adipate glycol based polyurethanes at varying soft segment contents [5]. It was reported that increasing soft segment content and length increased the gas permeability of polyurethanes. Knight and Lyman studied the effect of chemical structure and fabrication variables on the oxygen, nitrogen and carbon dioxide permeability of PEO and polypropylene oxide (PPO) based polyether-urethane and polyetherurethane-urea membranes [6]. It was found that PEO based polyetherurethanes were less permeable to oxygen, nitrogen and carbon dioxide gases than PPO based polyetherurethanes.

In this study, the effects of soft segment content, hydrophilicity of polyether soft segment and the presence of polydimethylsiloxane co-soft segment on oxygen and water vapor permeability of polyether-urethanes for biomedical applications were investigated.

EXPERIMENTAL

The polyetherurethane hard segment was formulated from 4,4'-diphenyl-methane diisocyanate (MDI) and 1,4-butanediol (BDO) extender. Polytetramethylene oxide (PTMO) glycol of 1000 molecular weight and polyethylene oxide (PEO) glycol of 1000 molecular weight were utilized as the soft segment. Dow Corning DC-3667 silicone fluid, polyethylene oxide glycol / polydimethylsiloxane (PDMS) / polyethylene oxide glycol triblock copolymer of 2000 molecular weight, was used as co-soft segment in silicone containing polyetherurethanes. Polyetherurethanes were synthesized using a one-shot bulk polymerization method. The polyethers were dried at 60-70 °C under vacuum (4-6 mm Hg) for 4-6 hours to remove moisture. Water content (Karl Fisher titration) and polyol hydroxyl number (phthalic anhydride pyridine method) were determined to adjust formulation stoichiometry. MDI was filtered and vacuum stripped (4-6 mm Hg) for

2-4 hours. Stoichiometric amounts of polyether macroglycol and extender (BDO) were placed in the polymerization vessel and degassed at 60 °C for 30 minutes. A stoichiometric amount of MDI (1.02 index) was then added and stirred vigorously until the polymerization temperature reached about 85-90 °C. The polymer was discharged and postcured at 125 °C for 30 minutes. Thin films were cast, from solutions of these polymers in a solvent mixture of N,N-dimethyl-formamide (DMF) / tetrahydrofuran (THF) (25/75, by weight), on silicone release paper. An adjustable film casting knife was utilized to achieve the desired film thickness (0.5-1.5 mil). Oxygen and water vapor transmission rates were determined by Mocon Modern Controls, Inc., Minneapolis, Minnesota, USA. The oxygen transmission rate was measured at 38 °C, 100% relative humidity (RH) and 4 mm Hg oxygen gradient, and extrapolated to 760 mm Hg oxygen gradient. The water vapor transmission rate measurement conditions were 40 °C and 80% RH test gradient.

RESULTS

Water vapor permeability of polyetherurethanes increased with increasing soft segment content (Figure 1). An increase of soft segment content by 10% increased the water vapor permeability of PTMO and PEO based polyurethanes respectively by 30% and 50%. However, the structure of the polyether soft segment had the most influence on the water vapor permeability of polyetherurethanes (Figure 1), as reported by Schneider et al. [4]. PEO based polyetherurethanes had approximately 3.7 times higher water vapor permeability than the PTMO based polyetherurethanes at the same soft segment content (60% by weight). This trend is in agreement with the water absorption behaviors of PEO and PTMO based polyetherurethanes (Figure 2). The water absorption of a PEO based polyetherurethane with a 60% soft segment content is approximately 59% by weight, whereas, that of a PTMO based polyurethane with a 60% soft segment content is less than 2% by weight. Utilization of PEO/PDMS/PEO glycol (DC-3667 silicone fluid) as co-soft segment substantially increased the water vapor permeability of PTMO based polyurethanes and slightly decreased that of PEO based polyurethanes (Figure 3). This effect is most likely related to the polyethylene oxide content of DC-3667 silicone fluid rather than the polydimethylsiloxane content. Replacement of some of the polyethylene oxide glycol with DC-3667 silicone fluid in a PEO based polyurethane formulation reduces its PEO content resulting in reduced water vapor permeability. In PTMO based polyurethanes, replacement of some of the PTMO glycol with DC-3667 increases its PEO content, therefore, its water vapor permeability increases.

Oxygen transmission rates of PEO and PTMO based polyurethanes were determined at 100% relative humidity and 38 °C to simulate actual biomedical application conditions. Samples were left for approximately 12 hours at these conditions to equilibrate. Oxygen permeability of PTMO based polyetherurethanes increased with increasing soft segment content (Figure 4) as previously reported [5]. A 5% increase in the soft segment content of the PTMO based polyetherurethane approximately doubled its oxygen transmission rate. The oxygen permeability of the PEO based polyetherurethanes was relatively low and did not significantly change with increasing soft segment content (Figure 4). The influence of PEO/PDMS/PEO glycol co-soft segment on the oxygen permeability of polyurethanes was the reverse of that of the water vapor permeability of polyetherurethanes (Figure 5). The oxygen permeability of PTMO based polyurethanes was slightly reduced by replacement of some of the PTMO glycol with DC-3667 silicone glycol in the polyurethane formulation. The oxygen permeability of the PEO based polyetherurethanes slightly increased with the addition of DC-3667 silicone fluid. The effect of PEO/PDMS/PEO co-soft segment on the oxygen permeability of polyurethanes is most likely due to its polydimethylsiloxane segment. Polydimethylsiloxane has one of the highest oxygen permeabilities of the known homogenous synthetic polymers.

DISCUSSION

In previously reported studies, oxygen permeability of polyether-urethanes was determined at dry conditions [5,6]. In this study, oxygen permeability of polyetherurethanes was determined at 100% RH test condition to simulate the actual biomedical application conditions. The absorbed water is known to cause plasticization in polyurethanes and alter most of their physical properties [7]. The low oxygen permeability of PEO based polyetherurethanes at dry conditions has been reported [6]. However, results of this study indicates that oxygen permeability of high soft segment PEO based polyetherurethanes at 100% RH is significantly lower than PTMO based polyurethanes.

Polytetramethylene oxide based polyetherurethanes have very desirable properties for biomedical applications such as: hydrolytic stability, relatively good resistance to autoxidation, excellent mechanical properties and good biocompatibility. High soft segment PTMO-PEUs also have good oxygen permeability for various biomedical applications. Their relatively poor water vapor permeability can be improved by adding some PEO or PEO/PDMS/PEO co-soft segment. Thus, a high oxygen and water vapor permeable biomaterial with excellent mechanical and bio-performance characteristics will be obtained. This material also affords a wide formulation latitude to tailor desired mechanical and physical properties.

CONCLUSIONS

Water vapor permeability of polyetherurethanes increases with increasing soft segment content. The water vapor permeability of PEO based polyurethanes is significantly higher than that of PTMO based polyurethanes. Addition of PEO/PDMS/PEO glycol co-soft segment increased the water vapor permeability of PTMO based polyetherurethane and decreased that of PEO based polyetherurethane.

The oxygen permeability of polyetherurethanes increase with increasing soft segment content. PTMO based polyetherurethanes have higher oxygen permeabilities than PEO based polyetherurethanes at 100% RH test condition. Addition of PEO/PDMS/PEO glycol co-soft segment reduces the oxygen permeability of PTMO based polyetherurethanes but increases the oxygen permeability of PEO based polyetherurethanes.

REFERENCES

1 Ulrich H and Bonk HW (1984) Emerging biomedical applications of polyurethane elastomers in: Planck H, Egbers G and Syre I (eds) Polyurethanes in Biomedical Engineering, Elsevier, Amsterdam
2 Ng HN, Allegrezza AE, Seymour RW and Cooper SL (1973) Effect of segment size and polydispersity on the properties of polyurethane block polymers, Polymer, 14: 255
3 Karakelle M and Zdrahala RJ (1987) Hydrophilic polyurethane membranes: Ionic permeability/structure relationship, in: Sedlack B and Kahovec J (eds) Synthetic Polymer Membranes, deGruyter, W Berlin
4 Schneider NS, Dusablon LV, Snell EW and Prosser RA (1969) Water vapor transport in structurally varied polyurethanes, J Macromol Sci, B3(4): 623
5 McBride JS, Massaro TA and Cooper SL (1979) Diffusion of gases through polyurethane block polymers, J Appl Poly Sci, 23: 201
6 Knight PM and Lyman DJ (1984) Gas permeability of various block copolyether-urethanes, J Memb Sci, 17:245
7 Zdrahala RJ, Spielvogel DE and Strand MA (1988) Softening of thermoplastic polyurethanes: A structure/property study, J Biomat Appl, 2:544

ACKNOWLEDGEMENTS

The permission by Becton Dickinson and Company to publish this work is greatly appreciated.

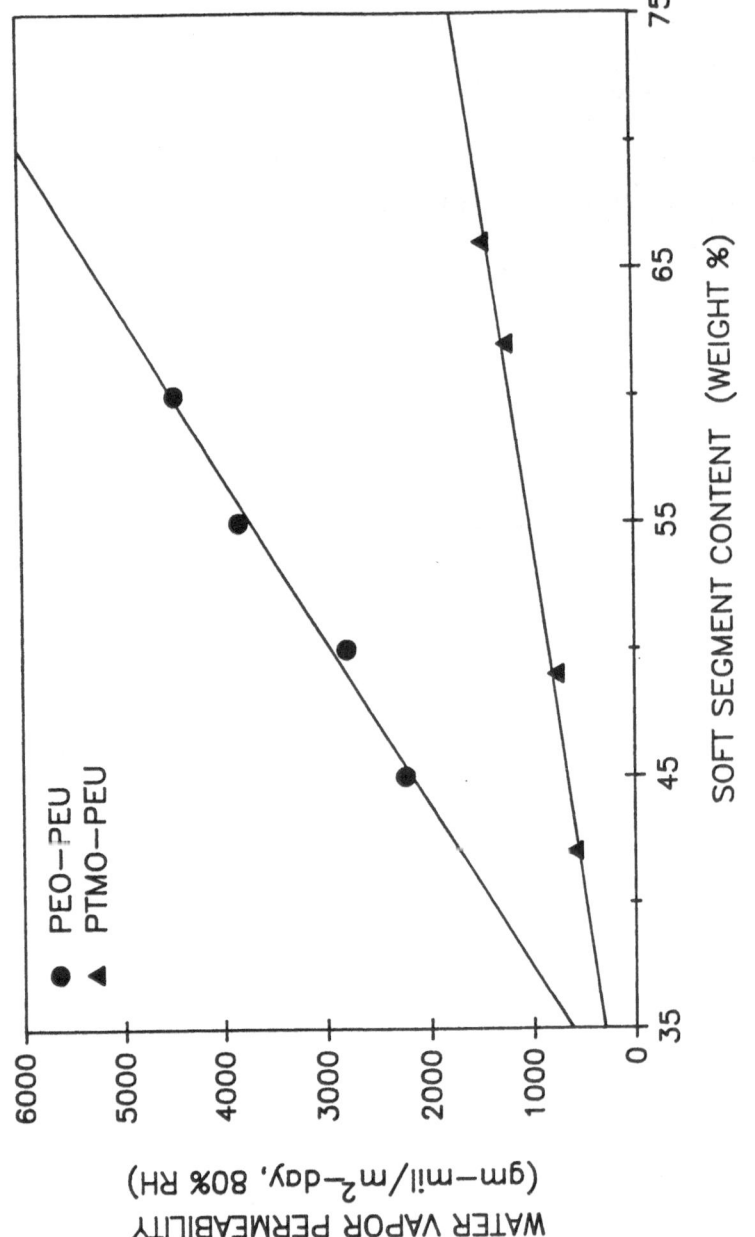

Figure 1. Polyetherurethane water vapor permeability dependence on soft segment content.

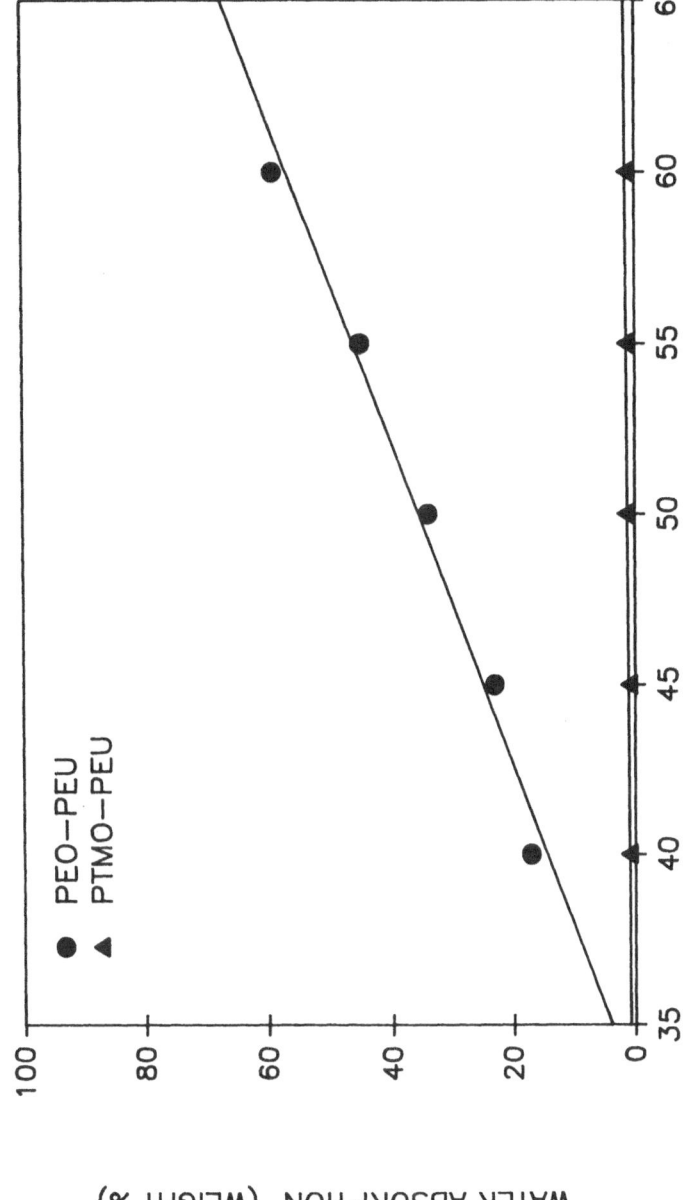

Figure 2. Polyetherurethane water absorption dependence on soft segment content.

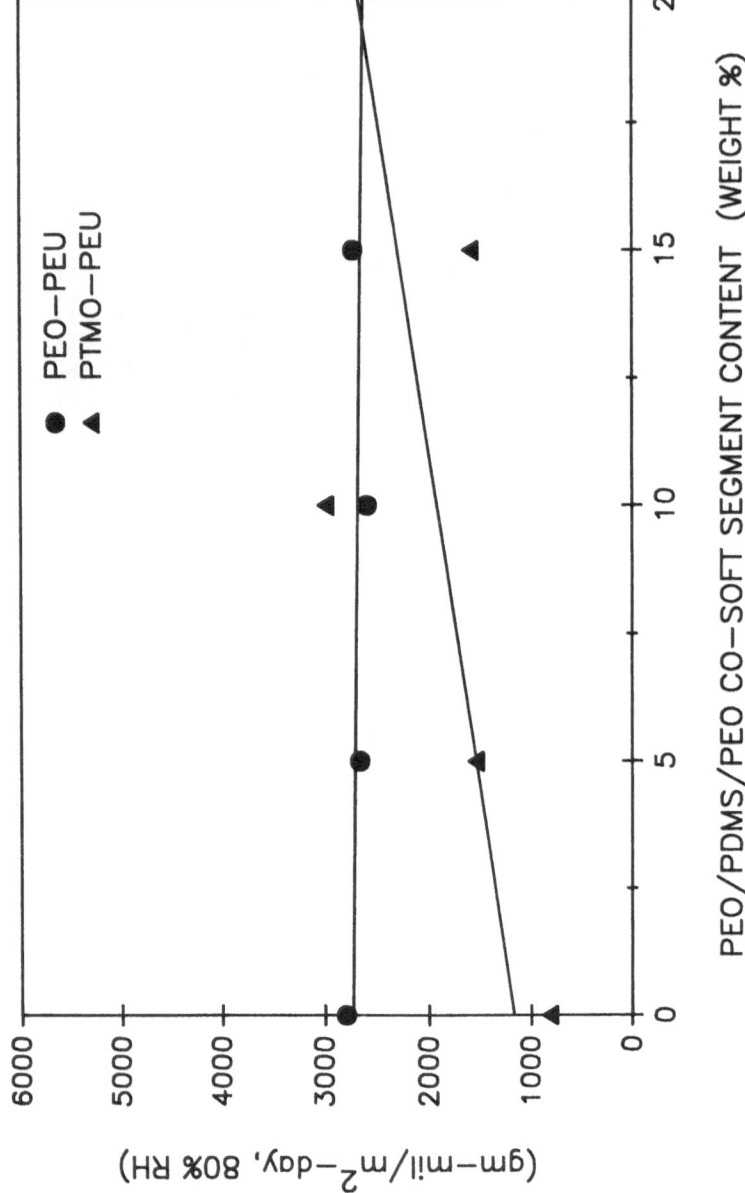

Figure 3. Polyetherurethane water vapor permeability dependence on silicone co-soft segment content.

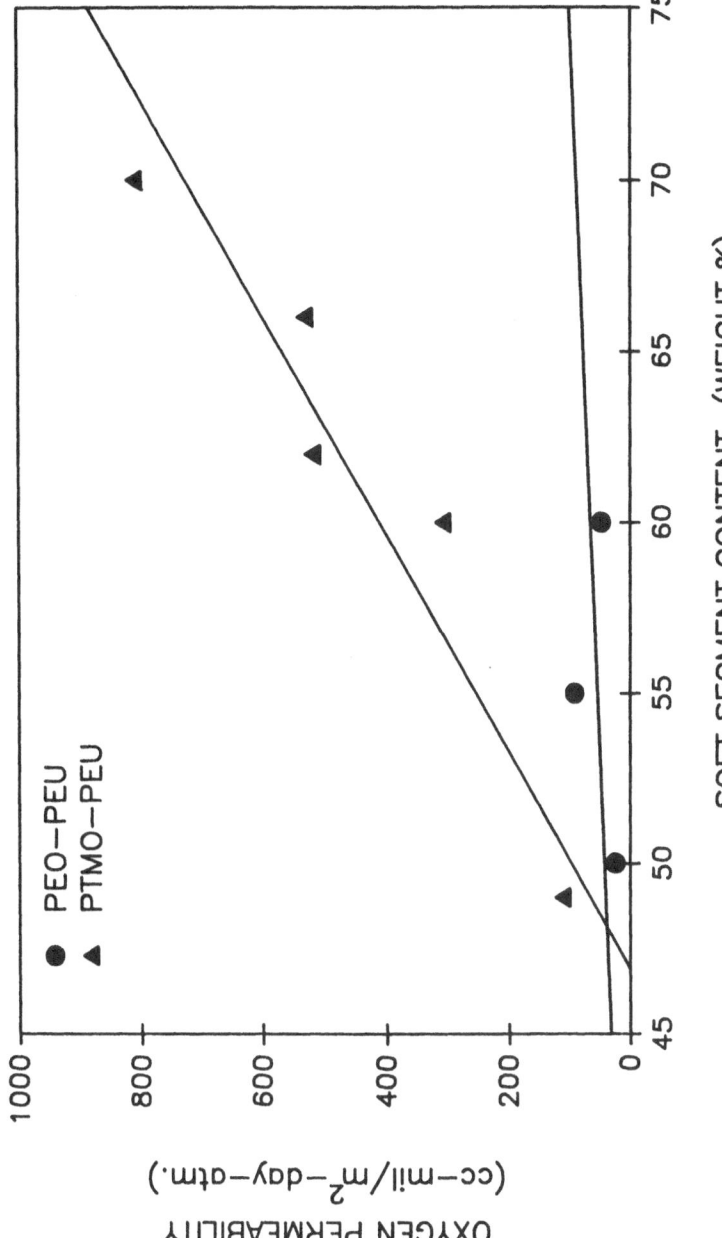

Figure 4. Polyetherurethane oxygen permeability dependence on soft segment content.

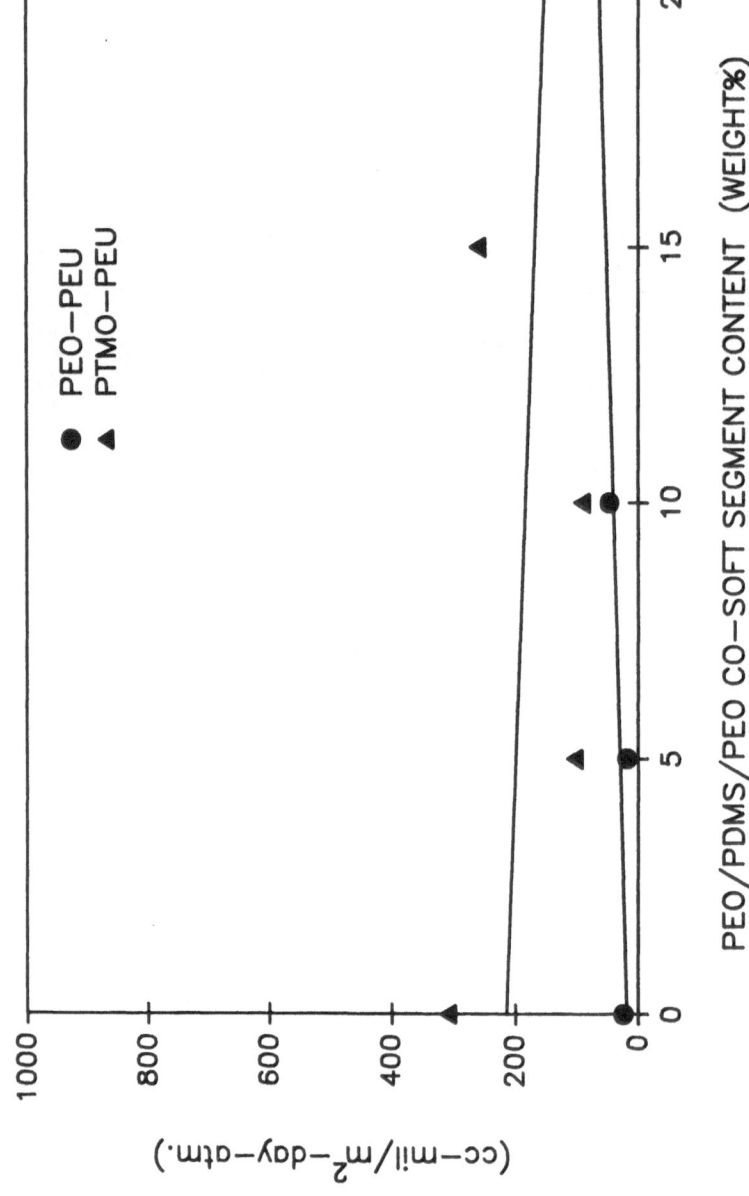

Figure 5. Polyetherurethane oxygen permeability dependence on silicone co-soft segment content.

Round-Table Discussion at the End of the Conference

Chairman: Prof. Dr. G. Egbers
Participants: Prof. Dr. C.R. Jerusalem
              Prof. Dr. A. von Recum
              Prof. Dr. H.J. Gerhardt
              Dr. H. Kniepkamp

Egbers: In the last three days we have heard a lot about the use of textiles for biomedical applications.
I would like to open the discussion asking you the most important factors coming up during this conference.
Prof. Jerusalem, could you start please with comments regarding the vascular grafts.

Jerusalem: The most impressive information during that conference was, that the human organism is able to attack by enzymes synthetic materials which are entirely resistant in vitro. That means, that the organism adaptively can form enzymes and break open fibers as the aramid fiber, one of the strongest manmade fibers.
Further, this conference has shown particulary in the field of prosthesiology that we do not have an ideal vascular substitute yet. It should be as strong as possible, dilatation and aneurysms should not occur. It should not be attacked by cells, not weakened by biochemical attacks of enzymes.
It has been shown, that a porous structure is advantageous, but the pores should not be so large, that preclotting is a problem because of secondary fibrinolysis. Very dense, nonporous prostheses may probably be of advantage in a lot of patients which are completely unable to react with the cellular activity. But that we don't know before the implantation. So I think, during the next years the combined effort of clinical research and of reserach of factories producing these vascular prostheses should be strengthened.

Egbers: Prof. von Recum, could you please summarize the role of histology as a key method for evaluating tests.

Von Recum: As you know the area of implantology started with naive implantation of available materials. During the sixties we learned to recognize specific responses of different tissues to different implant materials; it was more an observation of responses than a scientific approach of evalutation. In the seventies then, we learned some quantitative methods of differentiation between materials. The chemistry, the bulk chemistry and the surface chemistry, as well as the electro-chemical phenomena were considered to make the difference between materials and between their histological responses. In the eighties we had to learn, that this was still too superficial to understand the interplay between material and tissue.

Now we are at a stage where we are looking at this interplay at a cellular, subcellular and actually molecular level. And we are asthonished on one hand how difficile cells react to a material which is unknown to them; on the other hand we are asthonished to recognize that the bulk chemistry and the surface chemistry may not be - or not be alone - the major determinants of biocompatibility and that there are other factors that determine the type of cells at the interface and their reaction to the interface.

So with our increasing knowledge of the eigthies I'm convinced that eventually we will have materials that are generally biocompatible and useful for implantation.

But as we saw specifically at this meeting, the demands on the materials are manyfold and not combinable in one material. We know that for example hemocompatibility has different demands on a material than fibroblasts require for attachment.

Egbers: I think, we, the doctors and the mechanical engineers, live with the idea that here is a field where we can help people.

Prof. Gerhardt has agreed to sum up what we discussed today especially on artificial skin.

Gerhardt: My impression is, that meanwhile we have a very large scale of different materials in aspect of their biophysical properties, and on the other hand we have new biological techniques as the cell culture. In the next years we should try to combine these possibilities, to learn to play on this piano.

As Prof. von Recum just told us, we need not only one material with one special biophysical property but a combination. I'll try to make it clear in a seemingly very simple problem, but still it is solved unsufficiently by the plastic surgery: the replacement of an oricle. The oricle is an organ composed of a very thin skin on the anterior side (only at the eye-lid we have such a thin skin in the body), and behind there is a thicker skin of an elastic cartilage. It's the only site in the body where you have an elastic cartilage. When you try to take cartilage from the rip as most of the plastic surgeons now do, you have another structure. And, when you have a too thick skin on the anterior side, you don't get this very delicate profile as you have in the oricle. So all these artificial oricles made by the surgeons until now are plump.

So you have to combine, for instance, you can make the anterior side with keratinocytes as a very thin skin and you can make the skeleton of the oricle of polyurethanes. Some parts of it need not be permanent, they may weaken or they may biodegrade as a temporary support and then they can be replaced by the connective tissue. Other parts of the profile should stand, the rim for instance.

This may be a simple example, but perhaps it shows that the problem is not so easy. We have to combine different materials and techniques to copy what we see in the human body. That should be an aspect for the next years.

Egbers: Our modern society has created a lot of time for leisure and for sport activities. Not all of these are in favour for the human body. For instance, if you take a look at all these people lying in the sun, it creates a lot of problems with the skin. Here might be a large field for application of artificial skin in the future.

Another area where textiles seem to be a nearly natural product is the area of ligaments and tendons. Here again sport activities from soccer to skiing create a lot of problems not only at the knees. Here is another point where it is difficult to meet the complex requirements. Now Dr. Kniepkamp tries to summarize, what we've heard about that.

Kniepkamp: If you look at the development in the knee ligament replacement during the last years, there was an optimistic phase just in the beginning. The people thought the only thing, which is necessary is to have a very strong tenacity in the longitudinal direction. But unfortunately the ligament has a certain torsion and you must fix the ligament and therefore you need a multipurpose material. And I think we must combine different properties to have the best compromise for the patient.

Egbers: The task to create a structure that nature has developed over 100 million of years, this task has tempted at any time. I just remember a book from Bertrand about 1722, a childrens book. Amongst other topics there are a lot of pictures showing natural structures and manmade copies of these natural structures, like an outside structure of a coconut and the non woven, e.g..

Actually there is a research group at the University of Stuttgart trying to copy from the architectural design of the nature.

And here again we try to copy the nature regarding the use of textiles for medical applications. Sometimes I really wonder if this is the only approach we can take, but very often I think the human body reacts much better than we actually have expected and overcomes a lot of the deficiencies of the products. So we should try to create a functional design of an implant and not only copy the nature.

Planck: I think its necessary to look at the functions of a natural tissue and not only to imitate the structures. The structures of an implant may differ but the functions have to be the same.

Jerusalem: Looking at the functional performance of an implant you have to test the devices in a homostatic position. Testing materials only with subcutaneous implantation may lead to false conclusions.

I remember our investigation on the dura substitute. The material implanted subcutaneously will be replaced by the cheapest material of the body, by fat cells, and it looses tensile strength immediately. But inserted in a homostatic condition, that means inserted in the dura mater, it is remodelled with the tissue surrounding the implant, by strong and dense connective tissue. After several weeks it is almost no longer distinguishable from the autochthoneous dura mater.

Turnquist: I would like to make a comment to the presentation on the improving of test methods for suture materials. That's a very important work to use the engineering skills of an institute like yours to improve test methods and compare the commonly used methods. People working on product development find themselves disenchanted by test methods that are inadequate or not sensitive enough in many cases. But we don't have the time to develop the test methods so I really like to encourage such work. Perhaps it's little glamorous but it's very important.

Chawla: My comment is regarding the ultimate vascular research. So far the research seems to have been oriented towards the use of materials trying to minimize any reaction.

Picking up Prof. Gerhardts aspects perhaps we should try to use biodegradable matrices as a skeleton and build up cells on that skeleton and apply that. When the matrix has disappeared you'll have only the patient's own vascular cells. I know that may be a long way. But now there are growth factors available, epithelial growth factor, neuronal growth factor and vascular growth factor. Distributing those properly along the prosthesis you want to implant could speed up that process.

Planck: Prof. Gerhardt, do you think cell seeding techniques would be a method to avoid foreign body reaction to an implant?

Gerhardt: I don't know if they can be avoided generally. The material implanted must by highly biocompatible of course, but on the other hand the layer of autologous cells, that you can achieve by cell seeding speeds up the process of taking the prosthesis by the organism.

An example may be the replacement of an eyelid. You need a very thin layer made from an appropriate biomaterial. You have inside a layer from mucosa and outside from a very thin delicate skin. You cannot find any skin in the body with this specialities. But perhaps you can make a very thin skin by seeding keratinocytes so you nearly have no corneum on it.

Chu: I have some general thoughts:

First, I feel there is a strong need for the relationship of the textile structure and particulary the surface structure of the textile material to the clinical performance of the medical textile material. But most of the surgical implants are developed by clinicians alone or developed by a nontextile scientist. I haven't seen one single surgical implant research program involving the collaboration between textile technologists and the clinicians. I think textile materials are not very different from general polymer materials in terms of the chemical properties. However their surface structure is so different from that of conventional polymer materials. That's the part where we, the textile scientists, should focus on the differences regarding pore size, the distribution of the pores and the roughness of the surface. Some of these properties are very difficult to characterize. Nonwoven structures are a typical example.

My second point is, I would like to echo to Dr. Chawla's comment on the use of biodegradable materials for reconstruction. I think that will be the future direction for surgical implants. However the difficulty we have in this area is, we have not been able to develop a material that serves its function and then disappears as soon as possible. But all current biodegradable materials remain in the body far longer after the loss of their function. That would be one of the goals to design materials that degrade as fast as they can after loss of function and thus minimize the foreign body reactions.

Finally I feel its very important to utilize molecular biology in this field, particular some of these growth factors Dr. Chawla mentioned before and also sophisticate biochemical analysis. Additional to basical histological analysis, I think, we should go into some molecular biological study on the implant material.

Jerusalem: Nevertheless, we should not forget the enormous progress we have achieved during the last decade in the application of synthetic materials in human medicine.

Remember about 30 years ago, when a handmade tube was suitable to replace the human aorta. And when we remember, that someone took the nylon slips of his wife to produce a prosthesis, it was implanted in a human being and it was functioning, than we can imagine that we had a good progress during the last years. Unfortunately experimentation today is no longer as easy as previously. It is not only the question of money, but it is also the question of the permission for animal tests. It has been shown on this conference that it is impossible to draw definitive conclusions only from in vitro experiments, especially regarding the reaction or the ingrowth of tissue.

Here we do need the animal experiment, and I think the human being is too expensive to be used as an experimental animal.

KEYWORDS

General keywords: Biomaterials, Textiles

Aramid fiber	111, 123
Artificial skin	293, 305, 319, 333
Bioartificial pancreas	247, 253, 259, 265
Biocompatibility	35, 77, 219, 223, 293
Carbon fiber	129
Cell adherence	61
Cell culture	17, 77, 253
Cell orientation	193
Cell seeding	187, 193, 273, 285
Cellulose	35
Chemical structure	17
Clinical results	219, 319, 323
Coagulation system	35
Cytocompatibility	77, 187
Database	85, 93
Degradation	123, 231, 243, 329
Endothelial cell	187, 193
Epithelial cell	273, 285
Ex vivo study	35
Gas permeability	333
Hemodialysis	35
Immune system	35
Immunoisolation	247, 253
In vitro study	17, 77, 129, 137, 187, 193, 231, 243, 253, 259, 265, 285, 329
In vivo study	61, 111, 123, 265, 273, 293, 305, 329
Islets of Langerhans	253, 265
Ligament prosthesis	99, 111, 123, 129
Macrophage	17
Mechanical behaviour	99, 129, 137, 231, 329
Mechanical response	193
Membrane	1, 35, 193, 247, 253, 259, 293, 305, 319, 323, 333

Non wovens   1, 61
Patches   319, 323, 329
Polymers, comparison of   17, 35, 61, 85, 129, 223, 259,
                                    265, 329
Polyacrylate   253, 265
Polyacrylonitrile   35
Polyamide   17
Polydioxanone   129
Polyetherketone   259, 265
Polyethylene   17
Polyethyleneterephthalate,
        polyester   17, 129, 137
Polyglactin   329
Polyglycolic acid   231, 243, 329
Polyglyconate   243
Polylactic acid   329
Polypropylene   17, 129
Polystyrene   17
Polysulfone   17, 259, 265
Polytetrafluoroethylene,
        expanded   17, 61, 129, 219, 273, 285, 319, 323
Polyurethane   111, 187, 293, 333
Polyvinylidene fluoride   259
Pore size   1, 61, 123, 187, 247, 273, 293
Processing parameters   1, 99, 223, 293, 333
Protein coating   253, 259, 265
Silicone   17, 193
Smooth muscle cell   193
Surface structure   1, 17, 61, 187, 223, 231, 247,
                                    265, 285
Sutures   223, 231, 243
Textile structure   1, 61, 99, 129, 137, 223, 231, 329
Tissue ingrowth   61, 99, 123, 247, 265, 273, 305
Tracheal prosthesis   273, 285
Vascular graft   61, 137, 187, 219

The page numbers are referring to the first page of the paper
concerned

Author Index

Albert F.W., Abt. für Klinische Hämostaseologie und Transfusionsmedizin, Unikliniken, D-6650 Homburg/Saar (pp. 35-60)

Appel, M., Department of Textiles & Apparel, Martha Van Rensselaer Hall, Cornell University, Ithaca, NY 14853-4401 (pp. 17-34)

Bambauer, R., Abt. für Nephrologie, Universität des Saarlandes, D-6650 Homburg/Saar (pp. 93-98)

Barriuso C., Cardiovascular Surgery, Hospital Clínico y Provincal, University of Barcelona, Barcelona, Spain (pp. 319-322)

Bernier-Cardou, M., Biomaterials Institute, Room Fl-304, St. Francois d'Assise Hospital, Quebec City, Qc, G1L 3L5, Canada (pp. 137-186)

Betz, E., Institut für Physiologie, Universität Tübingen, Gmelinstr. 5, D-7400 Tübingen 1 (pp. 193-218)

Böhm F., Dermatologie, Humboldt-Universität Berlin, Charité, Schumannstr. 20/21, DDR-1040 Berlin (pp. 273-284, pp. 285-292)

Bohnert, G., Abt. für Klinische Hämostaseologie und Transfusionsmedizin, Unikliniken, D-6650 Homburg/Saar (pp. 93-98)

Boyer, D., Biomaterials Institute, Room Fl-304, St. Francois d'Assise Hospital, Quebec City, Qc, G1L 3L5, Canada (pp. 137-186)

Bretzel, R.G., Medizinische Klinik III und Poliklinik der Justus-Liebig-Universität, Rodthohl 6, D-6300 Gießen (pp. 253-258, pp. 265-272)

Cardou, A., Biomaterials Institute, Room Fl-304, St. Francois d'Assise Hospital, Quebec City, Qc, G1L 3L5, Canada (pp. 137-186)

Chaput, C., Biomaterials Institute, Room Fl-304, St. Francois d'Assise Hospital, Quebec City, Qc, G1L 3L5, Canada (pp. 137-186)

Charara, J., Biomaterials Institute, Room Fl-304, St. Francois d'Assise Hospital, Quebec City, Qc, G1L 3L5, Canada (pp. 137-186)

Chu, C.C., Department of Textiles & Apparel, Martha Van Rensselaer Hall, Cornell University, Ithaca, NY 14853-4401 (pp. 17-34)

Claes, L., Sektion für Unfallchirurgische Forschung und Biomechanik, Universität Ulm, Helmholtzstr. 14, D-7900 Ulm/Donau (pp. 129-136)

Dadgar, L., Biomaterials Institute, Room Fl-304, St. Francois d'Assise Hospital, Quebec City, Qc, G1L 3L5, Canada (pp. 137-186)

Dartsch, P.C., Institut für Physiologie, Universität Tübingen, Gmelinstr. 5, D-7400 Tübingen 1 (pp. 193-218)

Dauner, M., Institut für Textil- und Verfahrenstechnik, Körschtalstr. 26, Postfach 1155, D-7306 Denkendorf (pp. 99-110, pp. 111-122, pp. 123-128)

354

Debille, E., Biomaterials Institute, Room Fl-304, St. Francois d'Assise Hospital, Quebec City, Qc, G1L 3L5, Canada (pp. 137-186)

Dittel, K.-K., Marienhospital, Chirurgische Klinik, D-7000 Stuttgart 1 (pp. 99-110, pp. 111-122, 123-128)

Dürselen, L., Sektion für Unfallchirurgische Forschung und Biomechanik, Universität Ulm, Helmholtzstr. 14, D-7900 Ulm/Donau (pp. 129-136)

Elser, C., Institut für Textil- und Verfahrenstechnik, Körschtalstr. 26, Postfach 1155, D-7306 Denkendorf (pp. 231-242, 243-246)

Federlin, K., Medizinische Klinik III und Poliklinik der Justus-Liebig-Universität, Rodthohl 6, D-6300 Gießen (pp. 247-252, pp. 253-258, pp. 259-264, pp. 265-272)

Feijen, J., University of Twente, Department of Chemical Technology, P.O. Box 217, NL-7500 AE Enschede (pp. 293-304)

Gerhardt, H.J., HNO-Klinik, Humboldt-Universität Berlin, Charité, Schumannstr. 20/21, DDR-1040 Berlin (pp. 273-284, pp. 285-292)

Gerlach, J., Chirurgische Universitätsklinik, UKRV/Charlottenburg, Freie Universität Berlin, Spandauer Damm 130, D-1000 Berlin 19 (pp. 187-192)

Grothe-Pfautsch, U., Abt. für Klinische Hämostaseologie und Transfusionsmedizin, Unikliniken, D-6650 Homburg/Saar (pp. 35-60)

Guidoin, R., Biomaterials Institute, Room Fl-304, St. Francois d'Assise Hospital, Quebec City, Qc, G1L 3L5, Canada (pp. 137-186)

Haake, K., HNO-Klinik, Humboldt-Universität Berlin, Charité, Schumannstr. 20/21, DDR-1040 Berlin (pp. 273-284)

Hafemann, B., Klinik für Verbrennungs- und Plastische Wiederherstellungschirurgie, Klinikum der RWTH Aachen, Pauwelsstr., D-5100 Aachen (pp. 305-318)

Helmling, E., Abt. für Klinische Hämostaseologie und Transfusionsmedizin, Unikliniken, D-6650 Homburg/Saar (pp. 35-60)

Hepp W., Chirurgische Klinik und Poliklinik, Universitätsklinikum Rudolf Virchow, Standort Charlottenburg, Spandauer Damm 130, D-1000 Berlin 19 (pp. 219-222)

Hess, F., Department of Cell Biology, K.-University of Nijmegen, P.O. Box 9101, NL-6500 HB Nijmegen (pp. 61-76)

Hettich R., Klinik für Verbrennungs- und Plastische Wiederherstellungschirurgie, Klinikum der RWTH Aachen, Pauwelsstr., D-5100 Aachen (pp. 305-318)

Hinrichs, W.L.J., University of Twente, Department of Chemical Technology, P.O. Box 217, NL-7500 AE Enschede (pp. 293-304)

Hyon, S.-H., Research Center for Medical Polymers & Biomaterials, Experimental Surgery and Chemistry of Biomedical Polymers, Kyoto University, 53 Kawahara-cho, Shogoin, Sakyo-ku, Kyoto 606, Japan (pp. 329-333)

Ikada Y., Research Center for Medical Polymers & Biomaterials, Experimental Surgery and Chemistry of Biomedical Polymers, Kyoto University, 53 Kawahara-cho, Shogoin, Sakyo-ku, Kyoto 606, Japan (pp. 329-333)

Jerusalem, C.R., Department of Cell Biology, K.-University of Nijmegen, P.O. Box 9101, NL-6500 HB Nijmegen (pp. 61-76, pp. 123-128)

Jutzler, G., Abt. für Nephrologie, Universität des Saarlandes, D-6650 Homburg/Saar (pp. 93-98)

Gerhardt, H.J., HNO-Klinik, Humboldt-Universität Berlin, Charité, Schumannstr. 20/21, DDR-1040 Berlin (pp. 273-284)

Karakelle, M., Becton Dickinson, Polymer Research, P.O. Box 1285, Dayton, OH 45401, USA (pp. 333-350)

Kaschke, O., HNO-Klinik, Humboldt-Universität Berlin, Charité, Schumannstr. 20/21, DDR-1040 Berlin (pp. 273-284)

Kiesewetter H., Abt. für Klinische Hämostaseologie und Transfusionsmedizin, Unikliniken, D-6650 Homburg/Saar (pp. 93-98)

Kniepkamp, H., B. Braun Melsungen AG, D-3508 Melsungen (pp. 223-230)

Konle, S., Institut für Textil- und Verfahrenstechnik, Körschtalstr. 26, Postfach 1155, D-7306 Denkendorf (pp. 85-92)

Kuit J., University of Twente, Department of Chemical Technology, P.O. Box 217, NL-7500 AE Enschede (pp. 293-304)

Kunberger A., Institut für Textil- und Verfahrenstechnik, Körschtalstr. 26, Postfach 1155, D-7306 Denkendorf (pp. 85-92)

Lim, J.O., Department of Textiles & Apparel, Martha Van Rensselaeer Hall, Cornell University, Ithaca, NY 14853-4401 (pp. 17-34)

Lommen, E.J.C.M.P., University Hospital of Groningen, Department of Experimental Cardiopulmonary Surgery, P.O. Box 30001, NL-9700 RB Groningen (pp. 293-304)

Marceau, D., Biomaterials Institute, Room Fl-304, St. Francois d'Assise Hospital, Quebec City, Qc, G1L 3L5, Canada (pp. 137-186)

Mayr, K., Institut für Textil- und Verfahrenstechnik, Körschtalstr. 26, Postfach 1155, D-7306 Denkendorf (pp. 231-242)

Meinecke, H.J., Chirurgische Klinik und Poliklinik, Universitätsklinikum Rudolf Virchow, Standort Charlottenburg, Spandauer Damm 130, D-1000 Berlin 19 (pp. 219-222)

Mestres, C.A., Cardiovascular Surgery, Hospital Clínico y Provincal, University of Barcelona, Barcelona, Spain (pp. 319-322, pp. 323-328)

Milwich, M., Institut für Textil- und Verfahrenstechnik, Körschtalstr. 26, Postfach 1155, D-7306 Denkendorf (pp. 231-242)

Müller-Lierheim, W., Dr. Müller-Lierheim AG, Biological Laboratories, Behringstraße 6, D-8033 Planegg (pp. 77-84)

Mulet, J., Cardiovascular Surgery, Hospital Clínico y Provincal, University of Barcelona, Barcelona, Spain (pp. 319-322)

Nakamura, T., Research Center for Medical Polymers & Biomaterials, Experimental Surgery and Chemistry of Biomedical Polymers, Kyoto University, 53 Kawahara-cho, Shogoin, Sakyo-ku, Kyoto 606, Japan (pp. 329-333)

Ninot S., Cardiovascular Surgery, Hospital Clínico y Provincal, University of Barcelona, Barcelona, Spain (pp. 319-322)

Pallua, C., Chirurgische Klinik und Poliklinik, Universitätsklinikum Rudolf Virchow, Standort Charlottenburg, Spandauer Damm 130, D-1000 Berlin 19 (pp. 219-222)

Pindur, G., Abt. für Klinische Hämostaseologie und Transfusionsmedizin, Unikliniken, D-6650 Homburg/Saar (pp. 35-60, pp. 93-98)

Planck, H., Institut für Textil- und Verfahrenstechnik, Körschtalstr. 26, Postfach 1155, D-7306 Denkendorf (pp. 1-16, pp. 85-92, pp. 93-110, pp. 111-122, pp. 123-128, pp. 187-192, pp. 231-242, pp. 243-246, pp. 253-258, pp. 259-264, pp. 265-272)

Pomar, J.L., Cardiovascular Surgery, Hospital Clínico y Provincal, University of Barcelona, Barcelona, Spain (pp. 319-322, pp. 323-328)

Renardy, M., Institut für Textil- und Verfahrenstechnik, Körschtalstr. 26, Postfach 1155, D-7306 Denkendorf (pp. 231-242, pp. 243-246, pp. 253-258, pp. 259-264, pp. 265-272)

Sauren, B., Klinik für Verbrennungs- und Plastische Wiederherstellungschirurgie, Klinikum der RWTH Aachen, Pauwelsstr., D-5100 Aachen (pp. 305-318)

Schauwecker H.H., Chirurgische Universitätsklinik, UKRV/Charlottenburg, Freie Universität Berlin, Spandauer Damm 130, D-1000 Berlin 19 (pp. 187-192)

Seyfert, U.T., Abt. für Klinische Hämostaseologie und Transfusionsmedizin, Unikliniken, D-6650 Homburg/Saar (pp. 35-60, pp. 93-98)

Shimamoto, T., Research Center for Medical Polymers & Biomaterials, Experimental Surgery and Chemistry of Biomedical Polymers, Kyoto University, 53 Kawahara-cho, Shogoin, Sakyo-ku, Kyoto 606, Japan (pp. 329-333)

Shimizu, Y., Research Center for Medical Polymers & Biomaterials, Experimental Surgery and Chemistry of Biomedical Polymers, Kyoto University, 53 Kawahara-cho, Shogoin, Sakyo-ku, Kyoto 606, Japan (pp. 329-333)

Shiraki, K., Research Center for Medical Polymers & Biomaterials, Experimental Surgery and Chemistry of Biomedical Polymers, Kyoto University, 53 Kawahara-cho, Shogoin, Sakyo-ku, Kyoto 606, Japan (pp. 329-333)

Siebers, U., Medizinische Klinik III und Poliklinik der Justus-Liebig-Universität, Rodthohl 6, D-6300 Gießen (pp. 253-258, pp. 259-264, pp. 265-272)

Sturm R., Medizinische Klinik III und Poliklinik der Justus-Liebig-Universität, Rodthohl 6, D-6300 Gießen (pp. 265-272)

Suzuki, M., Research Center for Medical Polymers & Biomaterials, Experimental Surgery and Chemistry of Biomedical Polymers, Kyoto University, 53 Kawahara-cho, Shogoin, Sakyo-ku, Kyoto 606, Japan (pp. 329-333)

Syré, I., Institut für Textil- und Verfahrenstechnik, Körschtalstr. 26, Postfach 1155, D-7306 Denkendorf (pp. 99-110, pp. 111-122)

Taller R.A., Becton Dickinson, Polymer Research, P.O. Box 1285, Dayton, OH 45401, USA (pp. 333-350)

Torché, D., Biomaterials Institute, Room F1-304, St. Francois d'Assise Hospital, Quebec City, Qc, G1L 3L5, Canada (pp. 137-186)

Trauter, J., Institut für Textil- und Verfahrenstechnik, Körschtalstr. 26, Postfach 1155, D-7306 Denkendorf (pp. 259-264, pp. 265-272)

Umutyan, G., Institut für Textil- und Verfahrenstechnik, Körschtalstr. 26, Postfch 1155, D-7306 Denkendorf (pp. 85-92)

Watanabe, S., Research Center for Medical Polymers & Biomaterials, Experimental Surgery and Chemistry of Biomedical Polymers, Kyoto University, 53 Kawahara-cho, Shogoin, Sakyo-ku, Kyoto 606, Japan (pp. 329-333)

Weber, O., Institut für Textil- und Verfahrenstechnik, Körschtalstr. 26, Postfach 1155, D-7306 Denkendorf (pp. 231-242)

Wenzel E., Abt. für Klinische Hämostaseologie und Transfusionsmedizin, Unikliniken, D-6650 Homburg/Saar (pp. 35-60, pp. 93-98))

Wenzel M., Anatomie, Humboldt-Universität Berlin, Charité, Schumannstr. 20/21, DDR-1040 Berlin (pp. 273-284, pp. 285-292)

Wildevuur, C.R.H., University Hospital of Groningen, Department of Experimental Cardiopulmonary Surgery, P.O. Box 30001, NL-9700 RB Groningen (pp. 293-304)

Zekorn, T., Medizinische Klinik III und Poliklinik der Justus-Liebig-Universität, Rodthohl 6, D-6300 Gießen (pp. 247-252, pp. 253-258, pp. 259-264, pp. 265-272)

Zschocke, P., Institut für Textil- und Verfahrenstechnik, Körschtalstr. 26, Postfach 1155, D-7306 Denkendorf (pp. 259-264, pp. 265-272)